Harley-Davidson Sportsters Owners Workshop Manual

by Curt Choate,
Tom Schauwecker
and John H Haynes
Member of the Guild of Motoring Writers

Models covered:

XL, XLH, XLCH 883 Hugger, Sportster, Deluxe. 1970 to 1971,
   1986 to 1999
XL, XLX-61, XLH, XLCH, XLS 1000 Sportster, Roadster,
   Anniversary. 1972 to 1985
XL, XLH 1100 Sportster. 1986 to 1987
XL, XLH 1200 Sportster. 1988 to 1999
Note: XR-1000 engine information not included

(5B4 – 702)

Haynes Publishing
Sparkford Nr Yeovil
Somerset BA22 7JJ England

Haynes Publications, Inc
861 Lawrence Drive
Newbury Park
California 91320 USA

Acknowledgements

Thanks to Riders of Bridgwater who supplied the XLH 883 Hugger shown on the front cover.

© Haynes Publishing Group 1990, 1993, 1997, 1999

A book in the **Haynes Owners Workshop Manual Series**

Printed in the USA

All rights reserved. No part of this book may be reproduced or transmitted in any form or by any means, electronic or mechanical, including photocopying, recording or by any information storage or retrieval system, without permission in writing from the copyright holder.

ISBN 1 56392 344 0

Library of Congress Catalog Card Number 99-62154

British Library Cataloguing in Publication Data
A catalogue record for this book is available from the British Library.

We take great pride in the accuracy of information given in this manual, but motorcycle manufacturers make alterations and design changes during the production run of a particular motorcycle of which they do not inform us. No liability can be accepted by the authors or publishers for loss, damage or injury caused by any errors in, or omissions from, the information given.

Contents

Introductory pages
About this manual	6
Introduction to the Harley-Davidson Sportster	7
Identification numbers	8
Buying parts	10
Maintenance techniques, tools and working facilities	10
Motorcycle chemicals and lubricants	16
Safety first!	17
Troubleshooting	18

Chapter 1
Tune-up and routine maintenance — 26

Chapter 2
Engine, clutch and transmission — 50

Chapter 3
Fuel system — 116

Chapter 4
Ignition system — 132

Chapter 5
Frame and suspension — 138

Chapter 6
Wheels, brakes and tires — 150

Chapter 7
Electrical system — 171

Wiring diagrams — 184

English/American terminology — 194

Conversion factors — 195

Index — 197

1981 Harley-Davidson Sportster

Sportster engine/transmission (left side shown)

About this manual

Its purpose

The purpose of this manual is to help you maintain and repair your motorcycle. It can do so in several ways. It can help you decide what work must be done, even if you choose to have it done by a dealer service department or a repair shop, it provides information and procedures for routine maintenance and it offers diagnostic and repair procedures to follow when trouble occurs.

We hope you will use the manual to tackle the work yourself. For many simple jobs, doing it yourself may be quicker than arranging an appointment to get the machine into a shop and making the trips to leave it and pick it up. More importantly, a lot of money can be saved by avoiding the expense the shop must pass on to you to cover its labor and overhead costs. An added benefit is the sense of satisfaction and accomplishment you feel after doing the job yourself.

Using the manual

The manual is divided into several Chapters. Each Chapter is divided into numbered Sections which are headed in bold type between horizontal lines. Each Section consists of consecutively numbered Paragraphs (often referred to as "Steps" in the text).

The two types of illustrations used (line drawings and photographs) are all referenced by a number preceding each caption. The number denotes the Section and Paragraph the illustration is intended to clarify (i.e. illustration 3.4 means Section 3, Paragraph [or Step] 4). A "Refer to illustration . . ." entry under the Section head (or in some cases, a sub-head) will alert you to the illustrations which apply to the procedure you're following.

Procedures, once described in the text, are not normally repeated. When it's necessary to refer to another Chapter, the reference will be given as Chapter and Section number (i.e. Chapter 1, Section 16). Cross-references given without use of the word "Chapter" apply to Sections and/or Paragraphs in the same Chapter. For example, "see Section 8" means in the same Chapter.

Reference to the left or right side of the motorcycle is based on the assumption you're sitting on the seat, facing forward.

Even though extreme care has been taken during the preparation of this manual, neither the publisher nor the author can accept responsibility for any errors in, or omissions from, the information given.

NOTE

A **Note** provides information necessary to properly complete a procedure or information which will make the procedure easier to understand.

CAUTION

A **Caution** provides a special procedure or special steps which must be taken while completing the procedure where the **Caution** is found. Not heeding a **Caution** can result in damage to the assembly being worked on.

WARNING

A **Warning** provides a special procedure or special steps which must be taken while completing the procedure where the **Warning** is found. Not heeding a **Warning** can result in personal injury.

Introduction to the Harley-Davidson Sportster

William S. Harley and Arthur Davidson jointly designed and built their first motorcycle during 1903, constructed along the lines of a powered bicycle. The single cylinder, air-cooled engine had a bore and stroke of 2-1/8 x 2-7/8 inches and is alleged to have been one of several built for experimental purposes. The weaknesses of the powered bicycle approach soon became apparent and a larger capacity engine followed, with much heavier flywheels. With the new, more powerful engine, the bicycle was able to climb hills more easily but, in turn, weaknesses in the frame design became apparent. And so the design progressed until something usable and marketable was available. Land was purchased in Milwaukee for the erection of a simple, two story factory and the Harley-Davidson joined the ranks of the world's motorcycles. By the end of 1906, fifty machines had been assembled at Chestnut Street and the order book was full, portending well for the year to follow.

The year 1909 proved to be a milestone in the history of the company, for it was during that year the company launched its first V-twin. It was this model, and others that followed, that helped create the legend of the large capacity V-twin, a field in which Harley-Davidson ruled supreme for many years. In private hands it gave a standard of comfort, performance and reliability second to none, while on the racetrack it became a snarling, powerful beast, with the ability to perform incredible feats of speed and endurance that made the headlines.

Contrary to expectations, the American motorcycle boom didn't last long and by the 1930s' it had all but petered out. It was largely the more sporting side that kept things going, although motorcycles were also used by police for law enforcement purposes. The American "speed cop' was very much a part of life and a big, powerful V-twin gave him a certain amount of flexibility that wouldn't have been possible in a patrol car.

The year 1957 marked the introduction of the first OHV Sportster V-twin, based on the 55 cubic inch KH model. Others followed and by 1965 the Sportster had achieved a reputation all its own. Despite excessive weight, it was capable of speeds around 120 mph and had a very unique torque band. Superlatives appeared liberally in almost every road test report, leaving little doubt that Harley-Davidson had a real winner on its hands.

The development of the Evolution engine, which was introduced in 1986 models, ensures the Sportster mystique will continue, since the new engine is as reliable, oil-tight and serviceable as the engines in competitive motorcycles, without discarding the unique characteristics of previous Sportster engines. Because it's more powerful across the board, lighter, cooler running and cleaner, in terms of exhaust emissions, than its predecessor, it continues the trend toward refinement that should produce many more generations of satisfied Sportster riders.

Identification numbers

Harley-Davidson motorcycles have a Vehicle Identification Number (VIN) stamped into the frame on the right side of the steering head, as well as on the right side of the engine crankcase **(see illustrations)**. The VIN is made up of a model code, a serial number, the model year and the manufacturer's identification.

The VIN number should be recorded and kept in a safe place so it can be furnished to law enforcement officials in the event of theft. The VIN should also be available when purchasing or ordering parts for the machine. It's a good idea to write it on a card and keep it tucked away with your driver's license, then it'll be handy when you need it.

Frame VIN number location

Engine VIN number location

Identification numbers

| 1980 and Earlier |||||
Letters	Model No.	Serial No.	Mfr.	Year
XL or XLH	3A	10000 & up	H-1970's	0
XLS	4E	(5 digits)	J-1980's	(1980)
XLCH	4A	Harley-		
XLA	4D	Davidson		
XLT	2G			

VIN key for pre-1982 models (typical)

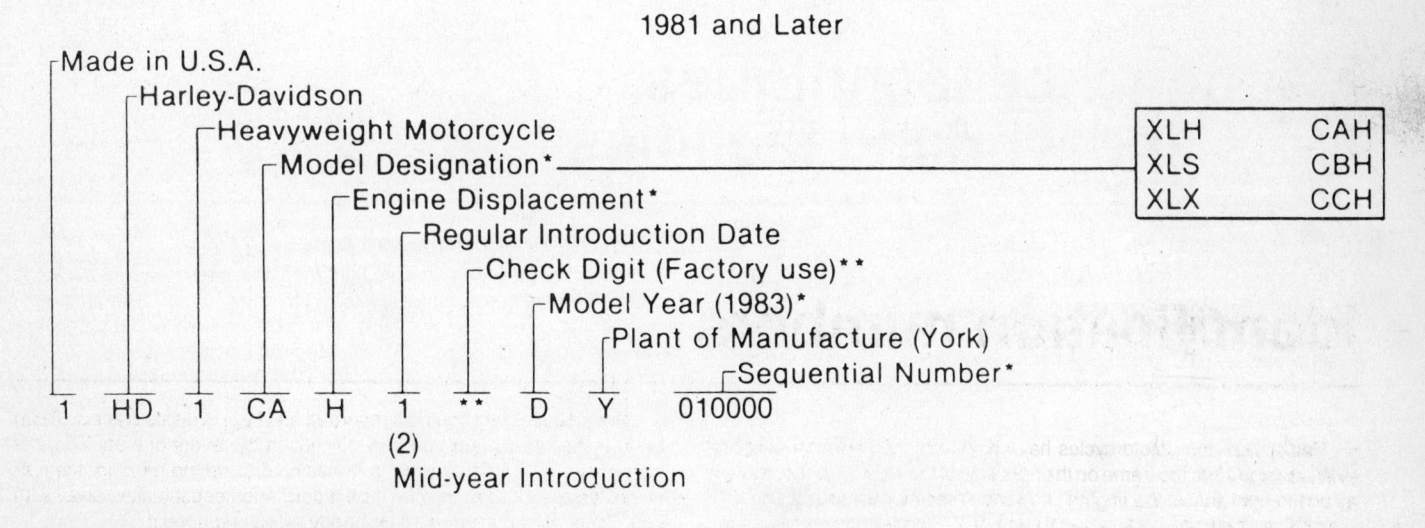

1981 and Later

```
┌Made in U.S.A.
│ ┌Harley-Davidson
│ │ ┌Heavyweight Motorcycle
│ │ │ ┌Model Designation*─────────────────────
│ │ │ │  ┌Engine Displacement*
│ │ │ │  │┌Regular Introduction Date
│ │ │ │  ││ ┌Check Digit (Factory use)**
│ │ │ │  ││ │┌Model Year (1983)*
│ │ │ │  ││ ││┌Plant of Manufacture (York)
│ │ │ │  ││ │││  ┌Sequential Number*
1  HD 1 CA H 1  ** D Y  010000
          (2)
         Mid-year Introduction
```

XLH	CAH
XLS	CBH
XLX	CCH

*Abbreviated VIN on Engine
**Varies - can be 0 thru 9 or X

Sample V.I.N. as it appears on the steering head - 1 HD 1CAH1 5 DY010234

Sample abbreviated V.I.N. as it appears on the engine - CAHD 010000

VIN key for 1982 and later models (typical)

Buying parts

Once you've located all the identification numbers, record them for reference when buying parts. Since manufacturers change specifications, parts and vendors (companies that manufacture various components on the machine), providing the ID numbers is the only way to be reasonably sure you're buying the correct parts.

Whenever possible, take the worn part to the dealer so direct comparison with the new component can be made. Along the trail from the manufacturer to the parts shelf, there are numerous places the part can end up with the wrong number or be listed incorrectly.

The two places to purchase new motorcycle parts – franchised dealers and independent accessory stores – differ in the type of parts they carry. While a dealer can obtain virtually every stock part on the motorcycle, as well as aftermarket items, the accessory dealer is usually – not always – limited to items such as shock absorbers, tune-up parts, engine gaskets, cables, brake parts, etc. Often, however, an accessory outlet will sell aftermarket suspension components, cylinders, transmission gears and other major components.

Used parts can be obtained for roughly half the price of new ones, but you can't always be sure of what you're getting. Once again, take the worn part to the salvage yard for direct comparison.

Whether buying new, used or rebuilt parts, it's a good idea to deal directly with someone who specializes in parts for Harley-Davidson motorcycles.

Maintenance techniques, tools and working facilities

Basic maintenance techniques

There are a number of techniques involved in motorcycle maintenance and repair that will be referred to throughout this manual. Application of these techniques will enable the amateur mechanic to be more efficient, better organized and capable of performing the various tasks properly, which will ensure the repairs are thorough and complete.

Fastening systems

Fasteners, basically, are nuts, bolts and screws used to hold two or more parts together. There are a few things to keep in mind when working with fasteners. Almost all of them require a locking device of some type (either a lock washer, locknut, lockplate, locking tab or thread adhesive). All threaded fasteners should be clean, straight, have undamaged threads and undamaged corners on the hex head where the wrench fits. Develop the habit of replacing all damaged nuts and bolts with new ones.

Rusted nuts and bolts should be treated with penetrating oil to make removal easier and prevent breakage. Some mechanics use turpentine in a spout type oil can, which works quite well. After applying the penetrating oil, let it "work" for a few minutes before trying to loosen the nut or bolt. Badly rusted fasteners may have to be chiseled off or removed with a hacksaw or special nut breaker, available at tool stores.

Flat washers and lock washers, when removed from an assembly, should always be replaced exactly as removed. Discard damaged washers and replace them with new ones. Always use a flat washer between a lock washer and any soft metal surface (such as aluminum), thin sheet metal or plastic. Special locknuts can only be used once or twice before they lose their locking ability and must be replaced.

If a bolt or stud breaks off in an assembly, it can be drilled out and removed with a special tool called an E-Z out. Most dealer service departments and motorcycle repair shops (and automotive machine shops) can perform this task, as well as others (such as the repair of threaded holes that have been stripped out).

Torquing sequences and procedures

When threaded fasteners are tightened, they are often tightened to a specific torque value (torque is basically a twisting force). Over-tightening the fastener can weaken it and cause it to break, while under-tightening can cause it to eventually come loose. Each bolt, depending on the material it's made of, the diameter of its shank and the material it's threaded into, has a specific torque value, which is noted in the Specifications Section at the beginning of each Chapter. Be sure to follow the torque recommendations closely. For fasteners not requiring a specific torque, use common sense when they're tightened.

Fasteners laid out in a pattern (such as cylinder head bolts, engine case bolts, etc.) must be loosened or tightened in a sequence to avoid warping the component. Initially, the bolts/nuts should go on finger-tight only. Next, they should be tightened one full turn each, in a criss-cross or diagonal pattern. After each one has been tightened one full turn, return to the first one and tighten them all one-half turn, following the same pattern. Finally, tighten each of them one-quarter turn at a time until each fastener has been tightened to the proper torque. To loosen and remove the fasteners the procedure would be reversed.

Disassembly sequence

Component disassembly should be done slowly and carefully to make sure the parts go back together properly during reassembly. Always keep track of the sequence in which parts are removed. Note special characteristics or marks on parts that can be installed more than one way (a good example would be the rocker arm shafts on the Evolution engine – they have a small cutout on only one end for the bolts to pass through). It's a good idea to lay the disassembled parts out on a clean surface in the order they were removed. It may also be helpful to make sketches or take instant photos of components before removal.

When removing fasteners from a component, keep track of their locations. Sometimes threading a bolt back into a part, or putting the washers and nut back on a stud, can prevent mix-ups later. If nuts and bolts can't

Maintenance techniques, tools and working facilities

be returned to their original locations, they should be kept in a compartmented box or a series of small boxes. A cupcake or muffin tin is ideal for this purpose, since each cavity can hold the bolts and nuts from a particular area (i.e. engine case bolts, rocker arm cover bolts, engine mount bolts, etc.). A pan of this type is especially helpful when working on assemblies with very small parts (such as the carburetor and the valve train). The cavities can be marked with paint or tape to identify the contents.

When wiring looms, harnesses or connectors are separated, it's a good idea to identify the two halves with numbered pieces of masking tape so they can be easily reconnected.

Gasket sealing surfaces

Gaskets are used to seal the mating surfaces between components and keep lubricants, fluids, vacuum or pressure contained in an assembly.

Many times these gaskets are coated with a liquid or paste-type gasket sealing compound before assembly. Age, heat and pressure can sometimes cause the two parts to stick together so tightly they're very difficult to separate. In most cases, the part can be loosened by striking it with a soft-face hammer near the mating surfaces. A regular hammer can be used if a block of wood is placed between the hammer and the part. Do not hammer on cast parts or parts that could be easily damaged. With any particularly stubborn part, always recheck to make sure all fasteners have been removed.

Avoid using a screwdriver or bar to pry components apart, as they can easily mar the gasket sealing surfaces of the parts (which must remain smooth). If prying is absolutely necessary, use a piece of wood, but keep in mind that extra clean-up will be necessary if the wood splinters.

After the parts are separated, the old gasket must be carefully scraped off and the gasket surfaces cleaned. Stubborn gasket material can be soaked with a gasket remover (available in aerosol cans) to soften it so it can be easily scraped off. Gasket scrapers are widely available – just be careful not to gouge the sealing surfaces if you use one. Some gaskets can be removed with a wire brush, but regardless of the method used, the mating surfaces must be left clean and smooth. If the gasket surface is scratched or gouged, then a gasket sealer thick enough to fill scratches will have to be used during reassembly of the components. For most applications, a non-drying (or semi-drying) gasket sealer is best.

Hose removal tips

Hose removal precautions closely parallel gasket removal precautions. Avoid scratching or gouging the hose fitting or the connection may leak. Because of various chemical reactions, the rubber in hoses can bond itself to the metal fitting it's installed on. To remove a hose, first loosen the clamp(s) securing it to the fitting. Next, using slip-joint pliers, grab the hose near the fitting and rotate it back-and-forth until it's completely free, then pull it off (silicone or other lubricants will ease removal if they can be applied between the hose and the fitting). Apply the same lubricant to the inside of the hose and the outside of the fitting before installation.

If a hose clamp is broken or damaged, don't reuse it. Also, don't reuse hoses that are cracked, hardened or chafed.

Tools

A selection of good tools is a basic requirement for anyone who plans to maintain and repair a motorcycle. For the owner who has few tools, if any, the initial investment might seem high, but when compared to the spiraling costs of routine maintenance and repair, it's a wise one.

To help the owner decide which tools are needed to perform the tasks detailed in this manual, the following tool lists are offered: *Maintenance and minor repair*, *Repair and overhaul* and *Special*. The newcomer to practical mechanics should start off with the *Maintenance and minor repair* tool kit, which is adequate for the simpler jobs. Then, as confidence and experience grow, the owner can tackle more difficult tasks, buying additional tools as they're needed. Eventually the basic kit will be built into the *Repair and overhaul* tool set. Over a period of time, the experienced do-it-yourselfer will assemble a tool set complete enough for most repair and overhaul procedures and will add tools from the *Special* category when it's felt the expense is justified by the frequency of use.

Maintenance and minor repair tool kit

The tools in this list (many of which are included in the tool kit supplied when the motorcycle is purchased new) should be considered the minimum required for performance of routine maintenance, servicing and minor repair work. We recommend the purchase of combination wrenches (box-end and open-end combined in one wrench); while more expensive than open-end ones, they offer the advantages of both types of wrench.

Combination wrench set
Adjustable wrench – 10-inch
Spark plug socket (with rubber insert)
Spark plug gap adjusting tool
Feeler gauge set
Standard screwdriver (5/16-inch x 6-inch)
Phillips screwdriver (no. 2 x 6-inch)
Combination (slip-joint) pliers – 6-inch
Hacksaw and assortment of blades
Tire pressure gauge
Control cable pressure luber
Grease gun
Oil can
Fine emery cloth
Wire brush
Hand impact screwdriver and bits
Funnel (medium size)
Safety goggles

A special wire-type gauge is needed for checking and adjusting spark plug gaps

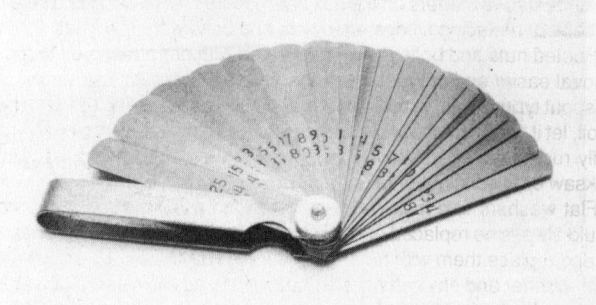

Feeler gauges are needed for measuring clearances

Maintenance techniques, tools and working facilities

Pressure lubricating adapters are used with aerosol cable lubricants

Many factory-tightened screws and bolts are hard to loosen without an impact driver (various size standard, Phillips and Allen-head bits are available)

Buy an oil filter wrench that fits over the end of the filter and accepts a 3/8-inch drive ratchet or breaker bar

Oil filter wrench
Drain pan
Spoke wrench
Battery hydrometer
Battery charger (1 amp)

Note: *Since basic ignition timing checks are a part of routine maintenance, you'll have to purchase a good quality, inductive pick-up timing light. Although it's included in the list of Special tools, it's mentioned here because ignition timing checks can't be made without it. You'll also need the special threaded clear plastic timing mark view plug (part no. HD-96295-65C) that's installed in the crankcase so the timing marks can be seen without oil spraying out of the view hole.*

Repair and overhaul tool set

These tools are essential for anyone who plans to perform major repairs and are intended to supplement those in the Maintenance and minor repair tool kit. Included is a comprehensive set of sockets which, though expensive, are invaluable because of their versatility (especially when various extensions and drives are available). We recommend 3/8-inch drive over 1/2-inch drive for general motorcycle maintenance and repair (ideally, a mechanic would have both).

Socket set(s)
Reversible ratchet
Extension – 6-inch
Universal joint
Torque wrench (same size drive as sockets)
Ball pein hammer – 12 oz
Soft-face hammer (plastic/rubber)
Standard screwdriver (1/4-inch x 6-inch)
Standard screwdriver (stubby – 5/16-inch)
Phillips screwdriver (no. 3 x 8-inch)
Phillips screwdriver (stubby – no. 2)
Pliers – vise-grip
Pliers – lineman's
Pliers – needle-nose
Pliers – snap-ring (internal and external)
Cold chisel – 1/2-inch
Scribe
Gasket scraper
Center punch
Pin punches (1/16, 1/8, 3/16-inch)
Steel rule/straightedge – 12-inch
Allen wrench set
Pin-type spanner wrench
A selection of files
A selection of brushes for cleaning small passages
Wire brush (large)
Work light with extension cord

Note: *Another tool which is often useful is an electric drill with a chuck capacity of 3/8-inch (and a set of good quality drill bits).*

Special tools

The tools in this list include those which aren't used regularly, are expensive to buy, or which need to be used in accordance with their manufacturer's instructions. Unless these tools will be used frequently, it's not very economical to purchase many of them. A consideration would be to split the cost and use between yourself and a friend or friends (such as members of a motorcycle club).

This list primarily contains tools and instruments widely available to the public, as well as some special tools produced by the vehicle manufacturer for distribution to dealer service departments. As a result, references

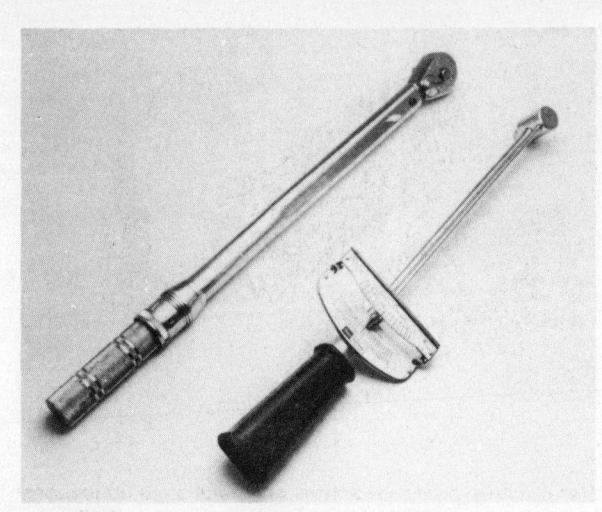

Maintenance techniques, tools and working facilities

Two types of torque wrenches are commonly available (left – click type; right – deflecting beam type

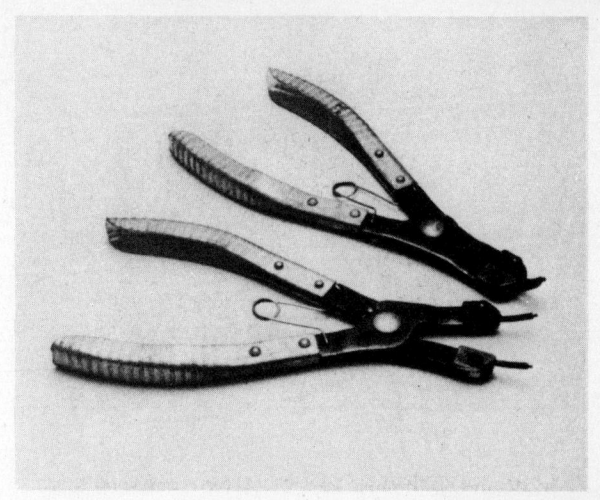

Snap-ring pliers are necessary for many disassembly and reassembly procedures

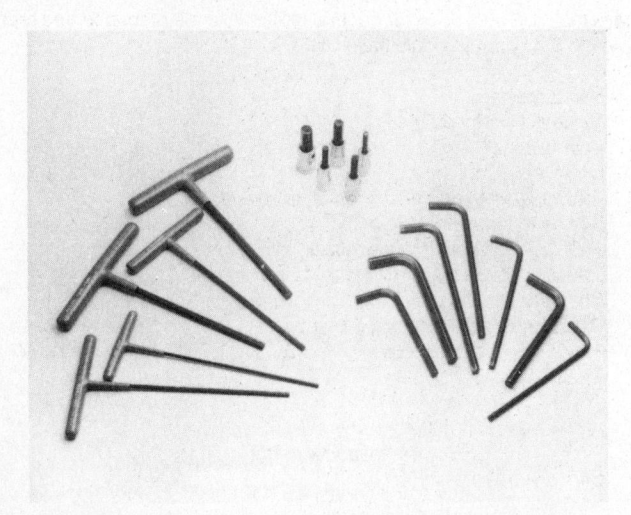

Hex (Allen) wrenches are available in a variety of styles

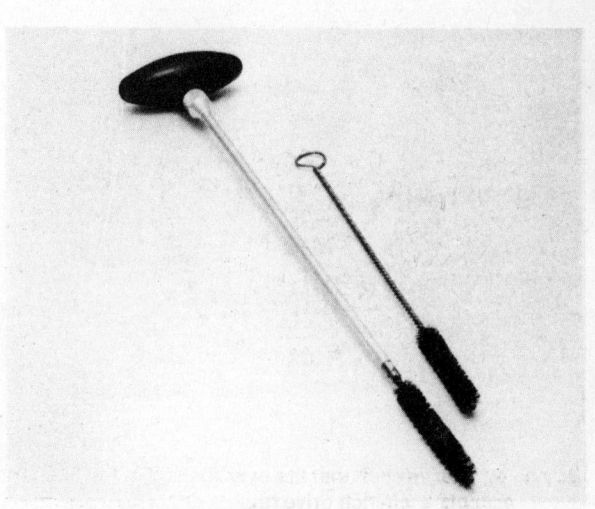

Small nylon or copper brushes can be used to clean internal passages in the engine and other components

to the manufacturer's special tools are occasionally included in the text of this manual. Generally, an alternative method of doing the job without the special tool is offered. However, sometimes there is no alternative to their use. Where this is the case, and the tool can't be purchased or borrowed, the work should be turned over to a dealer service department or motorcycle repair shop.

Valve spring compressor
Valve lapping tool
Piston ring removal and installation tool
Piston ring compressor
Telescoping gauges
Micrometer(s) and/or dial/Vernier calipers
Cylinder surfacing hone
Compression gauge
Dial indicator set
Multimeter
Tap and die set

Timing light and timing mark view plug
Small air compressor with blow gun and tire chuck

Buying tools

For the do-it-yourselfer just starting to get involved in motorcycle maintenance and repair, there are a number of options available when purchasing tools. If maintenance and minor repair is the extent of the work to be done, the purchase of individual tools is satisfactory. If, on the other hand, extensive work is planned, it would be a good idea to purchase a modest tool set from one of the large retail chain stores. A set can usually be bought at a substantial savings over the individual tool prices (and they often come with a tool box). As additional tools are needed, add-on sets, individual tools and a larger tool box can be purchased to expand the tool collection. Building a tool set gradually allows the cost of the tools to be spread over a longer period of time and gives the mechanic the freedom to choose only those tools that will actually be used.

Tool stores and motorcycle dealers will often be the only source of some of the special tools that are needed, but regardless of where tools

Maintenance techniques, tools and working facilities

A valve spring compressor is required if cylinder head work is needed

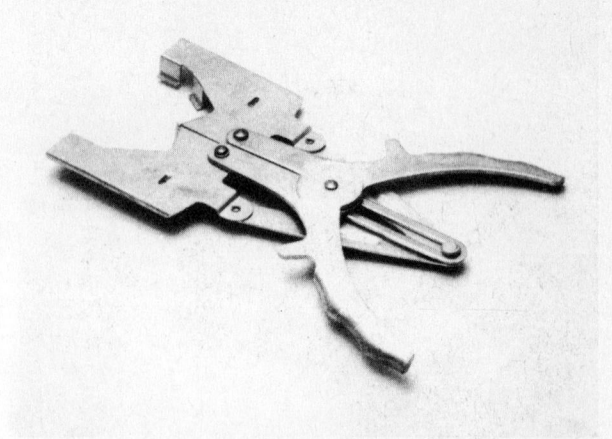

A piston ring removal and installation tool should be used to remove and install rings

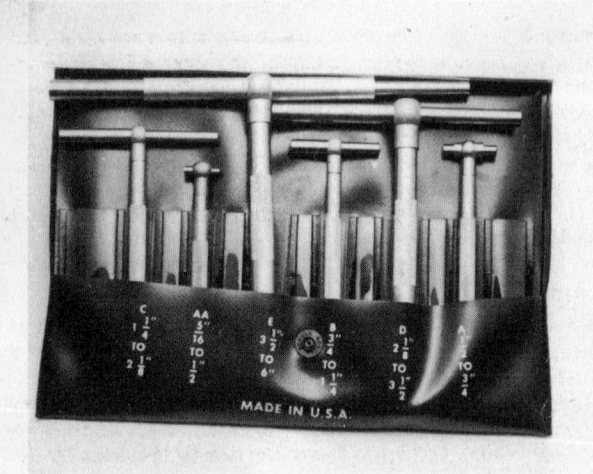

Telescoping gauges are used with micrometers to measure cylinder diameters and other internal dimensions

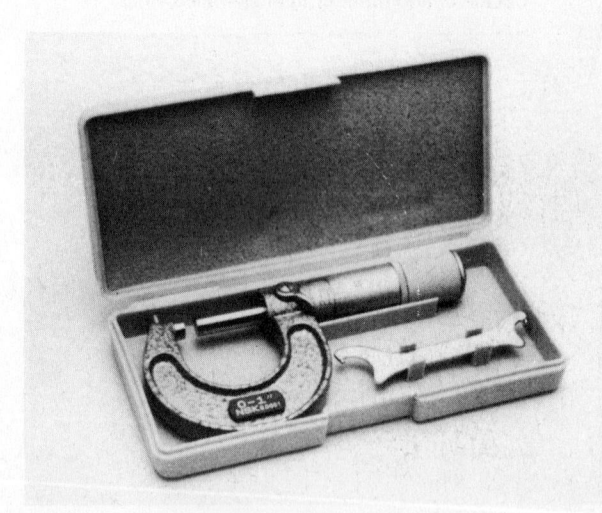

Micrometers are used to accurately measure shaft diameters and other outside dimensions

A hone is needed to resurface cylinders so new piston rings will seat properly

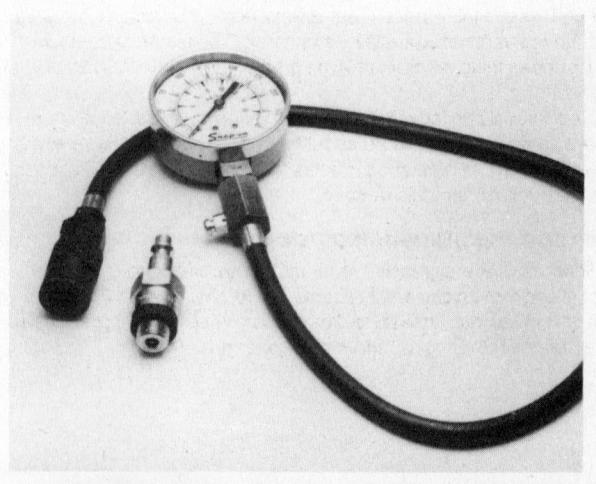

A compression gauge with a threaded fitting for the spark plug hole is preferred over the type that requires hand pressure to maintain the seal at the plug hole

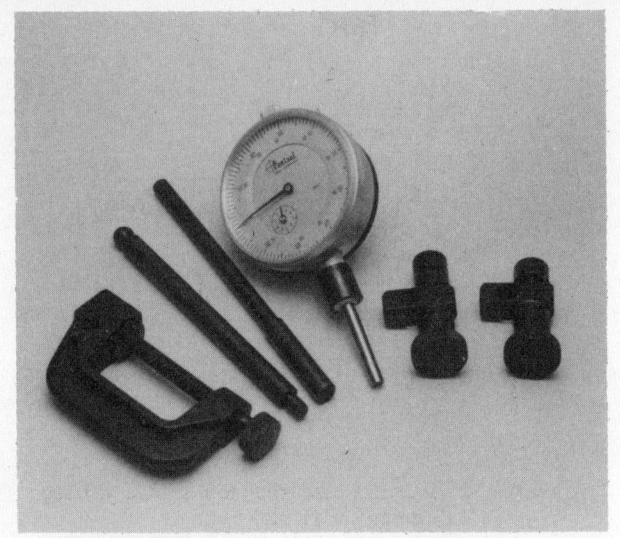

Maintenance techniques, tools and working facilities

A dial indicator is used to determine shaft runout, gear backlash and other critical specifications

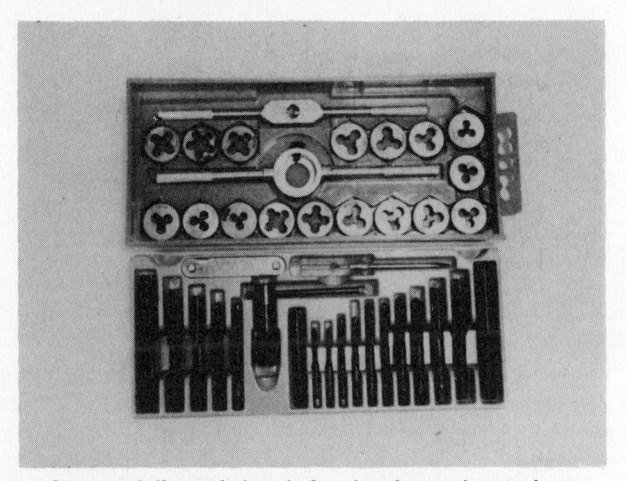

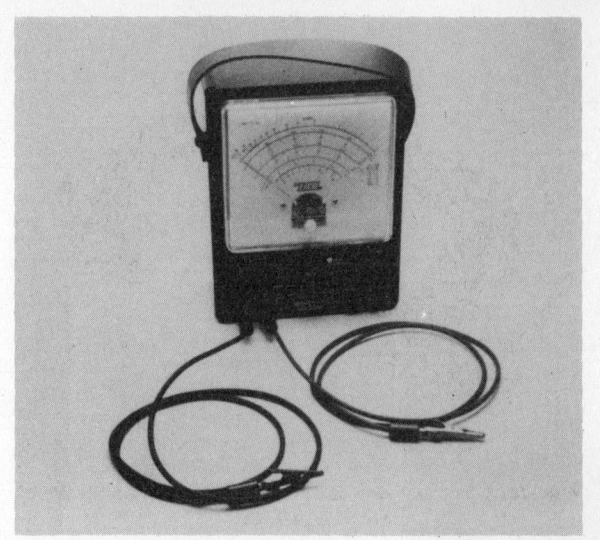

A multimeter (volt/ohm/ammeter) is needed if electrical system diagnosis is attempted

A tap and die set is handy for cleaning and restoring stripped or dirty threads

are bought, try to avoid cheap ones (especially when buying screwdrivers and sockets) because they won't last very long. There are plenty of tools around at reasonable prices, but always aim to purchase items which meet the relevant national safety standards. The expense involved in replacing cheap tools will eventually be greater than the initial cost of quality tools.

It is obviously not possible to cover the subject of tools fully here. For those who wish to learn more about tools and their use, there is a book entitled Motorcycle Workshop Practice Manual (Book no. 1454) available from the publishers of this manual.

Care and maintenance of tools

Good tools are expensive, so it makes sense to treat them with respect. Keep them clean and in usable condition and store them properly. Always wipe off dirt, grease and metal chips before putting them away. Never leave tools lying around in the work area.

Some tools, such as screwdrivers, pliers, wrenches and sockets, can be hung on a panel mounted on a garage or workshop wall, while others should be kept in a tool box or tray. Measuring instruments, gauges, meters, etc. must be carefully stored where they can't be damaged by weather or impact from other tools.

When tools are used with care and stored properly, they'll last a very long time. Even with the best of care, tools will wear out if used frequently. When a tool is damaged or worn out, replace it; subsequent jobs will be safer and more enjoyable if you do.

Working facilities

Not to be overlooked when discussing tools is the workshop. If anything more than routine maintenance is going to be done, some sort of special work area is essential.

It's understood, and appreciated, that many home mechanics don't have a good workshop or garage available and end up removing an engine or doing major repairs outside (the overhaul or repair should be completed under the cover of a roof).

A clean, flat workbench or table of comfortable working height is an absolute necessity. The workbench should be equipped with a vise that will open at least four inches.

As mentioned previously, some clean, dry storage space is also required for tools, as well as the lubricants, fluids, cleaning solvents, etc. which soon become necessary.

Sometimes waste oil and fluids, drained from the engine or transmission during normal maintenance or repairs, present a disposal problem. To avoid pouring them on the ground or into a sewage system, simply pour the used fluids into large containers, seal them with caps and take them to an authorized disposal site or service station. Plastic jugs (such as old antifreeze containers) are ideal for this purpose.

Always keep a supply of old newspapers and clean rags available. Old towels are excellent for mopping up spills. Many mechanics use rolls of paper towels for most work because they're readily available and disposable. To help keep the area under the motorcycle clean, a large cardboard box can be cut open and flattened to protect the garage or shop floor.

When working over a painted surface (such as the fuel tank) cover it with an old blanket or bedspread to protect the finish.

Motorcycle chemicals and lubricants

A number of chemicals and lubricants are available for use in motorcycle maintenance and repair. They include a wide variety of products ranging from cleaning solvents and degreasers to lubricants and protective sprays for rubber, plastic and vinyl.

Contact point/spark plug cleaner is a solvent used to clean oily film and dirt from points, grime from electrical connectors and oil deposits from spark plugs. It's oil free and leaves no residue. It can also be used to remove gum and varnish from carburetor jets and other orifices.

Carburetor cleaner is similar to contact point/spark plug cleaner but it usually has a stronger solvent and may leave a slight oily reside. It isn't recommended for cleaning electrical components or connections.

Brake system cleaner is used to remove grease or brake fluid from brake system components (where clean surfaces are absolutely necessary and petroleum-based solvents must not be used); it also leaves no residue.

Silicone-based lubricants are used to protect rubber parts such as hoses and grommets, and are used as lubricants for hinges and locks.

Multi-purpose grease is an all purpose lubricant used wherever grease is more practical than a liquid lubricant such as oil. Some multi-purpose grease is colored white and specially formulated to be more resistant to water than ordinary grease.

Gear oil (sometimes called gear lube) is a specially designed oil used in transmissions, as well as other areas where high friction, high-temperature lubrication is required. It's available in a number of viscosities (weights) for various applications.

Motor oil, of course, is the lubricant specially formulated for use in the engine. It normally contains a wide variety of additives to prevent corrosion and reduce foaming and wear. Motor oil comes in various weights (viscosity ratings) of from 5 to 80. The recommended weight of the oil depends on the seasonal temperature and the demands on the engine. Light oil is used in cold climates and under light load conditions; heavy oil is used in hot climates and where high loads are encountered. Multi-viscosity oils are designed to have characteristics of both light and heavy oils and are available in a number of weights from 5W-20 to 20W-50.

Gas additives perform several functions, depending on their chemical makeup. They usually contain solvents that help dissolve gum and varnish that build up on carburetor and intake parts. They also serve to break down carbon deposits that form on the inside surfaces of the combustion chambers. Some types contain upper cylinder lubricants for valves and piston rings.

Brake fluid is a specially formulated hydraulic fluid that can withstand the heat and pressure encountered in brake systems. Care must be taken not to get it on painted surfaces or plastics. An opened container should always be resealed to prevent contamination by water or dirt.

Chain lubricants are formulated especially for use on motorcycle final drive chains. A good chain lube should adhere well and have good penetrating qualities to be effective as a lubricant inside the chain and on the side plates, pins and rollers. Most chain lubes are either the foaming type or quick drying type and are usually marketed as sprays.

Degreasers are heavy duty solvents used to remove grease and grime that may accumulate on engine and frame components. They can be sprayed or brushed on and, depending on the type, are rinsed off with either water or solvent.

Solvents are used alone or in combination with degreasers to clean parts and assemblies during repair and overhaul. The home mechanic should use only solvents that are non-flammable and don't produce irritating fumes.

Gasket sealing compounds may be used in conjunction with gaskets, to improve their sealing capabilities, or alone, to seal metal-to-metal joints. Many gasket sealers can withstand extreme heat, some are impervious to gasoline and lubricants, while others are capable of filling and sealing large cavities. Depending on the intended use, gasket sealers either dry hard or stay relatively soft and pliable. They're usually applied by hand, with a brush, or are sprayed on the gasket sealing surfaces.

Thread adhesive is a liquid chemical locking compound that solidifies after application and prevents threaded fasteners from loosening because of vibration. It's available in a variety of types for different applications.

Moisture dispersants are usually sprays that can be used to dry out electrical components such as the fuse block and wiring connectors. Some types can also be used as treatment for rubber and as a lubricant for hinges, cables and locks.

Waxes and polishes are used to help protect painted and plated surfaces from the weather. Different types of paint may require the use of different types of wax polish. Some polishes utilize a chemical or abrasive cleaner to help remove the top layer of oxidized (dull) paint on older machines. In recent years, many non-wax polishes (that contain a wide variety of chemicals such as polymers and silicones) have been introduced. These non-wax polishes are usually easier to apply and last longer than conventional waxes and polishes.

Safety first!

Professional mechanics are trained in safe working procedures. Regardless of how eager you may be to start work, take the time to make sure your safety isn't jeopardized. A moment's lack of attention can result in an accident, as can failure to follow certain simple safety precautions. The possibility of an accident will always exist, and the following points aren't intended to be a comprehensive list of all dangers; they are intended, rather, to make you aware of the risks and to encourage a safety-conscious approach to all work you carry out on your machine.

Essential DOs and DON'Ts

DON'T start the engine without first checking to see if the transmission is in Neutral.

DON'T use gasoline for cleaning parts – ever!

DON'T attempt to drain the oil until you're sure its cooled to the point that it won't burn you.

DON'T touch any part of the engine or exhaust system until it has cooled down sufficiently to avoid burns.

DON'T siphon toxic liquids such as gasoline and brake fluid by mouth, or allow them to remain on your skin.

DON'T inhale brake lining dust – it's potentially hazardous.

DON'T allow spilled oil or grease to remain on the floor – wipe it up before someone slips on it.

DON'T use loose fitting wrenches or other tools which may slip and cause injury.

DON'T push on wrenches when loosening or tightening nuts or bolts. Always try to pull the wrench towards you. If the situation calls for pushing the wrench away, push with an open hand to avoid scraped knuckles if the wrench slips.

DON'T attempt to lift a heavy component which may be beyond your capability – get someone to help you.

DON'T rush or take unsafe shortcuts to finish a job.

DON'T allow children on or around the motorcycle when you're working on it.

DO wear eye protection when using power tools such as a drill, bench grinder, etc.

DO keep loose clothing and long hair well out of the way of moving parts.

DO make sure any hoist used has a safe working load rating adequate for the job.

DO make sure the machine is securely supported, especially when removing wheels.

DO get someone to check on you periodically when working alone.

DO carry out work in a logical sequence and make sure everything is correctly assembled and tightened.

DO keep chemicals and fluids tightly capped and out of the reach of children and pets.

DO remember that your motorcycle's safety affects both yourself and others. If in doubt on any point, get professional advice.

Asbestos

Brake linings, gaskets and clutch discs often contain asbestos, which is a health hazard. *Be extremely careful not to inhale dust from such products – if in doubt, assume that they do contain asbestos!*

Fire

Remember – gasoline is extremely flammable! Never smoke or have any kind of open flame or unshielded light bulbs around when working on your machine. But, the risk doesn't end there; a spark caused by an electrical short-circuit, by two metal surfaces contacting each other, or even static electricity built up in your body under certain conditions, can ignite gasoline vapors, which in a confined space are highly explosive. Do not, under any circumstances, use gasoline for cleaning parts; use an approved safety solvent only.

Always disconnect the battery ground cable before working on any part of the fuel system and never risk spilling fuel on a hot engine or exhaust pipe.

Keep a fire extinguisher suitable for use on fuel and electrical fires handy in the garage or workshop at all times. Never try to extinguish a fuel or electrical fire with water.

Fumes

Certain fumes are highly toxic and can quickly cause unconsciousness and even death if inhaled to any extent. Gasoline vapor falls into this category, as well as vapors from some cleaning solvents. Draining and pouring of such volatile fluids should be done in a well-ventilated area.

When using cleaning fluids and solvents, read the instructions on the container carefully. Never use materials from unmarked containers.

Never run the engine in an enclosed space such as a garage; exhaust fumes contain carbon monoxide which is extremely poisonous. If you need to run the engine, always do so in the open air or at least have the rear of the machine outside the work area.

The battery

Never create a spark or allow a bare light bulb near the battery vent hose. It'll normally be giving off a certain amount of hydrogen gas, which is highly explosive.

Always disconnect the battery ground cable before working on the fuel or electrical systems.

If possible, loosen the filler caps when charging the battery from an external source. Don't charge it at an excessive rate or the battery may burst.

Be careful when adding water and when carrying a battery. The electrolyte, even when diluted, is very corrosive and shouldn't be allowed to contact clothing or skin.

Always wear eye protection when cleaning the battery to prevent the caustic deposits from getting in your eyes.

Household current

When using an electric power tool, inspection light, etc., which operates on household current, always make sure the tool is correctly connected to its plug and that, where necessary, it's properly grounded. Don't use such items in damp conditions and, again, don't create a spark or apply excessive heat in the vicinity of fuel or fuel vapor.

Secondary ignition system voltage

A severe electric shock can result from touching certain parts of the ignition system (such as the spark plug wires) when the engine is running or being cranked, particularly if components are damp or the insulation is defective. Where an electronic ignition system is involved, the secondary system voltage is much higher and could prove fatal.

Troubleshooting

Contents

Symptom	Section
Engine doesn't start or is difficult to start	
Starter motor doesn't rotate	1
Starter motor rotates but engine doesn't turn over	2
Starter works but engine won't turn over (seized)	3
No fuel flow	4
Engine flooded	5
No spark or weak spark	6
Compression low	7
Stalls after starting	8
Rough idle	9
Runs poorly at low speed	
Spark weak	10
Fuel/air mixture incorrect	11
Compression low	12
Poor acceleration	13
Runs poorly or no power at high speed	
Firing incorrect	14
Fuel/air mixture incorrect	15
Compression low	16
Knocking or pinging	17
Miscellaneous causes	18
Overheating	
Firing incorrect	19
Fuel/air mixture incorrect	20
Compression too high	21
Engine load excessive	22
Lubrication inadequate	23
Miscellaneous causes	24
Clutch problems	
Clutch slipping	25
Clutch not disengaging completely	26
Gear shifting problems	
Doesn't go into gear or lever doesn't return	27
Jumps out of gear	28
Overshifts	29

Symptom	Section
Abnormal engine noise	
Knocking or pinging	30
Piston slap or rattling	31
Valve noise	32
Other noise	33
Abnormal driveline noise	
Clutch noise	34
Transmission noise	35
Abnormal frame and suspension noise	
Front end noise	36
Shock absorber noise	37
Disc brake noise	38
Oil pressure indicator light comes on	
Engine lubrication system	39
Electrical system	40
Excessive exhaust smoke	
White smoke	41
Black smoke	42
Brown smoke	43
Poor handling or stability	
Handlebar hard to turn	44
Handlebar shakes or vibrates excessively	45
Handlebar pulls to one side	46
Poor shock absorbing qualities	47
Braking problems	
Brakes are spongy, don't hold (hydraulic disc brakes only)	48
Brake lever or pedal pulsates	49
Brakes drag	50
Electrical problems	
Weak or dead battery	51
Battery overcharged	52
Alternator not charging	53
Generator not charging	54
Generator charging rate below normal	55

Engine doesn't start or is difficult to start

1 Starter motor doesn't rotate

1 Engine stop switch or ignition switch Off.
2 Battery voltage low or terminals loose or corroded. The battery must be in good condition and fully charged before troubleshooting the starter. Check and clean the terminals and/or recharge the battery (Chapter 7).
3 Starter solenoid or relay defective. Check the components and circuits (Chapter 7).
4 Defective starter motor. Make sure the wiring to the starter is secure, then check the starter motor (Chapter 7).
5 Starter button not contacting. The contacts could be wet, corroded or dirty. Disassemble and clean the switch (Chapter 7).
6 Wiring open or shorted. Check all wiring connections and wiring harnesses to make sure they're clean, dry and tight. Also check for broken or frayed wires that can cause a short to ground.
7 Ignition switch defective.
8 Engine stop switch defective. Check for wet, dirty or corroded contacts. Clean or replace the switch as necessary (Chapter 7).

Troubleshooting

2 Starter motor rotates but engine doesn't turn over

1 Starter motor clutch defective. Inspect and repair or replace (Chapter 2).
2 Damaged idler or starter gears. Inspect and replace the damaged parts (Chapter 2).
3 Splined teeth in bad condition. Starter motor should be removed and the splines cleaned.

3 Starter works but engine won't turn over (seized)

Seized engine caused by one or more internally damaged components. Failure due to wear, abuse or lack of lubrication. Damage can include seized valves, valve lifters, camshaft, pistons, crankshaft, connecting rod bearings, or transmission gears or bearings. Refer to Chapter 2 for engine disassembly.

4 No fuel flow

1 No fuel in tank.
2 Fuel valve turned off.
3 Tank cap air vent obstructed. Usually caused by dirt or water. Remove it and clean the cap vent hole.
4 Fuel valve clogged. Remove the valve and clean it and the filter (Chapter 1).
5 Fuel line clogged. Pull the fuel line loose and carefully blow through it.
6 Inlet needle valve clogged. For the valve to be clogged, either a very bad batch of fuel with an unusual additive has been used, or some other foreign object has entered the tank. Many times after a machine has been stored for many months without running, the fuel turns to a varnish-like liquid and forms deposits on the inlet needle valve and jets. The carburetor should be removed and overhauled if draining the float bowl doesn't alleviate the problem.

5 Engine flooded

1 Float level incorrect. Check and adjust as described in Chapter 3.
2 Inlet needle valve worn or stuck open. A piece of dirt, rust or other debris can cause the inlet needle to seat improperly, causing excess fuel to be admitted to the float bowl. In this case, the float bowl should be cleaned and the needle and seat inspected. If the needle and seat are worn, then the leak will persist and the parts should be replaced with new ones (Chapter 3).

6 No spark or weak spark

1 Ignition switch Off.
2 Engine stop switch turned to the Off position.
3 Battery voltage low. Check and recharge the battery as necessary (Chapter 7).
4 Spark plug dirty, defective or worn out. Locate reason for fouled plug(s) using spark plug condition photos and follow the plug maintenance procedures in Chapter 1.
5 Spark plug cap or plug wires defective. Check condition. Replace either or both components if cracks or deterioration are evident (Chapter 4).
6 Spark plug cap not making good contact. Make sure each plug cap fits snugly over the plug end.
7 Ignition module or sensor defective (Chapter 4).
8 Contact breaker points defective or improperly adjusted. Refer to Chapter 1.
9 Ignition coil defective. Check the coil as described in Chapter 4.
10 Ignition or stop switch shorted. This is usually caused by water, corrosion, damage or excessive wear. The switches can be disassembled and cleaned with electrical contact cleaner. If cleaning doesn't help, replace the switches (Chapter 7).
11 Wiring shorted or broken between:
   a) Ignition switch and engine stop switch
   b) Ignition module and engine stop switch
   c) Ignition module and ignition coil
   d) Ignition coil and plug
Make sure all wiring connections are clean, dry and tight. Look for chafed and broken wires (Chapters 4 and 7).
12 Faulty condenser (contact breaker point systems). Refer to Chapter 4.

7 Compression low

1 Spark plug loose. Remove the plug and inspect the threads. Reinstall and tighten to the specified torque (Chapter 1).
2 Cylinder head bolts loose. If the cylinder head is suspected of being loose, then there's a chance the gasket or head is damaged if the problem has persisted for any length of time. The head bolts should be tightened to the proper torque in the correct sequence (Chapter 2).
3 Incorrect valve clearance (Chapter 1 – pre-Evolution engine only). This means the valve isn't closing completely and compression pressure is leaking past it. Check the lifters (Evolution engine – Chapter 2).
4 Cylinder and/or piston worn. Excessive wear will cause compression pressure to leak past the rings. This is usually accompanied by worn rings as well. A top end overhaul is necessary (Chapter 2).
5 Piston rings worn, weak, broken, or sticking. Broken or sticking piston rings usually indicate a lubrication or carburetion problem that causes excess carbon deposits or seizures to form on the pistons and rings. Top end overhaul is necessary (Chapter 2).
6 Piston ring-to-groove clearance excessive. This is caused by excessive wear of the piston ring lands. Piston replacement is required (Chapter 2).
7 Cylinder head gasket damaged. If the head bolts loosen up, or if excessive carbon build-up on the piston crown and combustion chamber causes extremely high compression, the head gasket may leak. Retorquing the head is not always sufficient to restore the seal, so gasket replacement is necessary (Chapter 2).
8 Cylinder head warped. This is caused by overheating or improperly tightened head bolts. Machine shop resurfacing or head replacement is necessary (Chapter 2).
9 Valve spring broken or weak. Caused by component failure or wear; the spring(s) must be replaced (Chapter 2).
10 Valve not seating properly. This is caused by a bent valve (from over-revving or improper valve adjustment), burned valve or seat (improper carburetion) or an accumulation of carbon deposits on the seat (from carburetion, lubrication problems). The valves must be cleaned and/or replaced and the seats serviced if possible (Chapter 2).

8 Stalls after starting

1 Improper choke action. Make sure the choke rod is getting a full stroke and staying in the "out" position.
2 Ignition malfunction. See Chapter 4.
3 Carburetor malfunction. See Chapter 3.
4 Fuel contaminated. The fuel can be contaminated with either dirt or water, or can change chemically if the machine is allowed to sit for several months or more. Drain the tank and float bowl (Chapter 3).
5 Intake air leak. Check for loose carburetor-to-intake manifold connection or intake manifold-to-head connections (Chapter 4).
6 Idle speed incorrect. Turn idle speed screw until the engine idles at the specified rpm (Chapters 1 and 3).

9 Rough idle

1 Ignition malfunction. See Chapter 4.
2 Idle speed incorrect. See Chapter 1.
3 Carburetor malfunction. See Chapter 3.
4 Fuel contaminated. The fuel can be contaminated with either dirt or water, or can change chemically if the machine is allowed to sit for several months or more. Drain the tank and float bowl. If the problem is severe, a carburetor overhaul may be necessary (Chapters 1 and 3).
5 Intake air leak.
6 Air cleaner clogged. Service or replace air filter element (Chapter 1).

Runs poorly at low speed

10 Spark weak

1 Battery voltage low. Check and recharge battery (Chapter 7).
2 Spark plug fouled, defective or worn out. Refer to Chapter 1 for spark plug maintenance.
3 Spark plug cap or wires defective. Refer to Chapters 1 and 4 for details on the ignition system.
4 Spark plug cap not making contact.
5 Incorrect spark plug. Wrong type, heat range or cap configuration. Check and install correct plugs listed in Chapter 1. A cold plug or one with a recessed firing electrode won't function at low speeds without fouling.
6 Ignition module defective. See Chapter 4.
7 Contact breaker points defective or incorrectly gapped. See Chapter 1.
8 Ignition coil defective. See Chapter 4.
9 Condenser faulty (contact breaker point system). See Chapter 4.

11 Fuel/air mixture incorrect

1 Fuel/air mixture screw(s) out of adjustment (Chapter 3).
2 Jet or air passage clogged. Remove and overhaul the carburetor (Chapter 3).
3 Air bleed holes clogged. Remove carburetor and blow out all passages (Chapter 3).
4 Air cleaner clogged, poorly sealed or missing.
5 Air cleaner-to-carburetor joint poorly sealed. Look for cracks, holes and loose bolts or clamps and replace or repair defective parts.
6 Fuel level too high or too low. Adjust the float (Chapter 3).
7 Fuel tank air vent obstructed. Make sure the air vent passage in the filler cap is open.
8 Intake manifold(s) loose. Check for cracks, holes and loose clamps or bolts.

12 Compression low

1 Spark plug loose. Remove the plug and inspect the threads. Reinstall and tighten to the specified torque (Chapter 1).
2 Cylinder head not sufficiently tightened down. If the cylinder head bolts are loose, then there's a chance the gasket and head are damaged if the problem has persisted for any length of time. The head bolts should be tightened to the proper torque in the correct sequence (Chapter 2).
3 Improper valve clearance (Chapter 1 – pre-Evolution engine only). This means the valve isn't closing completely and compression pressure is leaking past the valve. Check the lifters (Evolution engine – Chapter 2).
4 Cylinder and/or piston worn. Excessive wear will cause compression pressure to leak past the rings. This is usually accompanied by worn rings as well. A top end overhaul is necessary (Chapter 2).
5 Piston rings worn, weak, broken, or sticking. Broken or sticking piston rings usually indicate a lubrication or carburetion problem that causes excess carbon deposits or seizures to form on the pistons and rings. Top end overhaul is necessary (Chapter 2).
6 Piston ring-to-groove clearance excessive. This is caused by excessive wear of the piston ring lands. Piston replacement is necessary (Chapter 2).
7 Cylinder head gasket damaged. If the head bolts loosen up, or if excessive carbon build-up on the piston crown and combustion chamber causes extremely high compression, the head gasket may leak. Retorquing the head isn't always sufficient to restore the seal, so gasket replacement is necessary (Chapter 2).
8 Cylinder head warped. This is caused by overheating or improperly tightened head bolts. Machine shop resurfacing or head replacement is necessary (Chapter 2).
9 Valve spring broken or weak. Caused by component failure or wear; the spring(s) must be replaced (Chapter 2).
10 Valve not seating properly. This is caused by a bent valve (from over-revving or improper valve adjustment), burned valve or seat (improper carburetion) or an accumulation of carbon deposits on the seat (from carburetion, lubrication problems). The valves must be cleaned and/or replaced and the seats serviced if possible (Chapter 2).

13 Poor acceleration

1 Timing not advancing. Check the mechanical advance mechanism for proper operation (Chapter 4). Check the VOES (Chapter 4).
2 Engine oil viscosity too high. Using a heavier oil than recommended in Chapter 1 can damage the oil pump or lubrication system and cause drag on the engine.
3 Brakes dragging. Usually caused by debris which has entered the brake piston sealing boot, or from a warped disc or bent axle. Repair as necessary (Chapter 6).

Runs poorly or no power at high speed

14 Firing incorrect

1 Timing not advancing.
2 Spark plug fouled, defective or worn out. See Chapter 1 for spark plug maintenance.
3 Spark plug cap or wire defective. See Chapters 1 and 4 for details on the ignition system.
4 Spark plug cap not making good contact. See Chapter 4.
5 Incorrect spark plug. Wrong type, heat range or cap configuration. Check and install correct plugs listed in Chapter 1. A cold plug or one with a recessed firing electrode won't function at low speeds without fouling.
6 Ignition module defective. See Chapter 4.
7 Ignition coil defective. See Chapter 4.

15 Fuel/air mixture incorrect

1 Main jet clogged. Dirt, water and other contaminants can clog the main jet. Clean the fuel valve filter, the float bowl area, and the jets and carburetor orifices (Chapter 3).
2 Main jet wrong size. The standard jetting is for sea level atmospheric pressure and oxygen content.
3 Throttle shaft-to-carburetor body clearance excessive. Refer to Chapter 3.
4 Air bleed holes clogged. Remove and overhaul carburetor (Chapter 3).
5 Air cleaner clogged, poorly sealed or missing.
6 Air cleaner-to-carburetor joint poorly sealed. Look for a torn gasket or a warped mating surface.
7 Fuel level too high or too low. Adjust the float (Chapter 3).
8 Fuel tank air vent obstructed. Make sure the air vent passage in the filler cap is open.

Troubleshooting

9   Carburetor intake manifold loose. Check for cracks, holes and loose clamps or bolts.
10 Fuel valve clogged. Remove the valve and clean it and the filter (Chapter 1).
11 Fuel line clogged. Pull the fuel line loose and carefully blow through it.

16 Compression low

1   Spark plug loose. Remove the plug and inspect the threads. Reinstall and tighten to the specified torque (Chapter 1).
2   Cylinder head not sufficiently tightened down. If the cylinder head bolts are loose, then there's a chance the gasket and head are damaged if the problem has persisted for any length of time. The head bolts should be tightened to the proper torque in the correct sequence (Chapter 2).
3   Improper valve clearance (Chapter 1 – pre-Evolution engine only). This means the valve isn't closing completely and compression pressure is leaking past the valve. Check the lifters (Evolution engine – Chapter 2).
4   Cylinder and/or piston worn. Excessive wear will cause compression pressure to leak past the rings. This is usually accompanied by worn rings as well. A top end overhaul is necessary (Chapter 2).
5   Piston rings worn, weak, broken, or sticking. Broken or sticking piston rings usually indicate a lubrication or carburetion problem that causes excess carbon deposits or seizures to form on the pistons and rings. Top end overhaul is necessary (Chapter 2).
6   Piston ring-to-groove clearance excessive. This is caused by excessive wear of the piston ring lands. Piston replacement is necessary (Chapter 2).
7   Cylinder head gasket damaged. If the head bolts loosen up, or if excessive carbon build-up on the piston crown and combustion chamber causes extremely high compression, the head gasket may leak. Retorquing the head isn't always sufficient to restore the seal, so gasket replacement is necessary (Chapter 2).
8   Cylinder head warped. This is caused by overheating or improperly tightened head bolts. Machine shop resurfacing or head replacement is necessary (Chapter 2).
9   Valve spring broken or weak. Caused by component failure or wear; the spring(s) must be replaced (Chapter 2).
10 Valve not seating properly. This is caused by a bent valve (from over-revving or improper valve adjustment), burned valve or seat (improper carburetion) or an accumulation of carbon deposits on the seat (from carburetion, lubrication problems). The valves must be cleaned and/or replaced and the seats serviced if possible (Chapter 2).

17 Knocking or pinging

1   Carbon build-up in combustion chamber. Use of a fuel additive that will dissolve the adhesive bonding the carbon particles to the piston crown and combustion chamber is the easiest way to remove the build-up. Otherwise, the cylinder head will have to be removed and decarbonized (Chapter 2).
2   Incorrect or poor quality fuel. Old or improper grades of gasoline can cause detonation. This causes a knocking or pinging sound. Drain old gas and always use the recommended fuel grade.
3   Spark plug heat range incorrect. Uncontrolled detonation indicates the plug heat range is too hot. The plug in effect becomes a glow plug, raising cylinder temperatures. Install the proper heat range plug (Chapter 1).
4   Improper fuel/air mixture. This will cause the engine to run hot, which leads to detonation. Clogged jets or an air leak can cause this imbalance. See Chapter 3.
5   Incorrect ignition timing (too far advanced) (Chapter 1).
6   Fuel octane rating too low.
7   Faulty vacuum operated electric switch (VOES) (Chapter 4).

18 Miscellaneous causes

1   Throttle valve doesn't open completely. Adjust the cable slack (Chapter 3).
2   Clutch slipping. Caused by a cable that's improperly adjusted, or snagging or by damaged, loose or worn clutch components. Refer to Chapters 1 and 2 for adjustment and overhaul procedures.
3   Ignition timing not advancing.
4   Engine oil viscosity too high. Using a heavier oil than recommended in Chapter 1 can damage the oil pump or lubrication system and cause drag on the engine.
5   Brakes dragging. Usually caused by debris which has entered the brake piston sealing boot, or from a warped disc or bent axle. Repair as necessary.

Overheating

19 Firing incorrect

1   Spark plug fouled, defective or worn out. See Chapter 1 for spark plug maintenance.
2   Incorrect spark plug.
3   Improper ignition timing. Timing that's retarded will cause high cylinder temperatures and lead to overheating (Chapter 1).

20 Fuel/air mixture incorrect

1   Main jet clogged. Dirt, water and other contaminants can clog the main jet. Clean the fuel valve filter, the float bowl area and the jets and carburetor orifices (Chapter 3).
2   Main jet wrong size. The standard jetting is for sea level atmospheric pressure and oxygen content.
3   Air cleaner poorly sealed or missing.
4   Air cleaner-to-carburetor joint poorly sealed. Look for a broken gasket or a warped mating surface.
5   Fuel level too low. Adjust the float (Chapter 3).
6   Fuel tank air vent obstructed. Make sure the air vent passage in the filler cap is open.
7   Carburetor intake manifold loose. Check for cracks, holes and loose clamps or bolts.

21 Compression too high

1   Carbon build-up in combustion chamber. Use of a fuel additive that will dissolve the adhesive bonding the carbon particles to the piston crown and combustion chamber is the easiest way to remove the build-up. Otherwise, the cylinder heads will have to be removed and decarbonized (Chapter 2).
2   Improperly machined head surface or installation of incorrect gasket during engine assembly. Check Specifications (Chapter 2).

22 Engine load excessive

1   Clutch slipping. Caused by an out of adjustment or snagging damaged, loose or worn clutch components. Refer to Chapter adjustment and overhaul procedures.
2   Engine oil viscosity too high. Using a heavier oil than recommended Chapter 1 can damage the oil pump or lubrication system and drag on the engine.
3   Brakes dragging. Usually caused by debris which brake piston sealing boot, or from a warped disc or bent necessary.

23 Lubrication inadequate

1   Engine oil level too low. Friction caused by intermittent lack of lubrication or from oil that is "overworked" can cause overheating. The oil provides a definite cooling function in the engine. Check the oil level (Chapter 1) and the oil pump (Chapter 2).
2   Poor quality engine oil or incorrect viscosity or type. Oil is rated not only according to viscosity but also according to type. Some oils aren't rated high enough for use in the engine. Check the Specifications Section and change to the correct oil (Chapter 1).

24 Miscellaneous causes

1   Engine fins packed with mud or dirt. The cooling fins can be blocked by a build-up of mud and cause a decrease in cooling. Clean the cylinder and head area.
2   Engine cooling fins on head and cylinder painted. Most paints actually seal in heat, so a cylinder/head paint should be one specifically designed for that use.
3   Valves not seating properly (being held off seat) or valve seats/faces in poor condition.
4   Carbon deposits.
5   Defective vacuum operated electric switch (VOES).
6   Modified exhaust system. Most aftermarket exhaust systems cause the engine to run leaner, which makes them run hotter. When installing an accessory exhaust system, always rejet the carburetor.

Clutch problems

25 Clutch slipping

1   No clutch lever play. Adjust clutch lever free play according to the procedure in Chapter 1.
2   Friction plates worn or warped. Overhaul the clutch assembly (Chapter 2).
3   Steel plates worn or warped (Chapter 2).
4   Clutch springs broken or weak. Old or heat-damaged (from slipping clutch) springs should be replaced with new ones (Chapter 2).
5   Clutch release not adjusted properly. See Chapter 1.
6   Clutch cable hanging up. Caused by a frayed cable or kinked housing. Replace the cable. Repair of a frayed cable isn't recommended.
7   Clutch release mechanism defective. Check the shaft, cam, actuating arm and pivot. Replace any defective parts (Chapter 2).
8   Clutch hub or drum unevenly worn. This causes improper engagement of the discs. Replace the damaged or worn parts (Chapter 2).

26 Clutch not disengaging completely

1   Clutch lever play excessive. adjust at lever or at engine (Chapter 1).
2   Clutch plates warped or damaged. This will cause clutch drag, which in turn causes the machine to creep. Overhaul the clutch (Chapter 2).
3   Clutch spring tension uneven. Usually caused by a sagged or broken spring. Check and replace the springs (Chapter 2).
4   Engine oil deteriorated. Old, thin, worn out oil will not provide proper lubrication for the discs, causing the clutch to drag. Replace the oil and filter (Chapter 1).
5   Engine oil viscosity too high. Using a heavier oil than recommended in Chapter 1 can cause the plates to stick together, putting a load on the engine. Change to the correct weight oil (Chapter 1).
6   Clutch drum seized on shaft. Lack of lubrication, severe wear or damage can cause the drum to seize on the shaft. Overhaul of the clutch, and perhaps transmission, may be necessary to repair damage (Chapter 2).
7   Clutch release mechanism defective. Worn or damaged release mechanism parts can stick and fail to apply force to the pressure plate. Overhaul the clutch cover components (Chapter 2).
8   Loose clutch hub nut. Causes drum and hub misalignment putting a load on the engine. Engagement adjustment continually varies. Overhaul the clutch (Chapter 2).

Gear shifting problems

27 Doesn't go into gear or lever doesn't return

1   Clutch not disengaging. See Section 26.
2   Shift fork(s) bent or seized. Often caused by dropping the machine or from lack of lubrication. Overhaul the transmission (Chapter 2).
3   Gear(s) stuck on shaft. Most often caused by a lack of lubrication or excessive wear in transmission bearings and bushings. Overhaul the transmission (Chapter 2).
4   Shift cam binding. Caused by lubrication failure or excessive wear. Replace the cam and bearings (Chapter 2).
5   Shift lever return spring weak or broken (Chapter 2).
6   Shift lever broken. Splines stripped out of lever or shaft, caused by allowing the lever to get loose or from dropping the machine. Replace parts as necessary (Chapter 2).
7   Shift mechanism pawl broken or worn. Full engagement and rotary movement of shift cam results. Replace shaft assembly (Chapter 2).
8   Pawl spring broken. Allows pawl to "float", causing sporadic shift operation. Replace spring (Chapter 2).

28 Jumps out of gear

1   Shift cam worn, damaged or adjusted incorrectly.
2   Shift fork(s) worn or out of adjustment. Overhaul the transmission (Chapter 2).
3   Gear groove(s) worn. Overhaul the transmission (Chapter 2).
4   Gear dogs or dog slots worn or damaged. The gears should be inspected and replaced. No attempt should be made to service the worn parts.
5   Damaged gears.

29 Overshifts

1   Pawl spring weak or broken (Chapter 2).
2   Shift cam follower not functioning (Chapter 2).

Abnormal engine noise

30 Knocking or pinging

1   Carbon build-up in combustion chamber. Use of a fuel additive that will dissolve the adhesive bonding the carbon particles to the piston crown and combustion chamber is the easiest way to remove the build-up. Otherwise, the cylinder head will have to be removed and decarbonized (Chapter 2).

Troubleshooting

2 Incorrect or poor quality fuel. Old or improper fuel can cause detonation. This causes the piston to rattle, thus the knocking or pinging sound. Drain the old gas and always use the recommended grade fuel (Chapter 3).
3 Spark plug heat range incorrect. Uncontrolled detonation indicates the plug heat range is too hot. The plug in effect becomes a glow plug, raising cylinder temperatures. Install the proper heat range plug (Chapter 1).
4 Improper fuel/air mixture. This will cause the engine to run hot and lead to detonation. Clogged jets or an air leak can cause this imbalance. See Chapter 3.
5 Incorrect ignition timing (Chapter 1).

31 Piston slap or rattling

1 Cylinder-to-piston clearance excessive. Caused by improper assembly. Inspect and overhaul top end parts (Chapter 2).
2 Connecting rod bent. Caused by over-revving, trying to start a badly flooded engine or from ingesting a foreign object into the combustion chamber. Replace the damaged parts (Chapter 2).
3 Piston pin or piston pin bore worn or seized from wear or lack of lubrication. Replace damaged parts (Chapter 2).
4 Piston ring(s) worn, broken or sticking. Overhaul the top end (Chapter 2).
5 Piston seizure damage. Usually from lack of lubrication or overheating. Replace the pistons and bore the cylinders, as necessary (Chapter 2).
6 Connecting rod big and/or small end clearance excessive. Caused by excessive wear or lack of lubrication. Replace worn parts.

32 Valve noise

1 Incorrect valve clearances (Chapter 1).
2 Valve spring broken or weak. Check and replace weak valve springs (Chapter 2).
3 Valve(s) sticking in guide(s) or rocker arm binding on shaft.
4 Hydraulic tappets malfunctioning (Evolution engine only).
5 Camshaft lobes or gears or valve train components worn or damaged. Lack of lubrication at high rpm is usually the cause of damage. Insufficient oil or failure to change the oil at the recommended intervals are the chief causes (Chapter 1).
6 Low oil pressure. Usually caused by oil pump not functioning properly.

33 Other noise

1 Cylinder head gasket leaking. This will usually produce a pop or "whisper" each time the affected cylinder reaches the compression stroke. It's usually accompanied by wetness around the leak. If the leak is into an oil return channel, then the crankcase will be pressurized and the oil will quickly be contaminated. Replace the head gasket and check the head for warpage (Chapter 2).
2 Exhaust pipe leaking at cylinder head connection. Caused by improper fit of pipe(s) or loose exhaust flange. Sometimes confused with a leaking head gasket. All exhaust fasteners should be tightened evenly and carefully. Failure to do this will lead to a leak (Chapter 3).
3 Crankshaft runout excessive. Caused by a bent crank (from overrevving) or damage from an upper cylinder component failure. Can also be attributed to dropping the machine on either of the crankshaft ends.
4 Engine mount bolts loose. Tighten all bolts to the specified torque (Chapter 2).
5 Crankshaft bearings worn (Chapter 2).
6 Primary chain tensioner out of adjustment or defective. Adjust or replace as necessary (Chapters 1 and 2).

Abnormal driveline noise

34 Clutch noise

1 Clutch drum/friction plate clearance excessive (Chapter 2).
2 Loose or damaged clutch pressure plate and/or bolts (Chapter 2).

35 Transmission noise

1 Bearings worn. Also includes the possibility that the shafts are worn. Overhaul the transmission (Chapter 2).
2 Gears worn or chipped (Chapter 2).
3 Metal chips jammed in gear teeth. Probably pieces from a broken clutch, gear or shift mechanism that were picked up by the gears. This will cause early bearing failure (Chapter 2).
4 Transmission oil level too low. Causes a howl from transmission. Also affects engine power and clutch operation (Chapter 1).

Abnormal frame and suspension noise

36 Front end noise

1 Low fork oil level or improper viscosity oil. This can result in a "spurting" sound and is usually accompanied by irregular fork action (Chapter 5).
2 Spring weak or broken. Makes a clicking or scraping sound. Fork oil, when drained, will have a lot of metal particles in it (Chapter 5).
3 Steering head bearings loose or damaged. Clicks when braking. Check and adjust or replace as necessary (Chapter 5).
4 Fork clamps loose. Make sure all fork clamp pinch bolts are tight (Chapter 5).
5 Fork tube bent. Good possibility if machine has been dropped. Replace tube with a new one (Chapter 5).
6 Front axle or axle clamp nuts loose. Tighten to the specified torque (Chapter 5).

37 Shock absorber noise

1 Fluid level incorrect. Indicates a leak caused by defective seal. Shock will be covered with oil. Replace shock (Chapter 6).
2 Defective shock absorber with internal damage. This is in the body of the shock and cannot be remedied. The shock must be replaced with a new one (Chapter 6).
3 Bent or damaged shock body. Replace the shock with a new one (Chapter 6).

38 Disc brake noise

1 Squeal caused by shim not installed or positioned correctly (Chapter 6).
2 Squeal caused by dust on brake pads. Usually found in combination with glazed pads. Clean with brake system solvent (Chapter 6).
3 Contamination of brake pads. Oil, brake fluid or dirt causing brake to chatter or squeal. Clean or replace pads (Chapter 6).
4 Pads glazed. Caused by excessive heat from prolonged use or from contamination. Do not use sandpaper, emery cloth or any other abrasive to roughen the pad surfaces as abrasives will stay in the pad material and damage the disc. A very fine flat file can be used, but pad replacement is suggested as a cure (Chapter 6).
5 Disc warped. Can cause a chattering, clicking or intermittent squeal. Usually accompanied by a pulsating lever and uneven braking. Resurface or replace the disc (Chapter 6).

Oil pressure indicator light comes on

39 Engine lubrication system

1 Engine oil pump defective (Chapter 2).
2 Engine oil level low. Inspect for leak or other problem causing low oil level and add recommended lubricant (Chapters 1 and 2).
3 Engine oil viscosity too low. Very old, thin oil or an improper weight of oil used in engine. Change to correct lubricant (Chapter 1).
4 Crankshaft and/or bearings worn. Check and replace crankshaft and bearings (Chapter 2).
5 Relief valve stuck open. Repair or replace the valve (Chapter 3).

40 Electrical system

1 Oil pressure switch defective. Replace it if it's defective.
2 Oil pressure indicator light system wiring defective. Check for pinched, shorted, disconnected or damaged wires (Chapter 7).

Excessive exhaust smoke

41 White smoke

1 Piston oil ring worn. The ring may be broken or damaged, causing oil from the crankcase to be pulled past the piston into the combustion chamber. Replace the rings with new ones (Chapter 2).
2 Cylinders worn, cracked, or scored. Caused by overheating or oil starvation. The cylinders will have to be rebored and new pistons installed.
3 Valve stem seal damaged or worn. Replace seals with new ones (Chapter 2).
4 Valve guide(s) worn. Perform a complete valve job (Chapter 2).
5 Abnormal crankcase pressurization, which forces oil past the rings. Clogged breather usually the cause (Chapter 1).
6 Engine oil level too high (Chapter 1).

42 Black smoke

1 Air cleaner clogged. Clean or replace the element (Chapter 1).
2 Main jet too large or loose. Compare the jet size to the Specifications (Chapter 3).
3 Fuel level too high. Check and adjust the float level as necessary (Chapter 3).
4 Inlet needle held off seat. Clean float bowl and fuel line and replace needle and seat if necessary (Chapter 3).

43 Brown smoke

1 Main jet too small or clogged. Lean condition caused by wrong size main jet or by a restricted orifice. Clean float bowl and jets (Chapter 3).
2 Fuel flow insufficient. Fuel inlet needle valve stuck closed due to chemical reaction with old gas. Float level incorrect. Restricted fuel line. Clean line and float bowl and adjust float as necessary (Chapter 3).
3 Intake manifold loose (Chapter 3).
4 Air cleaner poorly sealed or not installed (Chapter 1).

Poor handling or stability

44 Handlebar hard to turn

1 Steering stem locknut too tight (Chapter 5).

2 Bearings damaged. Roughness can be felt as the bars are turned from side-to-side. Replace bearings and races (Chapter 5).
3 Races dented or worn. Denting results from wear in only one position (i.e., straight ahead) from impacting an immovable object or hole or from dropping the machine. Replace races and bearings (Chapter 5).
4 Steering stem lubrication inadequate. Causes are grease getting hard from age or being washed out by high pressure car washes. Disassemble steering head and repack bearings (Chapter 5).
5 Steering stem bent. Caused by hitting a curb or hole or from dropping the machine. Replace damaged part. Don't try to straighten stem (Chapter 5).
6 Front tire air pressure too low (Chapter 1).

45 Handlebar shakes or vibrates excessively

1 Tires worn or out-of-balance (Chapter 6).
2 Swingarm bearings worn. Replace worn bearings by referring to Chapter 5.
3 Rim(s) warped or damaged. Inspect wheels for runout (Chapter 6).
4 Wheel bearings worn. Worn front or rear wheel bearings can cause poor tracking. Worn front bearings will cause wobble (Chapter 6).
5 Handlebar clamp bolts loose (Chapter 5).
6 Steering stem or fork clamps loose. Tighten them to the specified torque (Chapter 5).
7 Engine mount bolts loose. Will cause excessive vibration with increased engine rpm (Chapter 2).

46 Handlebar pulls to one side

1 Frame bent. Definitely suspect this if the machine has been dropped. May or may not be accompanied by cracks near the bend. Replace the frame (Chapter 5).
2 Wheel out of alignment. Caused by improper location of axle spacers or from bent steering stem or frame (Chapter 5).
3 Swingarm bent or twisted. Caused by age (metal fatigue) or impact damage. Replace the arm (Chapter 5).
4 Steering stem bent. Caused by impact damage or from dropping the motorcycle. Replace the steering stem (Chapter 5).
5 Fork leg bent. Disassemble the forks and replace the damaged parts (Chapter 5).
6 Fork oil level uneven.

47 Poor shock absorbing qualities

1 Too hard:
 a) Excessive fork oil (Chapter 5).
 b) Fork oil viscosity too high. Use a lighter oil.
 c) Fork tube bent. Causes a harsh, sticking feeling (Chapter 5).
 d) Shock shaft or body bent or damaged (Chapter 5).
 e) Fork internal damage (Chapter 5).
 f) Shock internal damage.
 g) Tire pressure too high (Chapter 1).
2 Too soft:
 a) Fork or shock oil insufficient and/or leaking (Chapter 5).
 b) Fork oil level too low (Chapter 5).
 c) Fork oil viscosity too light (Chapter 5).
 d) Fork springs weak or broken (Chapter 5).

Braking problems

48 Brakes are spongy, don't hold (hydraulic disc brakes only)

1 Air in brake line. Caused by inattention to master cylinder fluid level or by leakage. Repair problem and bleed brakes (Chapter 6).

Troubleshooting

2 Pads or disc worn (Chapters 1 and 6).
3 Brake fluid leak. See Paragraph 1.
4 Pads contaminated with oil, grease, brake fluid, etc. Clean or replace pads. Clean disc thoroughly with brake cleaner (Chapter 6).
5 Old, deteriorated or contaminated brake fluid. Drain system, replenish with new fluid and bleed the system (Chapter 6).
6 Master cylinder internal parts worn or damaged causing fluid to by-pass (Chapter 6).
7 Master cylinder bore scratched. From ingestion of foreign material or broken spring. Repair or replace master cylinder (Chapter 6).
8 Disc warped. Replace disc (Chapter 6).

49 Brake lever or pedal pulsates

1 Disc warped. Replace disc (Chapter 6).
2 Axle bent. Replace axle (Chapter 6).
3 Brake caliper bolts loose (Chapter 6).
4 Brake caliper pins damaged or sticking, causing caliper to bind. Lube the pins and/or replace them if they're corroded or bent (Chapter 6).
5 Wheel warped or otherwise damaged (Chapter 6).
6 Wheel bearings damaged or worn (Chapter 6).

50 Brakes drag

1 Master cylinder piston seized. Caused by wear or damage to piston or cylinder bore (Chapter 6).
2 Lever slow or stuck. Check pivot and lubricate as needed (Chapter 6).
3 Brake caliper binds. Caused by inadequate lubrication or damage to caliper pins (Chapter 6).
4 Brake caliper piston seized in bore. Caused by wear or ingestion of dirt past deteriorated seal (Chapter 6).
5 Brake pad damaged. Pad material separating from backing plate. Usually caused by faulty manufacturing process or from contact with chemicals. Replace pads (Chapter 6).
6 Pads improperly installed (Chapter 6).
7 Rear brake pedal free play insufficient.
8 Front brake lever free play insufficient.

Electrical problems

51 Weak or dead battery

1 Battery faulty. Caused by sulphated plates which are shorted through the sedimentation or low electrolyte level. Also, broken battery terminal making only occasional contact (Chapter 7).
2 Battery cables making poor contact (Chapter 7).
3 Load excessive. Caused by addition of high wattage lights or other electrical accessories.
4 Ignition switch defective. Switch either grounds internally or fails to shut off system. Replace the switch (Chapter 7).

5 Regulator/rectifier defective (Chapter 7).
6 Alternator stator coil open or shorted (Chapter 7).
7 Wiring faulty. Wiring grounded or connections loose in ignition, charging or lighting circuits (Chapter 7).

52 Battery overcharged

1 Regulator/rectifier defective. Overcharging is noticed when battery gets excessively warm or "boils" over (Chapter 7).
2 Battery defective. Replace battery with a new one (Chapter 7).
3 Battery amperage too low, wrong type or size. Install manufacturer's specified amp-hour battery to handle charging load (Chapter 7).

53 Alternator not charging

1 Faulty regulator-rectifier module (Chapter 7).
2 Faulty stator or rotor.
3 Weak or damaged battery.
4 Loose connections in wire harness.

54 Generator not charging

1 Defective armature. Have the commutator turned at an automotive electrical shop or replace the armature (Chapter 7).
2 Brushes worn excessively. Replace them with new ones as described in Chapter 7.
3 Commutator shorted out, dirty or oily. Remove the armature from the generator and clean it. If it's shorted, it must be replaced with a new one (Chapter 7).
4 Voltage regulator defective or not grounded properly. Test the regulator and replace it if necessary as described in Chapter 7.
5 Positive brush holder grounded or brushes sticking in the holder. Remove the brushes and clean the holder. Check the springs behind the brushes for cracks and distortion and replace the springs if needed. Repair any grounded wires. If they can't be repaired, replace the generator as described in Chapter 7.
6 Loose or broken wire in the battery/generator circuit.
7 Loose terminal or broken field coil wire (both sides) (Chapter 7).

55 Generator charging rate below normal

1 Armature worn or defective. Install a new armature or have the original reconditioned at an automotive electrical shop.
2 Loose terminal or broken field coil wire (one side) (Chapter 7).
3 Brushes dirty or sticking in holder. Remove the brushes as described in Chapter 7 and clean them. Inspect them for wear and damage and install new ones if necessary (Chapter 7).
4 Commutator oily or dirty. Remove the armature as described in Chapter 7 and clean it thoroughly.

Chapter 1 Tune-up and routine maintenance

Contents

Routine maintenance intervals	1
Introduction	2
Fluid levels – check	3
Air cleaner – servicing	4
Drive chain – check, adjustment and lubrication	5
Brake shoes/pads – wear check	6
Clutch – adjustment	7
Fuel system – check	8
Brake system – general check	9
Lubrication – general	10
Clutch and front brake cable – lubrication	11
Fasteners – check	12
Tires/wheels/spokes – general check	13
Valve clearance – adjustment (pre-Evolution engine only)	14
Idle speed – adjustment	15
Automatic drive chain oiler – maintenance and adjustment	16
Engine oil and filter – change	17
Spark plugs – check and replacement	18
Contact breaker points (1970 through 1978) – check and replacement	19
Ignition timing – check and adjustment	20
Primary chain – adjustment	21
Fuel filter – cleaning and replacement	22
Transmission oil – change	23
Throttle – check and lubrication	24
Suspension – inspection	25
Steering head bearings – check	26
Fork oil (except 1995-on XL 1200S) – change	27
Wheel bearings – repack	28
Cylinder compression – check	29
Headlight aim – adjustment	30
Sidestand – check and maintenance	31
Brake hydraulic systems – bleeding and fluid replacement	32
Drive belt – check and adjustment	33

Specifications

Engine

Recommended spark plugs
  1970 through 1979 .... Harley-Davidson no. 4 or 4R (resistor type)
  1980 and 1981 .... Harley-Davidson no. 4-5
  1980 through 1985 .... Harley-Davidson no. 4R5 (resistor type)
  1986-on .... Harley-Davidson no. 6R12 (no substitutes)

Spark plug gap
  No. 4 and 4R
    1970 through 1978 .... 0.025 to 0.030 in (0.635 to 0.762 mm)
    1979 .... 0.060 in (1.524 mm)
  No. 4-5, 4R5 and 6R12 .... 0.038 to 0.043 in (0.965 to 1.092 mm)

Contact breaker point gap .... 0.020 in (0.51 mm)
Valve clearance (pre-Evolution engine only) .... No lash (pushrods just free to rotate)

Cylinder compression pressure
  Minimum .... 120 psi (8.28 Bars)
  Maximum variation between cylinders .... 10 psi (0.69 Bars)

Idle speed
  1970 through 1978 .... 900 to 1100 rpm
  1979 through 1987 .... 900 to 950 rpm
  1988 through 1990 .... 1000 to 1050 rpm
  1991-on .... 950 to 1050 rpm

Chapter 1 Tune-up and routine maintenance

Engine (continued)
Fast idle speed
   1979 through 1985 .... 1500 rpm
   1986 and 1987 .... 1500 to 1550 rpm

Frame, drivetrain and suspension
Drive chain free play
   1970 through 1985 .... 1/2 in (12.70 mm)
   1986-on .... 1/4 in (6.350 mm)
Drive belt free play .... 9/16 to 11/16 inch (14 to 17 mm)
Primary chain free play
   1970 through early 1984
      Cold engine .... 5/8 to 7/8 in (15.88 to 22.23 mm)
      Hot engine .... 3/8 to 5/8 in (9.53 to 15.88 mm)
   Late 1984-on
      Cold engine .... 3/8 to 1/2 in (9.53 to 12.7 mm)
      Hot engine .... 1/4 to 3/8 in (6.35 to 9.53 mm)
Minimum brake lining thickness
   Drum brake shoes .... 0.080 in (2 mm)
   Disc brake pads .... 1/16 in (1.5 mm)
Tire pressures (COLD)
   1970 through 1978
      Front .... 24 psi (1.66 Bars)
      Rear .... 30 psi (2.07 Bars)
   1979 through 1985
      Front .... 26 psi (1.79 Bars)
      Rear .... 30 to 32 psi (2.07 to 2.21 Bars)
   1986-on
      883 cc engine through 1990
         Front .... 26 psi (1.79 Bars)
         Rear .... 30 to 32 psi (2.07 to 2.21 Bars)
      1200 cc engine and 1991-on 883 cc engine
         Front .... 30 psi (2.07 Bars)
         Rear .... 36 to 40 psi (2.48 to 2.76 Bars)
Tire tread depth (minimum)
   Front .... 1/16 in (1.5 mm)
   Rear .... 3/32 in (2.0 mm)

Torque specifications

	Ft-lbs (unless otherwise indicated)	Nm
Tappet adjusting locknut (pre-Evolution engine only)	72 to 132 in-lbs	8 to 15
Primary chaincase drain plug	14 to 21	19 to 28
Outer clutch cover Torx bolts (later models)	84 to 108 in-lbs	9 to 12
Spark plugs		
Pre-Evolution engine	20	27
Evolution engine	11 to 18	15 to 24
Axle nut		
Front		
1970 through 1987	50	68
1988-on	50 to 55	68 to 75
Rear	60 to 65	81 to 88
Axle pinch bolt		
1970 through 1972	132 in-lbs	15
1988-on	21 to 27	28 to 37
Handlebar clamp nuts/bolts		
1970 through 1983	20	27
1984-on	12 to 15	16 to 20
Brake caliper mounting bolt		
Front		
1973 through 1978	35	47
1979 through 1983	80 to 90 in-lbs	9 to 10
1984-on	25 to 30	34 to 41
Rear		
1979 through 1985	155 to 190 in-lbs	18 to 21
1986 through early 1987	13 to 16	18 to 22
Late 1987-on	15 to 20	20 to 27
Carburetor-to-intake manifold		
1970 through 1985 (nuts)	19	26
1986 and 1987 (bolts)	15 to 17	20 to 23
1988-on (hose clamp)	10 to 15 in-lbs	1.2 to 1.7

Recommended lubricants and fluids

Engine oil
Type .... Harley-Davidson or equivalent
Viscosity
  Pre-Evolution engine
    Normal (20 to 90-degrees F) .... SAE 20W50
    Below 40-degrees F .... SAE 30
    Above 40-degrees F .... SAE 40
    Severe operating conditions .... SAE 60
  Evolution engine
    Below 40-degrees F .... SAE 10W40
    Above 40-degrees F .... SAE 20W50
    Above 60-degrees F .... SAE 50
    Above 80-degrees F .... SAE 60
Capacity .... 3 US quarts (2.8 liters)

Transmission oil
Type
  1970 through early 1984 .... Harley-Davidson Power Blend Super Premium or equivalent (20W50 above 40-degrees F/60W below 40-degrees F)
  Late 1984 and 1985 .... Harley-Davidson Front Chaincase Lubricant (part no. 99887-84)
  1986-on .... Harley-Davidson Sport Trans Fluid
Capacity
  Models through 1990 .... 24 US oz (25 Imp oz, 710 cc)
  1991-on models .... 40 US oz (42 Imp oz, 1183 cc)
  1998-on .... 32 US oz (40 Imp oz, 946 cc)

Front forks
Type
  1970 through 1978 .... Harley-Davidson Type B
  1979-on .... Harley-Davidson Type E

Capacity	Drain and fill	After overhaul
1970 through 1983	5 US oz (5.2 Imp oz, 148 cc)	6 US oz (6.2 Imp oz, 177 cc)
1984 through 1987	5.4 US oz (5.6 Imp oz, 160 cc)	6.4 US oz (6.7 Imp oz, 189 cc)
1988-on		
All except 1992-on 883 Hugger and 1995-on XL 1200S	9 US oz (9.4 Imp oz, 266 cc)	10.2 US oz, (10.6 Imp oz, 302 cc)
1992-on 883 Hugger and 1995-on XL 1200S	10.7 US oz (11.1 Imp oz, 316 cc)	12.1 US oz (12.6 Imp oz, 358 cc)

Brake fluid .... DOT 3 (early models), Dot 5 (later models) **Warning:** *Do not mix fluid types. If not known, fully drain the fluid and have the system flushed before refilling. Refer to the manufacturer's fluid recommendation stamped on the reservoir cap or cover.*

Final drive chain
  Conventional chain .... Harley-Davidson Chain Spray (pt no. 99870-58) or High Performance Chain Lube Plus (pt no. 99865-81)
  O-ring chain .... Harley-Davidson High Performance Chain Lube Plus (pt no. 99865-81) or API GL-5 gear lube SAE 80 or 90

Cables
  Clutch/throttle/brake/choke .... Light machine oil or cable lube
  Speedometer/tachometer .... Lightweight grease
  General lubrication .... Light machine oil

Harley-Davidson Sportster Routine maintenance intervals

Note: *The pre-ride inspection outlined in your owner's manual covers checks and maintenance that should be carried out on a daily or pre-ride basis. It's condensed and included here to remind you of its importance. Always perform the pre-ride inspection at every maintenance interval (in addition to the procedures listed). The intervals listed below are the shortest intervals recommended by the manufacturer for each particular operation during the model years covered in this manual. Your owners manual may have different intervals for your model.*

Daily or before riding

Check the engine oil level in the oil tank
Check the fuel level in the tank and look for fuel leaks
Check the operation of both brakes – look for fluid leakage (hydraulic disc brakes only) and adjust brake lever free play if necessary
Check the tires for damage, the presence of foreign objects and correct air pressure
Check the throttle for smooth operation and correct free play
Check for proper operation of the headlight, taillight, brake light, turn signals, indicator lights, speedometer and horn

After the initial 500 miles (800 km)

Change the engine oil and oil filter
Check and adjust the valve clearances (pre-Evolution engine only)
Check the idle speed
Check the throttle and choke adjustments
Change the transmission oil and clean the magnetic drain plug
Check the brake fluid level (models equipped with disc brakes)
Grease the speedometer drive gear
Check the electrolyte level in the battery and clean the battery terminals
Inspect the oil lines and the brake hydraulic system for leaks
Check the brake adjustments
Test the operation of the electrical components and switches
Check the clutch adjustment
Inspect the condition and operation of the forks and shocks
Check the tightness and condition of all visible fasteners
Inspect the tires for wear and proper inflation
Check the tightness of the spokes (if applicable)
Inspect the brake disc(s) and brake pads for wear
Clean the fuel filter screen at the fuel tank valve
Change the fork oil
Inspect the fuel lines and fittings for leaks
Adjust the drive belt tension
Lubricate and adjust the drive chain
Lubricate the clutch, choke and throttle cables
Lubricate the clutch and front brake hand lever pivots

Every 300 miles (500 km)

Lubricate the drive chain

Every 1000 miles (1600 km)

Service the air cleaner
Adjust the drive chain
Adjust the brakes
Check the hydraulic brake fluid level
Inspect the brake disc(s) and pads
Check the clutch adjustment
Check the fuel lines and connections for leaks
Inspect the oil and hydraulic brake lines for leaks
Lubricate the clutch and front brake hand lever pivots
Lubricate the clutch, throttle and choke cables
Grease the speedometer drive gear
Check the tightness and condition of all visible fasteners
Inspect the tires for wear and proper inflation
Check the tightness of the spokes, if applicable

Every 1250 miles (2000 km)

Check the valve clearances (pre-Evolution engine only)
Check the idle speed
Check the throttle and choke adjustments
Test the operation of the electrical components and switches

Every 2000 miles (3200 km)

Clean and adjust the chain oiler (pre-1976 models only)
Change the engine oil and filter
Check the condition of the spark plugs
Check the condition of the contact breaker points (1970 through 1976 models only)
Check the ignition timing

Every 2500 miles (4000 km)

Adjust the primary chain
Replace the spark plugs with new ones
Clean the fuel filter screen on the fuel control valve
Check battery electrolyte level and clean terminals

Every 5000 miles (8000 km)

Change the transmission oil and clean the magnetic drain plug
Grease the internal spiral of the throttle sleeve and the speedometer and tachometer cables
Inspect the shock absorbers and bushings
Check the steering head and swingarm bearing free play
Inspect the generator brushes for wear
Change the fork oil
Check drive belt tension and condition

Every 10 000 miles (16 000 km)

Repack the wheel bearings
Repack the swingarm pivot bearings
Grease the steering head bearings (late 1992-on models)

Chapter 1 Tune-up and routine maintenance

2 Introduction

This Chapter covers in detail the checks and procedures necessary for the tune-up and routine maintenance of your motorcycle. Section 1 includes the routine maintenance schedule, which is designed to keep the machine in proper running condition and prevent possible problems. The remaining Sections contain detailed procedures for carrying out the items listed on the maintenance schedule, as well as additional maintenance information designed to increase reliability.

Since routine maintenance plays such an important part in keeping the motorcycle in safe condition and operating at its optimum, this Chapter should be used as a comprehensive check list. For the rider who does all his own maintenance, these lists outline the procedures and checks that should be done on a routine basis.

Deciding where to start or plug into the routine maintenance schedule depends on several factors. If you have a motorcycle whose warranty has recently expired, and if its been maintained according to the warranty standards, you may want to pick up routine maintenance as it coincides with the next mileage or calendar interval. If you have owned the machine for some time but have never performed any maintenance on it, then you may want to start at the nearest interval and include some additional procedures to ensure that nothing important is overlooked. If a major engine overhaul has just been done, then you may want to start the maintenance routine from the beginning. If you have a used machine and have no knowledge of its history or maintenance record, it may be a good idea to combine all the checks into one large service initially and then settle into the prescribed maintenance schedule.

Note that the procedures normally associated with ignition and carburetion tune-ups are included in the routine maintenance schedule. A regular tune-up will ensure good engine performance and help prevent engine damage due to improper carburetion and ignition timing.

The Sections detailing the maintenance and inspection procedures are written as step-by-step comprehensive guides to the actual performance of the work. References to additional information in applicable Chapters is also included and shouldn't be overlooked.

The first step of this or any maintenance plan is to prepare yourself before the actual work begins. Read through the appropriate Sections for all work to be done before you begin. Gather up all necessary parts and tools. If it appears that you could have a problem during a particular job, don't hesitate to ask advice from your local dealer parts or service department.

Before beginning any actual maintenance or repair, the machine should be cleaned thoroughly, especially around the oil filter, spark plugs, air cleaner, carburetor, etc. Cleaning will help ensure that dirt doesn't contaminate the engine and will allow you to detect wear and damage that could otherwise easily go unnoticed.

3 Fluid levels – check

Engine oil

Refer to illustrations 3.3 and 3.5

1 Check the engine oil level after taking the motorcycle on a short run. It must be positioned upright and on level ground for the reading to be accurate.
2 An oil cap is attached to the oil tank just below the seat on the right side.
3 Pull straight up on the oil filler cap with a twisting motion to remove it. A dipstick is connected to the oil cap **(see illustrations)**.
4 Wipe the dipstick clean and reinsert it into the oil tank. Withdraw the dipstick and check the oil level on the end.
5 If the oil level is near the lower mark, add enough oil of the recommended grade and type to bring the level up to the upper mark **(see illustration)** – don't overfill it. Reinstall the cap.

Brake fluid

Refer to illustrations 3.8a and 3.8b

6 In order to ensure proper operation of the hydraulic disc brakes, the fluid level in the master cylinder must be properly maintained.
7 Turn the handlebars until the top of the master cylinder is as level as possible. If necessary, loosen the brake lever clamp and rotate the master cylinder assembly slightly on the handlebar to make it level.
8 On later models the fluid level can be checked via the transparent sight glass in the reservoir, whereas on earlier models the cover or cap must be removed for inspection of the level **(see illustrations)**.
9 Before the front brake master cylinder cap is removed, cover the gas tank to protect it from brake fluid spills (which will damage the paint) and remove all dust and dirt from the area around the cap.
10 Remove the screws and lift off the cap and rubber diaphragm. **Caution:** *Do not operate the brake lever with the cap removed*. If the level is low, more fluid must be added.
11 Add new, clean brake fluid of the recommended type until the level is about 1/4-inch from the top of the reservoir. Do not mix different brands of brake fluid in the reservoir – they may not be compatible.
12 Replace the rubber diaphragm and the cover. Tighten the screws evenly, but don't over-tighten them.

3.3 Pull the dipstick out of the oil tank opening . . .

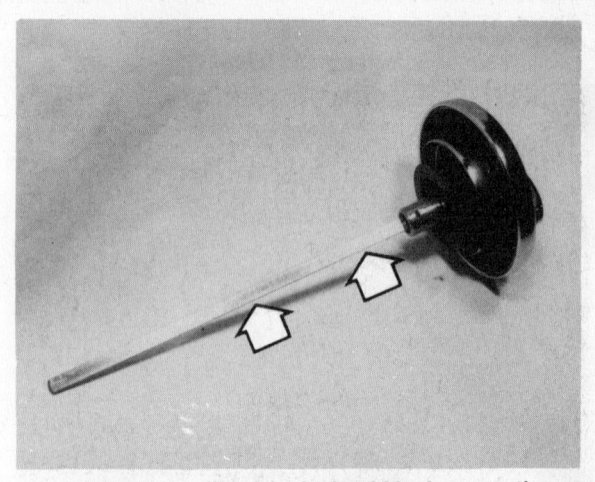

3.5 . . . and check the oil level – it should be between the upper and lower marks (arrows)

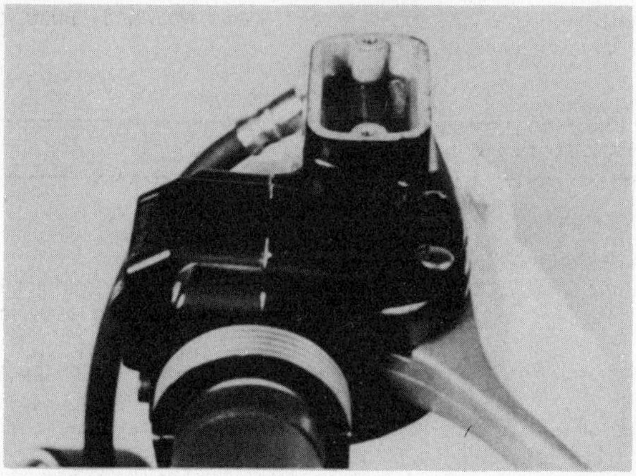

Chapter 1 Tune-up and routine maintenance

31

3.8a The fluid level in the front brake master cylinder on earlier models can only be checked after removing the screws and detaching the cover

3.8b Typical rear brake master cylinder – after cleaning it to avoid contamination of the brake system, remove the cover, or view the sightglass (arrow) to check the fluid level

13 If the brake fluid level was low, check the entire system for leaks. In the unlikely event the reservoir was completely empty, the system should be bled as described in Section 32.
14 Wipe any spilled fluid off the reservoir body and reposition and tighten the brake lever and master cylinder assembly if it was moved.

Battery electrolyte (through 1996)

Caution: *Be extremely careful when handling or working around the battery. The electrolyte is very caustic and an explosive gas is given off when the battery is charging.*

15 To check the electrolyte level in the battery, remove the side cover. The level should be between the upper and lower level marks printed on the outside of the battery case.
16 If the electrolyte level is at or below the lower mark, the battery must be removed in order to add more water. If necessary, refer to Chapter 7 for this procedure.
17 With the battery removed from the motorcycle, remove each cell cap and add enough distilled water to each cell to bring the level to the upper mark. Do not overfill. Also, do not use tap water, except in an emergency, as it will shorten the service life of the battery. The cell holes are quite small so it may help to use a plastic squeeze bottle with a small spout to add the water.
18 After filling, reinstall the battery. **Note:** *Be sure the vent hose is properly routed.*
19 The battery should periodically receive a thorough inspection, including a check of the electrolyte specific gravity. Refer to Chapter 7 for these procedures.

Transmission oil

Refer to illustration 3.21

20 Hold the motorcycle in an upright position throughout the inspection procedure.
21 Remove the oil fill plug and the level plug from the chaincase. On models through 1990 the level plug is positioned directly under the shift lever shaft, whereas from 1991-on it is at the rear of the chaincase **(see illustrations)**.
22 If the chaincase is equipped with a level plug, the oil level should be up to the bottom of the plug hole **(see illustrations)**. On later models without an oil level plug, remove the outer clutch cover; the oil level should be up to the bottom of the clutch diaphragm spring.
23 If necessary, add enough oil of the recommended type to fill the chaincase (transmission) to the proper level – don't overfill it.
24 Tighten the oil fill plug (and level plug, if equipped) securely. If the bike has an outer clutch cover, tighten its bolts to the torque listed in this Chapter's Specifications.

4 Air cleaner – servicing

Refer to illustration 4.1

1 In order to gain access to the air cleaner element, remove the plated cover/trim plate attached to the air cleaner assembly by screws or Allen

3.21a The transmission oil level should be at the bottom of the level plug opening with the motorcycle upright – models through 1990

1 Oil fill plug   2 Oil level plug

3.21b Transmission oil fill plug (1), level plug (2) and drain plug (3) – 1991-on models

4.1 The air filter element is accessible after removing the outer cover

4.5 Circular holes in element rear surface must locate over bolt heads – 1991-on models

head bolt(s). Some models have a baffle plate or seal band inside the cover. This will expose the filter element and screen **(see illustration)**.

2   The air cleaner element used on 1970 and 1971 models is made of metal mesh. It should be removed, washed in a non-flammable solvent and saturated with clean engine oil after its been allowed to dry. This type of service should be performed at least every 1000 miles and more frequently if the machine is used in very dusty conditions.

3   Most later models (1972 through 1989) have a foam-type filter. If a film of dirt has built up on the outer surface or if light spots appear on the surface, the filter should be cleaned and re-oiled. If the filter element is cracked, torn or distorted so it doesn't fit the screen, install a new one.

4   On 1972 through 1989 models, remove the foam element from the screen and wash it with hot water and soap. After it's dry, apply 1-1/4 tablespoons of engine oil to it with an atomizer or work it into the filter by hand. Squeeze out any excess oil, then reinstall the filter and other components. Tighten the screws/bolts securely.

5   On 1990-on models, wash the element with warm, soapy water to remove carbon and dirt. Soak it for 30-minutes to remove stubborn deposits (it can be considered clean when light is visible through the filter material when holding it up to a strong light source). Dry the element with low-pressure compressed air (32 psi maximum), then let it air dry for several hours. On 1991-on models the crankcase breather system vents vapor to the element via the air cleaner bolts which screw into the cylinder head; ensure that the circular holes in the back of the element fit over the bolt heads **(see illustration)**.

6   Never run the machine without the air cleaner attached or without the air filter element in place. The carburetion is set up to take into account the presence of the air cleaner and will be affected if the settings aren't changed to compensate.

7   A clogged or split air cleaner element will also have an adverse effect on carburetion and engine performance. It's better to give the air cleaner more frequent attention than necessary rather than to neglect it altogether. It's an item that shouldn't be ignored during the normal service routine.

5   Drive chain – check, adjustment and lubrication

Check

1   A neglected drive chain won't last long and can quickly damage the countershaft and rear wheel sprockets. Routine chain adjustment and lubrication isn't difficult and will ensure maximum chain and sprocket life.

2   To check the chain, place the motorcycle upright with a rider sitting on it and shift the transmission into Neutral. Make sure the ignition switch is Off.

3   Check for the specified free play (slack) at the lower chain run, midway between the sprockets. Chains usually don't wear evenly, so rotate the rear wheel and check the free play in a number of places. As wear occurs, the chain will actually get longer, which means that adjustment usually involves removing some slack from the chain. In some cases where

5.6 Loosen the axle nut (1) and tighten the chain adjusters (2) to take slack out of the drive chain

lubrication has been neglected, corrosion and galling may cause the links to bind and kink, which effectively shortens the chain"s length. If the chain is tight between the sprockets, rusty or kinked, it's time to replace it with a new one.

4   After checking the slack, grasp the chain where it wraps round the rear sprocket and try to pull it away from the sprocket. If more than 1/4-inch of play is evident, the chain is excessively worn and should be replaced with a new one.

Adjustment

Refer to illustration 5.6

5   Rotate the rear wheel until the chain is positioned where the least amount of slack is present.

6   Loosen the axle nut **(see illustration)**. On pre-1978 models, loosen the brake anchor bolt also.

7   Turn the axle adjusting nuts on both sides of the rear wheel until the proper chain tension is attained. Be sure to turn both adjusting nuts the same amount to keep the rear wheel in alignment. If the adjusting nuts reach the end of their travel, the chain is probably excessively worn and should be replaced with a new one. An accurate method of checking the alignment of the rear wheel is to measure the center-to-center distance between the swingarm pivot bolt and the rear axle on both sides of the machine. When the distances are equal, the rear wheel (and thus the chain and sprockets) should be properly aligned.

8   Tighten the axle nut and the anchor bolt (where applicable). Recheck the chain tension.

Lubrication

Refer to illustration 5.15

9   Pre-1977 models are equipped with an automatic chain oiler; refer to Section 16 for maintenance and adjustment procedures.

Chapter 1 Tune-up and routine maintenance

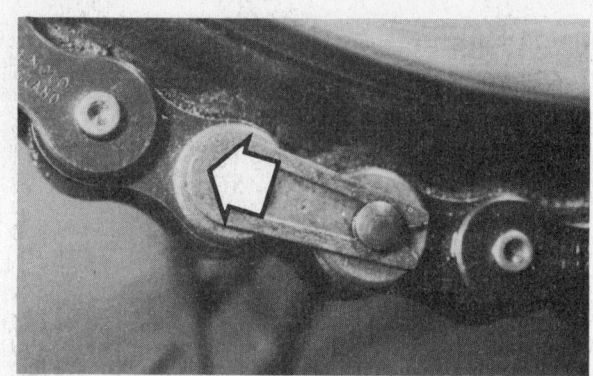

5.17 The closed end of the spring clip MUST face the direction of chain travel (arrow)

Conventional chain

10 The best time to lubricate the chain is after the motorcycle has been ridden. When the chain is warm, the lubricant will penetrate the joints between the side plates, pins, bushings and rollers to provide lubrication of the internal load bearing areas. Use a good quality chain lubricant and apply it to the area where the side plates overlap – not the middle of the rollers. After applying the lubricant, let it soak in for a few minutes before wiping off any excess.

11 If the chain is extremely dirty, it should be removed and cleaned before it's lubricated. Remove the master link retaining clip with pliers. Be careful not to bend or twist it. Slide out the master link and remove the chain from the sprockets. Clean the chain and master link thoroughly with solvent. Use a small brush to remove caked-on dirt. Wipe off the solvent, hang up the chain and allow it to dry.

O-ring chain

12 Later models (circa 1992) are fitted with an O-ring chain as standard equipment. Lubricant is sealed in the rollers by O-rings and thus, lubrication is only required on its outer working surfaces. Take care to use only a lubricant marked as being suitable for O-ring chains – other lubricants will cause damage to the O-rings.

13 If the chain is excessively dirty, it can be detached from its sprockets and cleaned in kerosene – don't use any strong solvents or gasoline for cleaning, otherwise the O-rings will deteriorate. When dry apply fresh lubricant to the outer rollers and sideplates.

Both chain types

14 Inspect the chain for wear and damage. Look for cracked rollers and side plates and check for excessive looseness between the links. To check for overall wear, lay the chain out on a clean, flat surface in a straight line. Push the ends together to take up all the slack between the links, then measure the overall length. Pull the chain ends apart as far as possible and measure the overall length again. Subtract the two measurements to determine the difference in the compressed and stretched lengths. If the difference, which is an indication of wear, is equal to or greater than 3-percent of the chain's nominal length, it's excessively worn and should be replaced with a new one.

15 Check the master link, especially the clip, for damage. A new master link should be used whenever the chain is reassembled.

16 Check the sprockets for wear also. If the teeth have a hooked shape, or are excessively worn or damaged, replace the sprockets with new ones. Never put a new chain on worn sprockets or a worn chain on new sprockets. Both chain and sprockets must be in good condition or the new parts will wear rapidly. Refer to Chapter 6 for rear wheel sprocket removal and installation procedures.

17 Reposition the chain on the sprockets and insert the master link. This should be done with both ends of the chain adjacent to each other on the back side of the rear wheel sprocket. On O-ring type chains, take care to position the O-rings correctly when assembling the master link. **Note:** *Make sure the closed end of the master link clip points in the direction of chain travel* **(see illustration)**.

18 Lubricate and adjust the chain as previously described.

6 Brake shoes/pads – wear check

1 The brake shoes and pads should be checked at the recommended intervals and replaced when worn beyond the specified limits.

Front drum brake

2 The front brake assembly can be withdrawn from the front hub after the axle has been pulled out and the wheel removed from the forks. Refer to Chapter 6 for the front wheel removal procedure.

3 Examine the brake shoe linings. If they're thin or worn unevenly, they should be replaced with new ones.

Front disc brake

Refer to illustrations 6.4 and 6.5

4 The front brake pads can be examined for wear by looking through the opening at the rear of the caliper **(see illustration)**. If in doubt about the condition of the brake pads, remove the caliper(s) and measure the thickness of the pad lining material. Refer to Chapter 6 for the caliper removal procedure.

5 Check the pads for wear, damage and looseness. Make sure the metal backing pad is flat and not distorted. If one pad has worn until the lining is less than 1/16-inch (1.5 mm) thick, replace both pads with new ones **(see illustration)**.

Rear drum brake

Refer to illustration 6.6

6 Some drum brake models have an inspection hole in the backing plate – remove the plug **(see illustration)** and check the brake shoe lining

6.4 The brake pad lining thickness (arrows) can be checked without removing the caliper

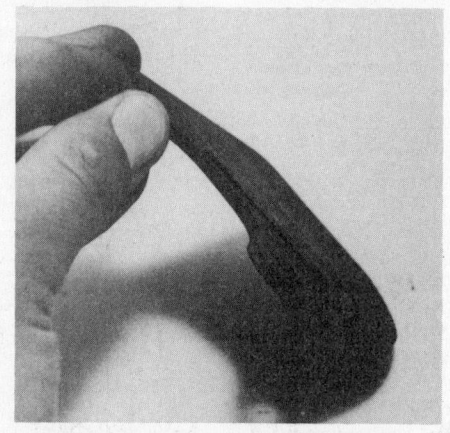

6.5 If the pads are allowed to wear to this extent, you risk damage to the disc(s)

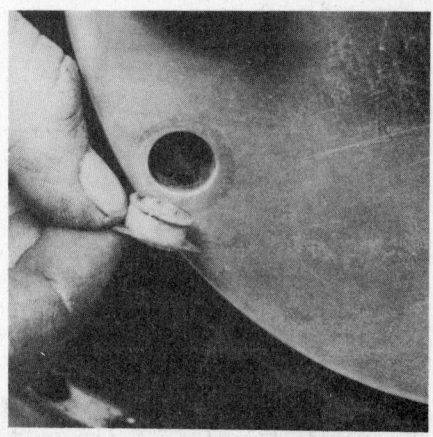

6.6 The rear drum brake backing plate has a small plug that can be removed to check the thickness of the brake shoe linings

Chapter 1 Tune-up and routine maintenance

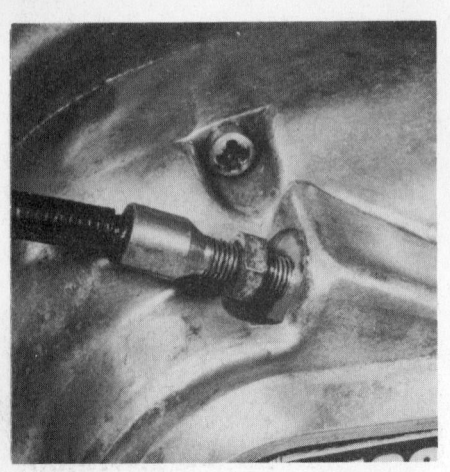

7.5 Loosen the clutch cable adjuster locknut at the primary chaincase

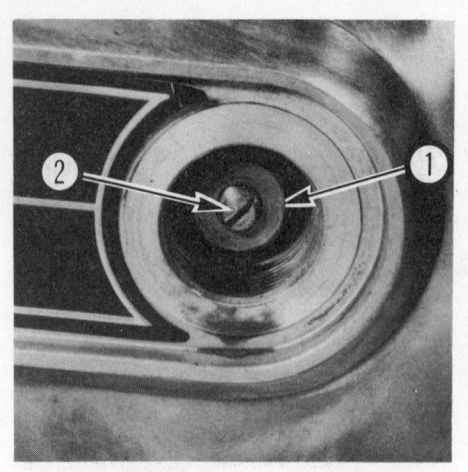

7.6 Remove the access plug from the primary chaincase to adjust the clutch
1 Locknut   2 Adjusting screw

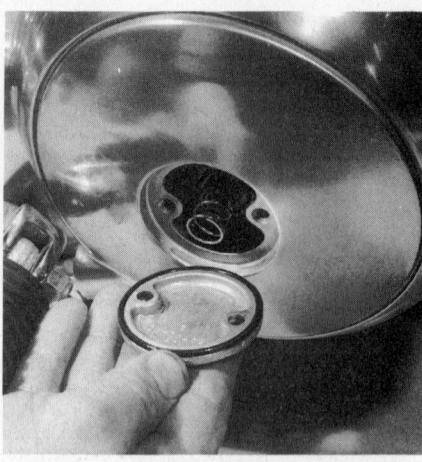

7.12a Late 1984-on – remove the access plug and lift out the spring . . .

thickness by looking through the hole. However, it's a good idea to remove the rear wheel to do a thorough inspection. Refer to Chapter 6 for the wheel removal procedure. If the linings are worn unevenly or worn to the specified service limit, they should be replaced with new ones.

Rear disc brake

7 On 1979 through 1981 models, refer to Steps 4 and 5 above (the calipers are very similar). On 1982 and later models, remove the caliper to examine the brake pads (see Chapter 6).

7 Clutch – adjustment

1970 models

1 Slacken the release mechanism locknut located in the center of the sprocket cover and back off (counterclockwise) the adjuster screw (see Chapter 2).
2 The release lever inside the cover should be heard to contact its stop when the handlebar lever is in the 'at-rest' position. Using the cable adjuster at the handlebar end, adjust so that the release lever does not quite return against its stop.
3 Turn the adjuster screw in the sprocket cover inwards (clockwise) until there is 1/8 inch (3 mm) free play at the lever before the clutch comes into operation. At this point hold the adjuster screw steady whilst its locknut is tightened.
4 It is possible to adjust the spring tension of the clutch itself – refer to a Harley-Davidson dealer for details.

1971 through early 1984

Refer to illustrations 7.5 and 7.6

5 Loosen the cable adjusting locknut where it enters the primary chaincase **(see illustration)**. Turn the cable adjuster in until the lever on the handlebar has plenty of free play.
6 Remove the clutch access plug from the primary chaincase, just behind the left foot peg **(see illustration)**.
7 Working through the opening in the chaincase, loosen the locknut and turn the adjusting screw in (clockwise) until it becomes difficult to turn. From this point, turn the screw in two more turns to be sure the clutch is disengaged.
8 Turn the cable adjuster out of the case until there's no free play in the cable. Don't put any tension on the cable. When there's no play at the hand lever, tighten the cable adjuster locknut.
9 Back off the clutch adjusting screw until it begins to turn easier (clutch is being engaged). Turn the screw back in until there's no free play. Back the adjusting screw out 1/8-to-1/4 turn, then tighten the locknut while holding the adjusting screw stationary. This can be done with a Harley-Davidson special tool or by inserting a screwdriver through a deep socket with a hex head on its upper end.
10 There should be 1/16-inch of free play between the hand lever and the mounting bracket. If necessary, adjust the cable until the desired free play at the lever is obtained.

Late 1984-on

Refer to illustrations 7.12a and 7.12b

11 On models through 1987 loosen the cable adjusting locknut where it enters the primary chaincase **(see illustration 7.5)**. Turn the adjuster in until the handlebar lever has plenty of play. The requirement is the same for 1988-on models, but use the in-line adjuster mid-way along the cable's length.
12 On models through 1993, remove the circular access plug from the chaincase and lift out the spring and adjusting screw lockplate **(see illustrations)**. The procedure is the same on 1994 and later models except that access is obtained by removing the four Torx screws that secure the outer clutch cover to the primary chaincase. Take the cover off without dislodging its O-ring.
13 Turn the adjusting screw out (counterclockwise) until it starts to turn hard (all free play removed).
14 Turn the adjusting screw in 1/4-turn (clockwise), then install the lockplate and spring. If the lockplate hex doesn't match up with the recess in the chaincase cover, turn the adjusting screw clockwise slightly until it does. Install the access plug or outer clutch cover, making sure the O-ring is in its groove. Tighten the clutch cover Torx screws evenly, in a criss-cross pattern, to the torque listed in this Chapter's Specifications.
15 Use the cable adjuster previously slackened, to obtain 1/16-to-1/8 inch (1.6 to 3 mm) free play between the cable ferrule and the lever butt at the handlebar.

8 Fuel system – check

Refer to illustrations 8.2a and 8.2b

1 The condition of the fuel system components, hoses and connections should be checked periodically to reduce the likelihood of a fuel leak developing. If the smell of gasoline is noticed while riding or after the motorcycle has been parked, the system should be checked immediately. **Warning:** *If a fuel odor is detected, be sure to work in a well-ventilated area and don't allow open flames (cigarettes, appliance pilot lights, etc.) or unshielded light bulbs in or near the work area.*
2 Inspect the area around the fuel tank, fuel valve and underneath the

Chapter 1 Tune-up and routine maintenance

7.12b ... then withdraw the adjusting screw lockplate

8.2a Check the fuel lines to make sure they're secure and in good condition – replace leaking and deteriorated ones immediately!

8.2b Original equipment hose clamps must be cut off and discarded

carburetor for evidence of leaks and damage. Carefully inspect all fuel lines to make sure they're tightly connected to the fittings and not cracked or otherwise deteriorated. If so, they should be replaced immediately with new ones **(see illustrations)**. Check all hose clamps to make sure they're tight.

3   If there's leakage from the fuel valve, make sure it's securely attached to the fuel tank. If a leak persists between the valve and tank, drain the tank (referring to Section 22, if necessary), remove the valve and apply thread sealing tape to the threads of the valve. If the leak is coming from the body of the valve, it'll have to be replaced with a new one.

4   If leakage is occurring at the carburetor, it indicates defective carburetor gaskets. The carburetor should be removed and disassembled as described in Chapter 3 to locate the problem.

5   Check all evaporative emission system hoses and components for damage and deterioration (later California models only).

9 Brake system – general check

1   A routine general check of the brakes will ensure that any problems are discovered and remedied before the rider's safety is jeopardized.

2   On early models, check the brake shoes for excessive wear and the lever, brake cable or rod and pedal for loose connections, excessive play, distortion and damage. Replace any damaged parts with new ones. Refer to Section 6 for the brake shoe wear check.

3   On disc brake systems, carefully examine the master cylinder, hoses and caliper(s) for brake fluid leaks. Pay particular attention to the hoses. If they're cracked, abraded, or otherwise damaged, replace them with new ones. If leaks are evident at the master cylinder or caliper(s) they should be rebuilt by referring to the appropriate Sections in Chapter 6.

4   Check the disc brake lever/pedal for proper operation. It should feel firm and return to its original position when released. If it feels spongy, or if travel is excessive, the system may have air trapped in it. Refer to Section 32 and bleed the brakes.

5   Check the brake pads for excessive wear by referring to Section 6.

6   Examine the brake disc(s) for cracks and score marks. Measure the thickness of each disc and compare it to the Specifications in Chapter 6. Any disc worn beyond the allowable limit must be replaced with a new one.

7   If the brake lever or pedal pulsates when the brakes are applied during operation of the machine, the disc(s) may be warped. Attach a dial indicator setup to the fork slider or swingarm and check the disc runout. If the runout is greater than specified, replace the disc with a new one. If a dial indicator isn't available, a dealer service department or motorcycle repair shop can make this check for you.

8   Make sure both brake light switches operate properly.

10 Lubrication – general

1   Since the controls, cables and various other components of a motorcycle are exposed to the elements, they should be lubricated periodically to ensure proper operation.

2   The clutch and brake lever pivots should be lubricated with light oil. Don't apply too much oil to the pivots, as the oil will attract dirt, which could cause the controls to bind.

3   The throttle, clutch, front brake and choke cables should be treated with a commercially available cable lubricant, which is specially formulated for use on motorcycle control cables. Small adapters for pressure lubricating the cables with spray can lubricants are available and work very well. **Caution:** *DO NOT lubricate the enrichener cable for the CV carburetor used on 1988 and later models.*

4   Speedometer and tachometer cables should be removed from their housings and lubricated with a very light grease or cable lubricant.

5   If the throttle operates roughly, a light coat of grease should be applied to the inside of it, as described in Section 24.

6   Finally, the pivot points for the rear brake pedal, gear shift lever, footpegs and sidestand should all be lubricated with multi-purpose grease.

11 Clutch and front brake cable – lubrication

1   Lubrication of the clutch cable and the front brake cable (on earlier motorcycles without disc brakes) is a simple operation which will help prevent premature wear of the cables.

2   Both cables can be serviced in the same way. Loosen the locknut on the cable adjuster, then loosen the cable adjuster as far as possible by turning it in a clockwise direction (refer to Section 7 if necessary).

3   Pull the cable housing out of the lever mount and position the cable so the end can be removed from the lever.

4   Spray cable lubricant into the cable housing until it comes out the other end. Special adapters are available for pressure lubricating the cables and can be purchased at most motorcycle shops.

5   When the cables are lubricated, attach the end to the lever and pull the cable housing into position in the mount.

6   Adjust the cables for the specified amount of free play, then secure the adjustment by tightening the locknut.

12 Fasteners – check

1   Since engine vibration tends to loosen fasteners, all nuts, bolts, screws, etc. should be periodically checked.

2   If a torque wrench is available, use it along with the torque Specifications at the beginning of each Chapter.

14.3 Push down on the spring retainer (1) and pull out the keeper (2) . . .

14.5 . . . so the pushrod cover can be lifted up to get at the tappets (pre-Evolution engine only)

13 Tires/wheels/spokes – general check

1 Routine tire and wheel checks should be made with the realization that your safety depends to a great extent on their condition.
2 Check the tires carefully for cuts, tears, embedded nails or other sharp objects and excessive wear. Operation of the motorcycle with excessively worn tires is extremely hazardous, as traction and handling are directly affected. Check the tread depth at the center of the tire and replace worn tires with new ones when the tread is worn excessively.
3 Repair or replace punctured tires and tubes as soon as damage is noted. Don't try to patch a torn tire, as wheel balance and tire reliability may be impaired.
4 Check the tire pressures when the tires are cold and keep them properly inflated. Proper air pressure will increase tire life and provide maximum stability and ride comfort. Keep in mind that low tire pressures may cause the tire to slip on the rim or come off, while high tire pressures will cause abnormal tread wear and unsafe handling.
5 The cast alloy wheels used on some models are virtually maintenance free, but they should be kept clean and checked periodically for cracks, dented rims and other damage. Never attempt to repair damaged cast wheels – they must be replaced with new ones.
6 On machines equipped with wire spoked wheels, periodic checks are extremely important. Inspect the rims for dents and cracks. Check all spokes for damage (such as cracks and distortion) and make sure they're tight. To check spoke tightness, strike each one lightly with a screwdriver or other metal tool and listen to the sound that's produced. A crisp, ringing sound indicates a properly tensioned spoke. If a dull thud is produced, the spoke is loose.
7 A spoke wrench of the proper size should be used to tighten the loose spokes. Also, lubricate the nipples (at the point where the spokes thread into them and at the rim) with light oil before tightening the spokes. Don't over-tighten them, as wheel concentricity, roundness and side play will be affected.
8 Refer to Chapter 6 for the procedures to follow for checking wheels.

14 Valve clearance – adjustment (pre-Evolution engine only)

Refer to illustrations 14.3 and 14.5

1 The valve clearances must be checked and adjusted with the engine cold.
2 Remove the spark plugs so the engine is easier to turn over.
3 Push down on the pushrod cover spring retainer and remove the keeper from the upper end to gain access to the tappets **(see illustration)**.
4 The valve clearances are checked with the valves closed. This can be determined by turning the engine over by hand. Put the transmission in gear and turn the rear wheel forward while watching the tappets/pushrods for one particular cylinder – when the tappets move down in the bores and remain there, the valves for that cylinder are closed.
5 Hold the lower tappet cover up, out of the way **(see illustration)**.
6 Loosen the locknut on the tappet adjusting screw.
7 Turn the adjusting screw into the tappet body until the pushrod can be moved up-and-down.
8 Slowly turn the adjusting screw out until the play between the adjusting screw and pushrod is nearly gone.
9 Tighten the locknut on the adjusting screw securely and check the pushrod. There should be no noticeable up-and-down play, but you should be able to rotate it without any binding.
10 Adjust the remaining valve clearances in the same manner.
11 When all of the clearances are correct, install the pushrod covers. Installation is the reverse of removal – be sure the ends of the covers are seated properly against the washers.

15 Idle speed – adjustment

Refer to illustration 15.3

1 The idle speed should be checked and adjusted when it's obviously too high or too low. Before adjusting the idle speed, be sure the ignition timing and spark plug gaps are correct.
2 The engine should be at normal operating temperature, which is usually reached after 10 or 15 minutes of stop and go riding. Place the motorcycle on the kickstand and make sure the transmission is in Neutral.
3 Turn the throttle stop screw until the specified speed is obtained **(see illustration)**.

15.3 The throttle stop screw (arrow) is turned to change the idle speed (a stubby screwdriver will be required to reach it)

Chapter 1 Tune-up and routine maintenance

16.1 Automatic drive chain oiler (arrow)

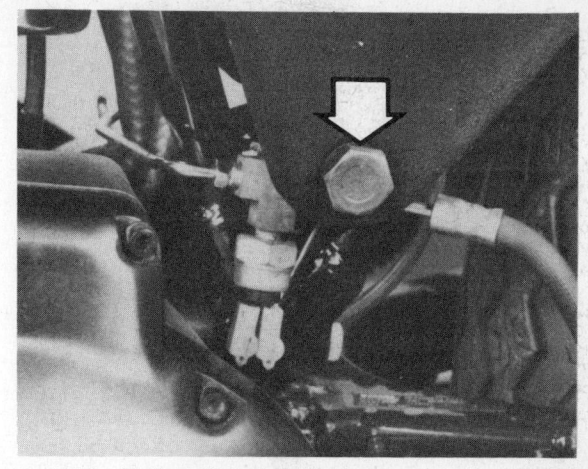

17.2 Typical oil tank drain plug location (arrow)

4 If a smooth, steady idle cannot be obtained, the fuel/air mixture may be incorrect. Refer to Chapter 3 for fuel/air mixture adjustment procedures.

16 Automatic drive chain oiler – maintenance and adjustment

Refer to illustration 16.1

1 Many models (1970 through 1976) are equipped with an automatic rear drive chain oiler **(see illustration)**. The oiler is exposed to the elements so it must be kept clean and properly adjusted.
2 Loosen the locknut and turn the oiler adjusting screw in until it bottoms on the seat. **Note:** *Keep track of the number of turns required to bottom the adjuster.*
3 Completely unscrew the adjuster and blow the orifice out with compressed air.
4 Install the adjusting screw and turn it in until it bottoms, then back it out the required number of turns to its original position (the normal setting is 1/4-turn open). Tighten the locknut.
5 The oiler should release two or three drops of oil per minute. Turn the adjusting screw in if less oil is desired; turn it out if more oil is needed.

17 Engine oil and filter – change

Refer to illustrations 17.2, 17.4, 17.5a, 17.5b and 17.6

1 Consistent routine oil and filter changes are the single most important maintenance procedure you can perform on a motorcycle. The oil not only lubricates the internal parts of the engine, but it also acts as a coolant, a cleaner, a sealant and a protectant. Because of these demands, the oil takes a terrific amount of abuse and should be replaced often with new oil of the recommended grade and type. Saving a little money on the difference in cost between good oil and cheap oil won't pay off if the engine is damaged.
2 Before changing the oil, warm up the engine so the oil will drain easily. Be careful when draining the oil – the exhaust pipes, the engine and the oil itself can cause severe burns. The drain plug is at the bottom of the oil tank **(see illustration)**.
3 Place a container below the oil tank to drain the old oil into. If required, a chute made out of sheet metal or cardboard can be used to guide the oil into the container. This will prevent oil from spilling on the motorcycle. It isn't necessary to drain the crankcase.
4 Remove the oil filler cap and the dipstick **(see illustration)**, followed by the oil tank drain plug.
5 On pre-1979 models, lift the oil filter out of the tank after the oil is drained **(see illustration)**. The filter is housed in a cartridge and can be removed after detaching the filter clip and sealing washer from the upper end of the cartridge tube **(see illustration)**. The felt filter element should be replaced with a new one every time the oil is changed. When replacing the element, make sure the O-ring is positioned correctly on the cartridge tube flange. The correct order of assembly within the cartridge tube is: tube seal, spring, lower filter retainer, filter element, sealing washer and filter clip.

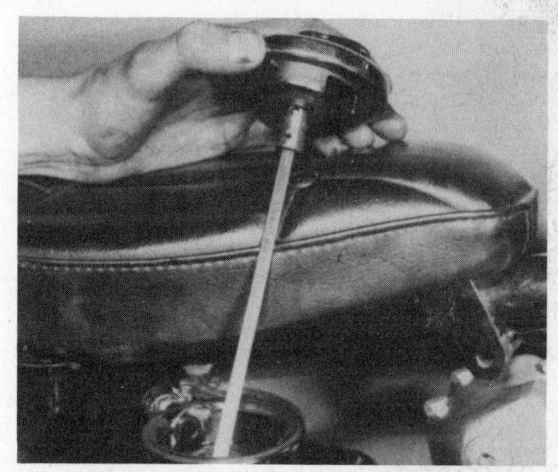

17.4 Remove the engine oil filler cap and dipstick, . . .

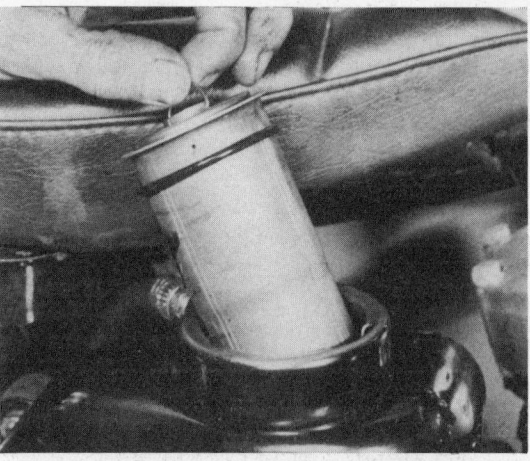

17.5a . . . then lift the oil filter out of the tank

Chapter 1 Tune-up and routine maintenance

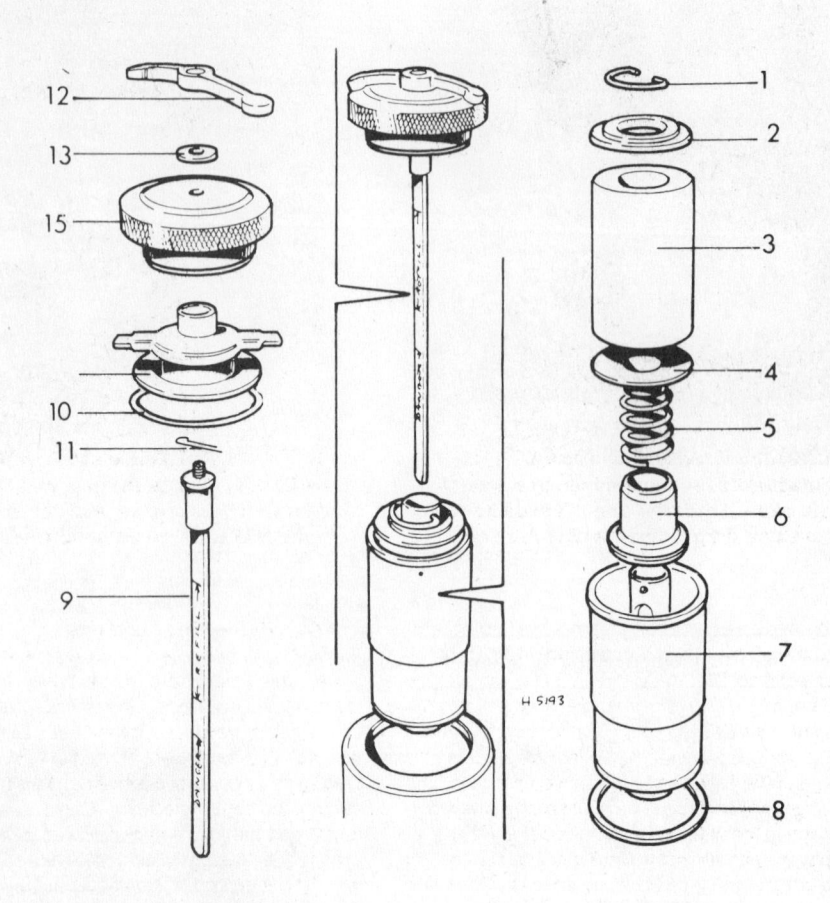

17.5b Oil tank filter unit (pre-1979 models) – exploded view

1 Filter clip	5 Spring	9 Dipstick	13 Washer
2 Sealing washer	6 Tube seal	10 Cap gasket	14 Not used
3 Filter element	7 Cartridge tube	11 Cotter pin	15 Cap
4 Lower filter retainer	8 O-ring	12 Wing nut	

17.6 The disposable oil filter is attached to the frame or the front of the engine, depending on the model year (1980/1981 model shown)

18.2 Detach the plug wires (A) to remove the spark plugs

Chapter 1 Tune-up and routine maintenance

18.6a Spark plug manufacturers recommend using a wire-type gauge when checking the gap – if the wire doesn't slide between the electrodes with a slight drag, adjustment is required

18.6b To change the gap, bend the side electrode only, as indicated by the arrows, and be very careful not to crack or chip the porcelain insulator surrounding the center electrode

6   1980 and 1981 models have an external oil filter mounted on a bracket between the engine and the oil tank **(see illustration)**. 1982 through early 1984 models have an external filter mounted on the lower left front engine bracket. 1985 and later models have an external filter mounted directly on the front of the engine. Use a filter wrench that fits over the end of the filter and is turned with a 3/8-inch drive ratchet to remove the filter. If a filter wrench isn't available and the filter can't be removed by hand, one last-ditch method of removing the filter is to pierce it with a long screwdriver and twist it off. The damage to the filter doesn't matter, since it'll be replaced with a new one anyway. If additional maintenance is planned for this time, check or service another component while the oil is allowed to drain completely.

7   Prior to installing the new spin-on filter, coat the rubber seal with clean engine oil, then screw it into place. Tighten the filter by hand an additional 1/4 to 1/2-turn after the seal first makes contact with the mounting surface.

8   The oil tank should be flushed at least every other oil change. To flush it, install the drain plug and pour approximately one quart of kerosene into the tank. Agitate the kerosene by rocking the motorcycle from side-to-side. Remove the drain plug and allow the kerosene and displaced sludge to drain out.

9   Be sure the seal is in good condition before final installation of the drain plug.

10   Pour the recommended grade of oil into the oil tank until the level is about one inch from the top. Insert the oil filter canister, on models so equipped, and replace the filler cap.

11   There's also a wire mesh screen-type filter in the timing cover. This filter prevents small pieces of foreign material, suspended in the oil, from finding their way into the scavenge pump. The filter should be cleaned whenever the timing cover is removed (see Chapter 2).

12   Start the engine and allow it to run for one minute at fast idle. **Caution:** *Make sure the oil pressure light goes out within one minute with the engine running at fast idle. If it doesn't, loosen the oil pressure switch (at the front of the oil pump) and the oil tank filler cap enough to allow about two ounces of oil to run out past the pressure switch threads while the engine is running. Retighten the switch and the filler cap.*

13   Check the oil filter and drain plug(s) for leaks. Check the oil level on the dipstick and, if necessary, add more oil.

18 Spark plugs – check and replacement

Refer to illustrations 18.2, 18.6a and 18.6b

1   Make sure your spark plug wrench or socket is the correct size before attempting to remove the plugs.

2   Disconnect the spark plug wires from the plugs **(see illustration)**. Clean any dirt from around the base of the plugs with compressed air, a damp cloth or a brush, then remove the plugs.

3   Inspect the electrodes for wear. Both the center and side electrodes should have square edges and the side electrode should be of uniform thickness. Look for excessive deposits and a cracked or chipped insulator around the center electrode. Compare the spark plugs to the spark plug photos on the inside back cover of this manual. Check the threads, the washer and the porcelain insulator body for cracks and other damage.

4   If the electrodes aren't excessively worn, and if the deposits can be easily removed with a wire brush, the plugs can be regapped and reused (if no cracks or chips are visible in the insulator). If in doubt concerning the condition of the plugs, replace them with new ones, as the expense is minimal.

5   Cleaning spark plugs by sandblasting isn't recommended, since grit from the sandblasting process may remain in the plug and be dislodged after it's installed in the engine, which obviously can cause damage and increased wear.

6   Before installing new plugs, make sure they're the correct type and heat range. Check the gap between the electrodes – they're not pre-set. For best results, use a wire-type gauge rather than a flat gauge to check the gap **(see illustration)**. If the gap must be adjusted, bend the side electrode only and be very careful not to chip or crack the insulator nose **(see illustration)**.

7   Thread the plug into the head by hand. Tighten the plugs finger-tight (until the washers bottom on the cylinder head) then use a wrench to tighten them to the torque listed in this Chapter's Specifications. DO NOT over-tighten them.

8   Reconnect the spark plug wires.

19 Contact breaker points (1970 through 1978) – check and replacement

Refer to illustrations 19.2, 19.4, 19.7 and 19.8

1   If the contact breaker points are badly burned, pitted or worn, they should be replaced with a new set. This also applies if the fiber heel that rides on the breaker cam is badly worn.

2   Detach the point cover **(see illustration)**. Prior to removal, mark the base plate in relation to the distributor body or engine side cover with a scribe or permanent felt-tip marker so the plate can be reinstalled in the same position. This will eliminate the need to retime the ignition after reassembly.

19.2 The breaker point cover is retained by two screws

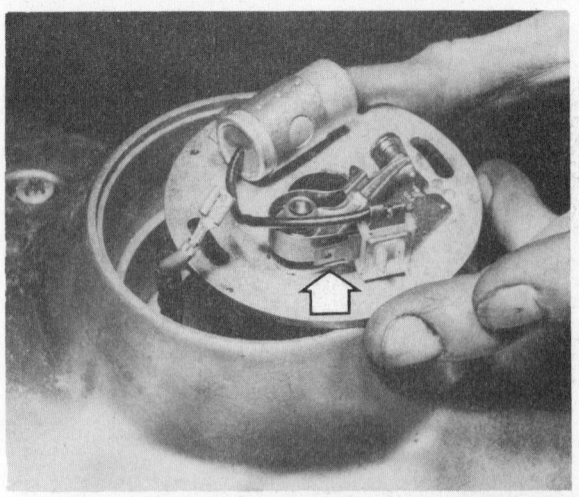

19.4 Detach the primary wire from the terminal (arrow), then remove the base plate with the points attached

19.7 Check the point gap with a feeler gauge – if the gap is correct, the feeler gauge will just slide between the two contacts with a slight amount of drag

19.8 Lock screw (1) and breaker point adjusting slot (2) locations

3 On 1970 models only, remove the two screws that secure the base plate to the distributor body **(see illustration 11.2 in Chapter 4)**. When the condenser wire and the external primary wire have been disconnected, the moving contact point will no longer be attached at the far end by its return spring so it can be lifted out of position. The fixed point can be detached by removing the screw that holds it to the base plate. Note the arrangement of the various washers and insulators; if they're installed wrong, the points could short out, causing failure of the ignition system.
4 On 1971 and later models, remove the base plate mounting screws, lift out the base plate and disconnect the primary wire **(see illustration)**.
5 Pull the condenser wire off the terminal post, unhooking the moving contact point return spring at the same time. Lift off the moving contact point and release the fixed contact by removing the single retaining screw through the base. Note the arrangement of the insulators and other washers to prevent them from being replaced in the wrong order.
6 Install the points in the reverse order of removal. Make sure the insulators are installed in the correct positions. It's a good idea to place a small amount of distributor cam lube on the pivot pin prior to installation of the moving contact arm.
7 Check and adjust the point gap with a feeler gauge when the points are completely opened by one of the cam lobes **(see illustration)**.
8 Loosen the lock screw at the base of the fixed contact point and move the point by inserting a screwdriver into the adjusting slot and turning it **(see illustration)**. Adjust the points until the specified gap is obtained, then retighten the lock screw and recheck the gap.
9 Turn the engine over until the points are completely opened by the other cam lobe and check the gap. The gap should be exactly the same for both cam lobes. If it isn't, the cam is defective and must be replaced with a new one.

20 Ignition timing – check and adjustment

Refer to illustrations 20.1a, 20.1b, 20.2a, 20.2b, 20.4a, 20.4b and 20.5

1 A timing light is the best and most accurate way to check the ignition timing, since it's done with the engine running. The timing light leads should be attached to the front spark plug wire (initially) and the battery terminals. It's a good idea to replace the plug in the left (models through 1990) or right (1991-on models) crankcase with the special clear plastic

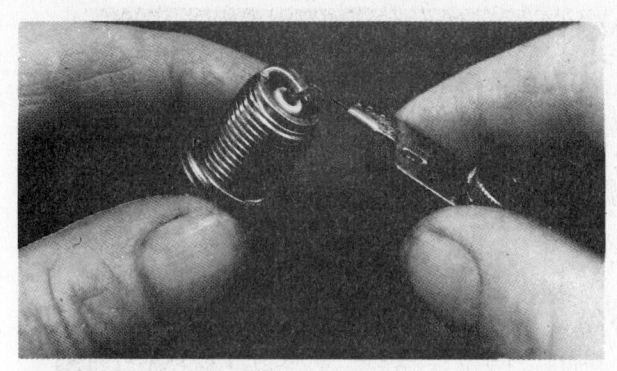

Spark plug maintenance: Checking plug gap with feeler gauges

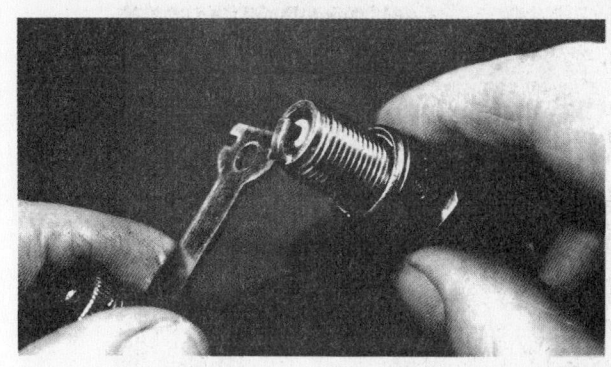

Altering the plug gap. Note use of correct tool

Spark plug conditions: A brown, tan or grey firing end is indicative of correct engine running conditions and the selection of the appropriate heat rating plug

White deposits have accumulated from excessive amounts of oil in the combustion chamber or through the use of low quality oil. Remove deposits or a hot spot may form

Black sooty deposits indicate an over-rich fuel/air mixture, or a malfunctioning ignition system. If no improvement is obtained, try one grade hotter plug

Wet, oily carbon deposits form an electrical leakage path along the insulator nose, resulting in a misfire. The cause may be a badly worn engine or a malfunctioning ignition system

A blistered white insulator or melted electrode indicates over-advanced ignition timing or a malfunctioning cooling system. If correction does not prove effective, try a colder grade plug

A worn spark plug not only wastes fuel but also overloads the whole ignition system because the increased gap requires higher voltage to initiate the spark. This condition can also affect air pollution

Chapter 1 Tune-up and routine maintenance

20.1a Install the special clear plastic plug in the crankcase inspection hole (arrow) before running the engine to check the ignition timing – models through 1990 shown

20.1b Timing inspection hole (arrow) is on right-hand side of engine from 1991-on

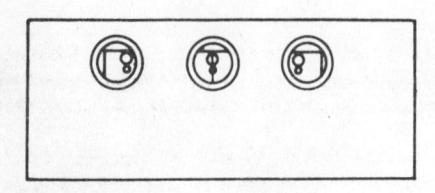

20.2a On 1970 through 1978 models, this is what the timing marks look like – ideally, they should be centered in the opening

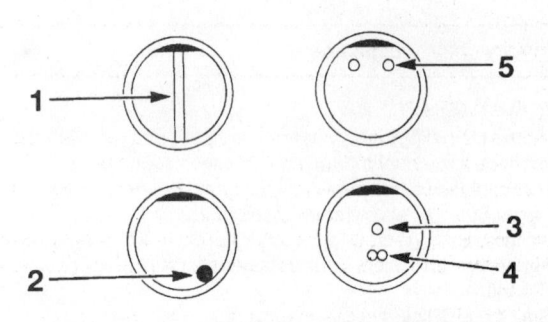

20.2b Timing marks for late 1979-on models

1. Front cylinder advance mark (late 1979 through early 1980) or front cylinder TDC mark (late 1980-on)
2. Front cylinder TDC mark (late 1979 through early 1980)
3. Front cylinder advance mark (late 1980 through 1994 all; 1995 domestic)
4. "Lazy 8" rear cylinder advance mark (some models)
5. Front cylinder advance mark (1995 International; 1996-on all)

factory Timing Mark View Plug (part no. HD-96295-65C for models through 1990, H-D-96295-65D for 1991-on models) **(see illustrations)**, otherwise oil spray will be a problem when the engine is running. Make sure the plug doesn't touch the flywheel.

2 Run the engine at approximately 2000 rpm (1300 rpm on 1983 through 1985 models; 1650 to 1950 rpm on 1986 through 1994 models and 1995 US models; 1050 to 1500 rpm on 1995 International and all 1996 and later models) and observe the timing marks through the crankcase opening **(see illustrations)**. The front cylinder timing mark should appear stationary in the opening. As the engine is revved up, the advance mark should move into view.

3 On 1970 models, the timing can be changed by turning the distributor body slightly after loosening the clamp. On other models with contact points, remove the point cover as described in Section 19 to get at the breaker point base plate.

4 On later models with electronic ignition, drill out the rivet heads **(see illustration)** and detach the outer cover, then remove the screws and take off the inner cover and gasket to get at the sensor plate **(see illustration)**. Mark the sensor plate and cover with a felt-tip pen or a scribe to ensure the sensor plate (ignition timing) can be returned to its original position if desired.

5 Adjustments can be made by loosening the contact breaker point base plate or electronic ignition sensor plate screws and rotating the plate very carefully, in small increments, with a screwdriver inserted in the slot provided **(see illustration)**. The ignition timing will change as the plate is moved. **Note:** *When checking the ignition advance on later models, be sure to check the Vacuum Operated Electric Switch (VOES) also. With the engine idling, unplug the VOES hose from the carburetor and plug the carburetor fitting. The timing should retard – the engine speed should decrease. When the hose is reattached to the carburetor, the engine speed should increase (the timing should advance). If it doesn't, check the VOES wire connection at the ignition module and the VOES ground wire connection. If they appear to be okay, the VOES may be defective and should be replaced with a new one.*

20.4a On later models with electronic ignition, drill out the rivet heads and remove the outer cover, . . .

Chapter 1 Tune-up and routine maintenance

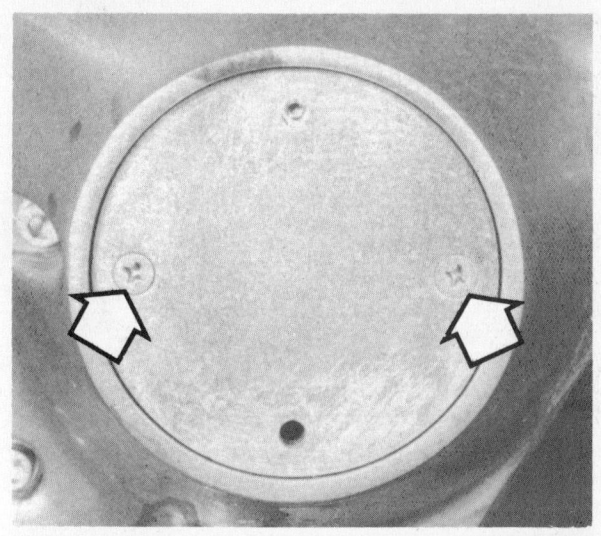

20.4b . . . then remove the screws (arrows) and detach the inner cover and gasket to get at the sensor plate

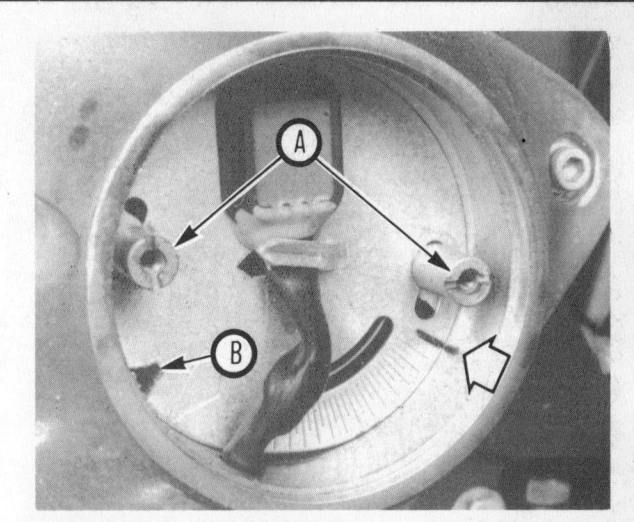

20.5 Mark the sensor plate (arrow), then loosen the screws (A) and insert a screwdriver into the slot (B) to change the position of the plate – move the plate a little at a time and recheck the timing

21 Primary chain – adjustment

Refer to illustrations 21.1 and 21.3

1   Remove the oil fill plug from the primary chaincase **(see illustration)**.
2   Turn the engine over to find the tightest point on the chain.
3   Loosen the adjusting screw locknut at the bottom of the chaincase, just in front of the kickstand **(see illustration)**.
4   Turn the adjusting screw IN to tighten the chain and OUT to loosen the chain. When the correct tension is obtained (see the Specifications) tighten the locknut.
5   Install the oil fill plug in the chaincase.
6   If the chain can't be adjusted tight enough, it's worn out or the adjuster is defective. Replace the faulty component.

22 Fuel filter – cleaning and replacement

Refer to illustration 22.2

Warning: *Gasoline is extremely flammable and highly explosive under certain conditions – safety precautions must be followed when working on any part of the fuel system! Don't smoke or allow open flames or un-shielded light bulbs in or near the work area. Don't do this procedure in a garage with a natural gas appliance (such as a water heater or clothes dryer). Also, before starting work, disconnect the negative battery cable from the battery.*

1   Make sure the fuel control valve is in the Off position. Remove the air cleaner assembly, detach the hose from the carburetor fitting and drain the contents of the fuel tank into a gasoline container.
2   Unscrew the valve from the bottom of the fuel tank. The screen-type filter is attached to the fuel control valve **(see illustration)**.
3   Thoroughly clean the filter. It can be removed from the valve to be cleaned or replaced if it's damaged.
4   If the valve itself leaks, it's not practical to attempt to repair it. The complete unit should be replaced with a new one.
5   Apply thread sealant to the threads before installing the valve in the fuel tank.

23 Transmission oil – change

Refer to illustration 23.1

1   Run the engine so that the transmission oil reaches normal operating temperature, then switch it off. Remove the drain plug from the bottom of

21.1 Remove oil fill plug to check primary chain tension

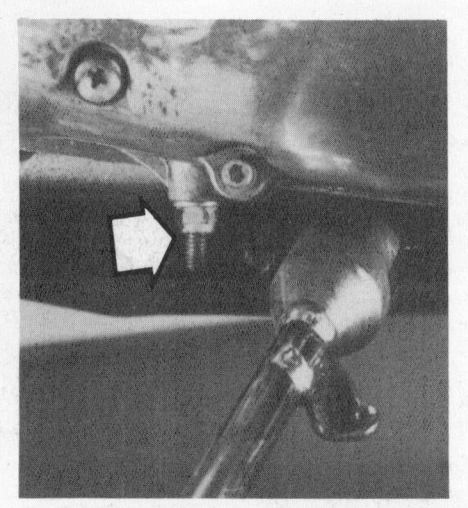

21.3 Primary chain adjusting screw location (arrow)

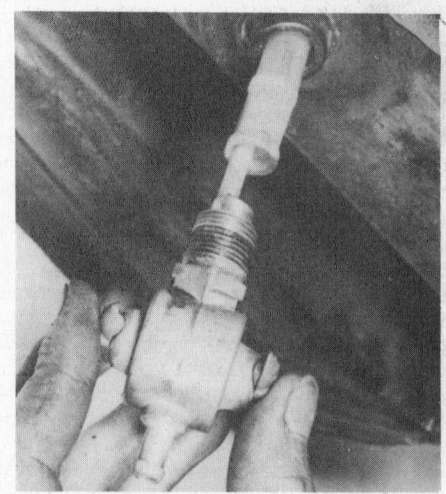

22.2 Unscrew the fuel valve from the tank to gain access to the screen-type filter for cleaning

Chapter 1 Tune-up and routine maintenance

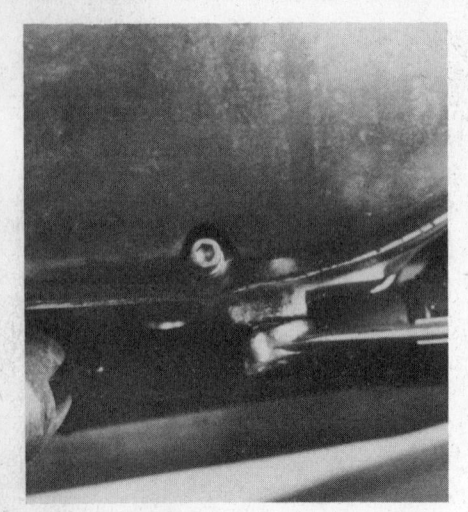

23.1 Remove the drain plug from the bottom of the primary chaincase to drain the transmission oil

24.2 Remove the screws securing the two halves of the throttle together, then lift off the top section to lubricate the throttle cable(s)

26.5 Steering grease nipple location – later models

the chaincase **(see illustration)** and allow the oil to drain completely into a shallow container.

2   Clean the magnetic drain plug and inspect the old oil for excessive metal chips and/or fiber-like clutch material. If either metal or fiber particles are found in the oil, something is wearing in the engine, transmission or clutch.

3   When all of the oil is drained, reinstall the drain plug. Be careful not to over-tighten it.

4   Remove the oil fill plug and level plug from the primary chaincase.

5   Hold the motorcycle in an upright position while filling the chaincase. Pour in the specified grade of oil until it just starts to flow out the level plug hole.

6   Install the oil fill and level plugs.

24 Throttle – check and lubrication

Refer to illustration 24.2

1   Make sure the throttle grip rotates easily from fully closed to fully open with the front wheel turned at various angles. The grip should return automatically from fully open to fully closed when released. If the throttle sticks, check the throttle cable for cracks or kinks in the housing. Make sure the inner cable is clean and well-lubricated and the adjustment screw located on the bottom of the lower throttle clamp isn't too tight.

2   On models with a drum-type throttle, the throttle cable(s) can be lubricated by first removing the screws that attach the top and bottom sections of the throttle assembly. Position the top section out of the way to gain access to the ends of the throttle cable(s) **(see illustration)**. The cable(s) can be lubricated with a spray lubricant specially made for motorcycle cables. As you apply the lubricant, twist the throttle back-and-forth to help move the lubricant through the cable(s). Reinstall the throttle assembly attaching screws.

3   On earlier models with spiral-type throttles, remove the handlebar end screw, screw spring and grip sleeve. Remove the roller pin and the rollers from the throttle cable plunger. Disconnect the throttle cable from the carburetor, then pull the plunger from the end of the handlebar with the cable still attached to it. Spray cable lubricant into the cable housing until it comes out of the carburetor end.

4   Assemble the throttle in the reverse order of disassembly.

5   Another possible cause of a sticking throttle grip is lack of lubrication between the throttle and the handlebar. To lubricate the throttle grip, first remove it from the handlebar. Apply a thin coat of multi-purpose grease to the handlebar area that's covered by the throttle, then reinstall the throttle.

25 Suspension – inspection

1   It's important that the suspension systems of the motorcycle be maintained in top operating condition to ensure rider safety. Loose, worn or damaged suspension parts decrease the vehicle's stability and control.

2   While standing alongside the motorcycle, lock the front brake and push on the handlebars to compress the forks several times. See if they move up-and-down smoothly without binding. If binding is felt, the forks should be disassembled and inspected as described in Chapter 5.

3   Carefully inspect the area around the fork seals for fork oil leakage. If leakage is evident, the seals must be replaced as described in Chapter 5.

4   Check the tightness of all suspension nuts and bolts to be sure they haven't worked loose.

5   Inspect the shocks for fluid leakage and make sure the mounting nuts are tight. If leakage is found, the shocks should be rebuilt or replaced.

6   Carefully raise the motorcycle and support it securely with the rear wheel off the ground. Make sure it can't fall over to either side. Grab the swingarm on both sides, just ahead of the axle. Rock the swingarm from side-to-side. There should be no discernible movement at the rear. If there's a little movement or a slight clicking can be heard, make sure the pivot bolt/locknut is correctly tightened. If the pivot is tight, but movement is still noticeable, then the swingarm will have to be removed and the bearings replaced, repacked and adjusted as described in Chapter 5.

7   Check the tightness of all rear suspension nuts and bolts.

26 Steering head bearings – check

Refer to illustration 26.5

1   Sportsters from 1975-on are equipped with tapered roller-type steering head bearings, which seldom require servicing. Earlier models are equipped with uncaged ball bearings, which may become dented, rough or loose during normal use of the machine. In extreme cases, worn or loose steering head bearings can cause steering wobble that's potentially dangerous.

2   To check the bearings, support the machine so the front wheel is in the air.

3   Point the wheel straight ahead and slowly move the handlebars from side-to-side. Dents or roughness in the bearing races (bearing cups) will be felt and the bars won't move smoothly.

4   Next, grasp the fork legs and try to move the wheel forward and backward. Any looseness in the steering head bearings will be felt. Refer to Chapter 5 for bearing maintenance and repair procedures.

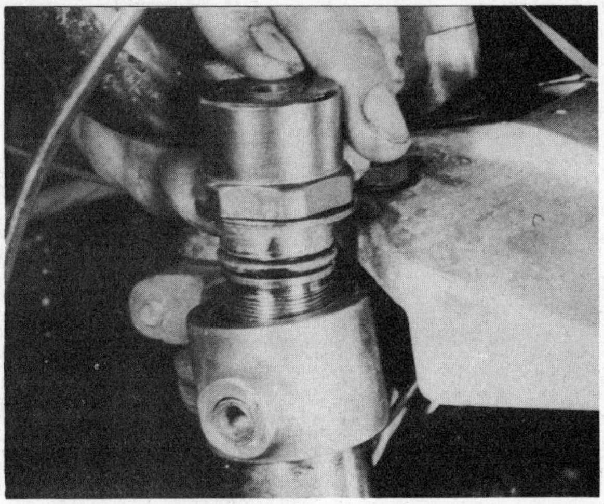

Chapter 1 Tune-up and routine maintenance

27.2 Unscrew the cap from the top of the fork leg

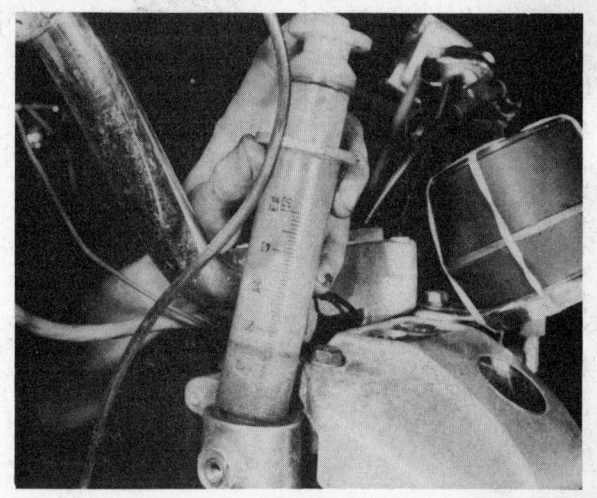

27.6 Fill the fork legs with the specified amount of fork oil (a baby bottle or measuring cup will make it easier to accurately determine the number of fluid ounces added)

5 Late 1992 models are equipped with a grease nipple in the left side of the steering head **(see illustration)**. Apply grease at the specified intervals. On earlier models it will be necessary to dismantle the steering head for greasing of the bearings, and although not specified at regular intervals, it is advised if the machine has covered a high mileage.

27 Fork oil (except 1995-on XL 1200S) – change

Refer to illustrations 27.2 and 27.6

1 Support the motorcycle securely with the front wheel off the ground.
2 Remove the cap from the top of one of the fork legs **(see illustration)**.
3 Remove the drain screw from the bottom of the fork leg (near the axle) that the cap was removed from.
4 Allow the oil to drain for a few minutes, then pump the forks up-and-down to force the remainder of the oil out.
5 Install the drain screw in the bottom of the fork leg.
6 Fill the fork with the specified amount of the recommended fork oil **(see illustration)**. If the fork has been disassembled, it will require additional fork oil.
7 Check the condition of the seal around the fork cap. If necessary replace the seal with a new one **(see illustration 27.2)**. Attach the fork cap to the top of the fork leg.
8 Repeat Steps 2 through 7 for the other fork leg.

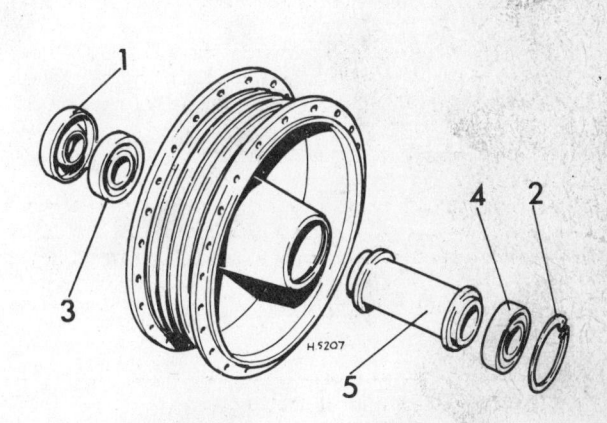

28 Wheel bearings – repack

Front wheel

Drum brake models

Refer to illustration 28.2

1 Remove the front wheel as described in Chapter 6.
2 Pry the seal out of the left side of the hub **(see illustration)**.
3 Lift the brake backing plate and shoe assembly out to expose the right-hand bearing. Remove the snap-ring that secures the bearing in the hub.
4 Drive the left-hand bearing into the hub as far as possible with a punch or a large socket. This will force the right-hand bearing out of position, along with the spacer that separates the bearings.
5 Remove the spacer, then drive the left bearing out from inside the hub.

Disc brake models

Refer to illustration 28.6

6 Models equipped with disc brakes have tapered roller bearings. The

28.2 Front wheel hub components – exploded view (drum brake type – 1970 through 1972)

1 Seal	4 Ball bearing
2 Snap-ring	(brake side)
3 Ball bearing	5 Spacer

seal must be pried out to remove the bearings and spacer **(see illustration)**.
7 If the bearings must be replaced, you'll have to replace the bearing race also. This requires a special tool to eliminate the risk of damaging the hub.

All models

Refer to illustrations 28.10, 28.11a and 28.11b

8 Remove all of the old grease from the hub and bearings, then clean the bearings with solvent. Check them for excessive play and of roughness as they're rotated. If there's any doubt about the condition of the bearings, they should be replaced with new ones.
9 Pack the wheel bearings with grease. On non-tapered bearings, pack the grease in from the open side of the bearing. Apply a liberal amount of grease to the inside of the hub and insert the bearings.
10 After the left bearing is in place, be sure to install the spacer separating the bearings **(see illustration)**. On mid-1991 on models, install the spacer shim, followed by the shouldered spacer; the smaller OD of the spacer should face the bearing on the right side of the hub.

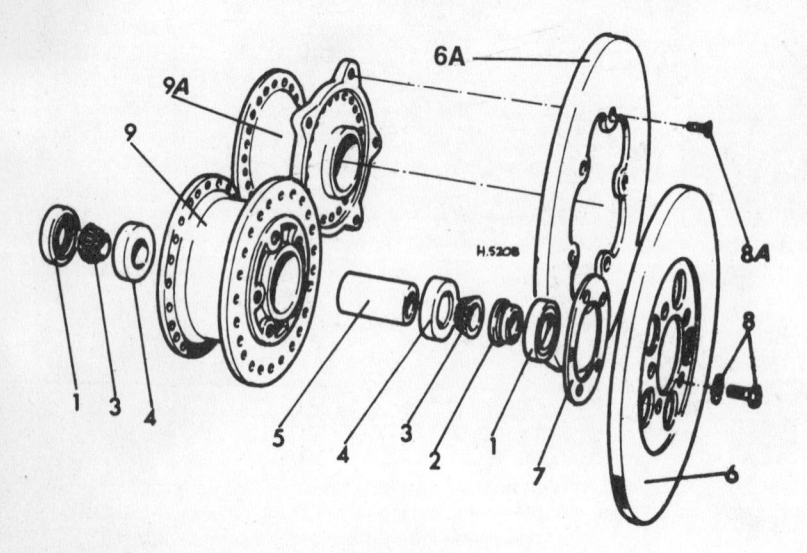

28.6 Front wheel hub components – exploded view (1973 through 1983 disc brake type shown; later models similar)

1. Seal
2. Spacer
3. Tapered roller bearing
4. Bearing race
5. Bearing spacer
6. Brake disc (1973)
6A. Brake disc (1974-on)
7. Brake disc spacer (1973)
8. Bolt and lock washer (1973)
8A. Allen-head bolt (1974-on)
9. Hub (1973)
9A. Hub (1974-on)

28.10 Install the spacer, ...

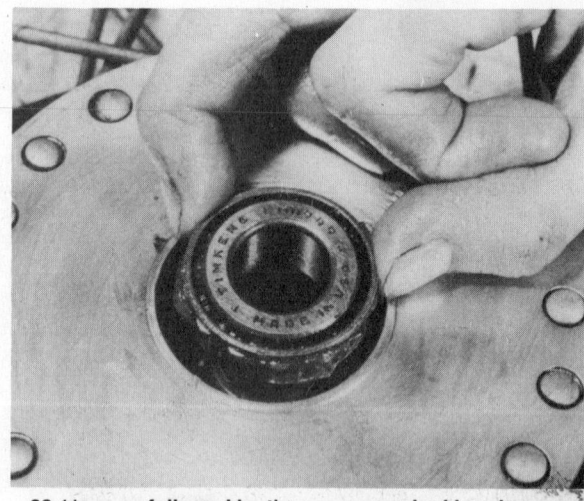

28.11a ... followed by the grease-packed bearing ...

28.11b ... and the seal in each side of the hub

11 Install the right side bearing and seal (**see illustrations**). When replacing the snap-ring on drum brake models, be sure the sharp edge faces away from the bearing.

Rear wheel

1970 through 1978

Refer to illustrations 28.13, 28.14a, 28.14b and 28.14c

12 Remove the rear wheel as described in Chapter 6.
13 Loosen the bearing locknut on the left side of the hub (**see illustration**). It's staked into place and may be difficult to turn initially.
14 Remove the locknut, seal and outer spacer from the hub (**see illustration**). Next, drive out the left bearing and remove the washer and bearing spacer (**see illustrations**).
15 Flip the wheel over and, working from the inside of the hub, drive the right bearing out. Remove the washer.
16 Service and install the bearings as described in Steps 9 through 11. Don't forget the spacer and washers. After it's tightened securely, stake the locknut with a hammer and punch.

1979 through 1983

17 Remove the snap-rings and washers from both sides of the hub.

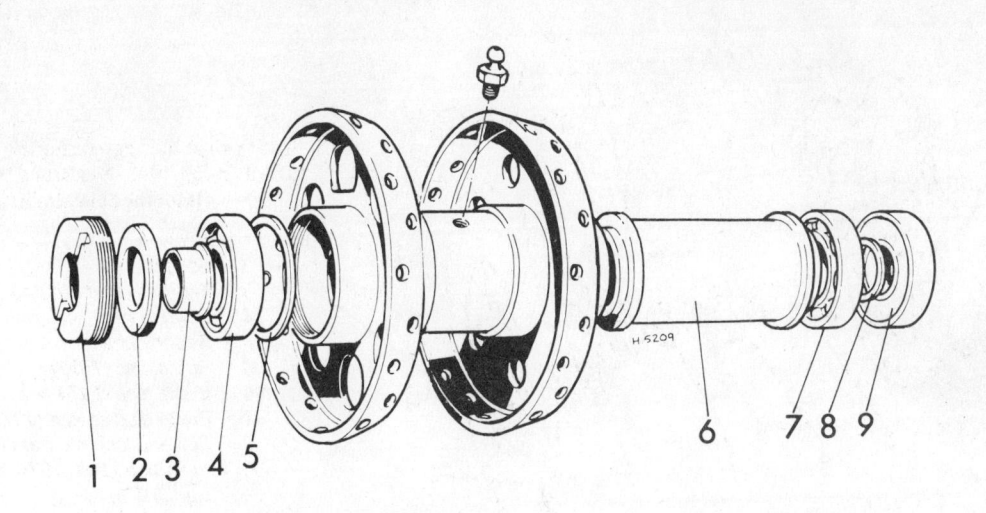

Chapter 1 Tune-up and routine maintenance

28.13 Rear wheel hub components – exploded view (1970 through 1978 models)

1 Bearing locknut
2 Seal
3 Outer spacer
4 Ball bearing
5 Washer
6 Spacer
7 Ball bearing
8 Washer
9 Sealed ball bearing

28.14a Remove the rear wheel bearing locknut, seal and outer spacer, . . .

28.14b . . . then drive out the bearing . . .

28.14c . . . and remove the center spacer

18 Carefully pry out both seals and remove the spacers. Note that one spacer is longer than the other.
19 Remove the bearings and the spacer from the center of the hub. Refer to Steps 8 and 9 above. If replacement is necessary, the races must be pressed out of the hub. The bearings and races must be replaced as matched sets.
20 Install the spacer in the center of the hub, then pack the bearings full of grease and install them. The seals must be installed so they are 3/16 to 1/4-inch below the edge of the hub. Apply grease to the outer edge of the seals before they're pressed into the hub.
21 The short spacer should be installed on the disc side of the hub; the long spacer on the other side.
22 Install the washers and snap-rings (be sure the sharp edge of each snap-ring faces out).

1984-on

23 Later models are very similar to 1979 through 1983 models, except they don't have snap-rings holding the bearings in the hub.

30.3a Pry out the plug . . .

30.3b . . . to get at the headlight adjusting locknut (arrow)

24 From mid-1991 on, a spacer shim and shouldered spacer were fitted. Make a note of which side of the hub they reside in before removal. When installing the shouldered spacer, note that its smaller OD must face the bearing.

29 Cylinder compression – check

1 Among other things, poor engine performance may be caused by leaking valves, incorrect valve clearances, a leaking head gasket or worn pistons, rings and/or cylinder walls. A cylinder compression check will help pinpoint these conditions and can also indicate the presence of excessive carbon deposits in the cylinder head(s).
2 The only tools required are a compression gauge and a spark plug wrench. Depending on the results of the initial test, a squirt-type oil can may also be needed. A compression gauge that screws into the spark plug hole is preferred over the type that requires hand pressure to maintain the seal at the plug hole.
3 Warm up the engine to normal operating temperature (ten or fifteen minutes of stop-and-go riding should be sufficient). Park the motorcycle and remove any dirt around the spark plugs with compressed air or a small brush, then remove the plugs. Work carefully, don't strip the spark plug hole threads and don't burn your hands. Use jumper wires to ground the spark plug wires on the cylinder heads.
4 Install the compression gauge in one of the spark plug holes. Make sure the choke is open and hold or block the throttle wide open. The transmission must be in Neutral.
5 Crank the engine over a minimum of five to seven revolutions and note the initial movement of the compression gauge needle as well as the final total gauge reading (write down the results). Repeat the procedure on the remaining cylinder and check the Specifications in this Chapter.
6 If the compression in both cylinders built up quickly and evenly to the specified amount, you can assume the engine upper end is in reasonably good mechanical condition. Worn or sticking piston rings and worn cylinders will produce very little initial movement of the gauge needle, but compression will tend to build up gradually as the engine spins over. Valve and valve seat leakage, or head gasket leakage, is indicated by low initial compression, which doesn't build up.
7 To further confirm your findings, add about 1/2-ounce of engine oil to each cylinder by inserting the nozzle of a squirt-type oil can through the spark plug holes. The oil will tend to seal the piston rings if they're leaking. Repeat the test on both cylinders.
8 If the compression increases significantly after the addition of oil, the piston rings and/or cylinders are definitely worn. If the compression doesn't increase, the pressure is leaking past the valves or the head gasket. Leakage past the valves may be caused by burned or cracked valve seats or faces, warped or bent valves or insufficient valve clearances.
9 If the compression readings are considerably higher than specified, the combustion chambers are probably coated with excessive carbon deposits. It's possible for carbon deposits to raise the compression enough to compensate for the effects of leakage past rings or valves. Refer to Chapter 2 to remove the cylinder heads and carefully decarbonize the combustion chambers.

30 Headlight aim – adjustment

Refer to illustrations 30.3a and 30.3b
1 An improperly adjusted headlight may cause problems for oncoming traffic or provide poor, unsafe illumination of the road ahead. Before adjusting the headlight, be sure to consult local traffic laws and regulations.
2 To set up the headlight, the machine should be placed on level ground at least 25-feet from a wall in its normal position (off the stand and with a rider – also a passenger if that's usually the case). On high beam, the top of the beam on the wall should be the same as the ground-to-headlight distance. This will ensure that oncoming drivers/riders won't be blinded.
3 To adjust the beam, pry the decorative plug out, loosen the headlight adjusting locknut and twist the headlight assembly in the desired direction until the beam is aimed properly **(see illustrations)**.
4 Tighten the locknut and recheck the aim.
5 Install the decorative plug.

31 Sidestand – check and maintenance

1 The sidestand should be checked periodically for smooth operation and wear. Clean the stand and inspect it carefully for cracks and other damage. Also, be sure it's not bent.
2 The sidestand pivot must be in good condition and well lubricated so the stand will retract when the machine is raised to an upright position. This applies to the return spring also.
3 To lubricate the sidestand, first disengage the spring, then remove the sidestand pivot bolt. Apply a coat of grease to the pivot, then reinstall the stand.

32 Brake hydraulic systems – bleeding and fluid replacement

1 Because the hydraulic systems used with both the front and rear disc brakes are virtually the same, the following procedures apply to both systems.

Bleeding

Refer to illustration 32.3
2 Every time any part of the brake hydraulic system is disassembled or develops a leak, or when the fluid in the master cylinder reservoir runs low, air will enter the system and cause a decrease in performance. To eliminate the air, the system must be bled using the following procedure.

Chapter 1 Tune-up and routine maintenance

32.3 Remove the rubber cap from the bleeder valve and attach the tubing to the valve fitting – open the bleeder valve slightly with a wrench as pressure is applied to the brake lever or pedal – close the valve before releasing the brake

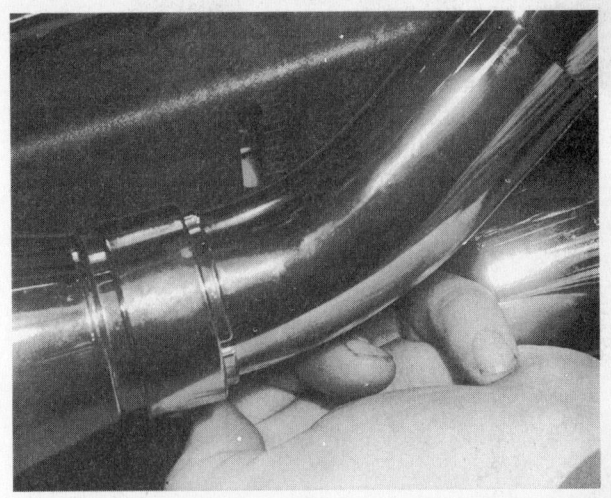

33.3 Belt tension window – apply force midway between the sprockets on the bottom run

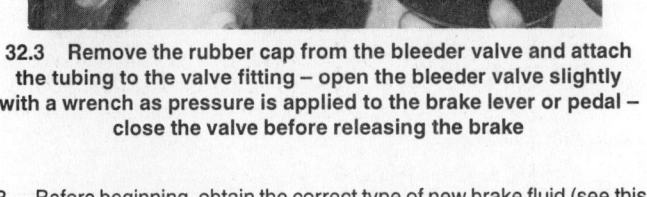

3 Before beginning, obtain the correct type of new brake fluid (see this Chapter's Specifications. Have an assistant on hand, as well as an empty clear plastic container, a length of plastic, rubber or vinyl tubing to fit over the bleeder valve and a wrench to open and close the bleeder valve **(see illustration)**.

4 Make sure the master cylinder reservoir is full of fluid and be sure to keep it at least half full during the entire operation. If, at any point, the reservoir runs low on fluid, the entire bleeding procedure must be repeated.

5 Loosen the bleeder valve slightly, then tighten it to a point where it's snug but can still be loosened quickly and easily.

6 Place one end of the tubing over the bleeder valve fitting and submerge the other end in brake fluid in the container **(see illustration 32.3)**.

7 Have your assistant apply the brakes a few times to get pressure in the system. On the last pump, have him hold the lever or pedal firmly depressed.

8 With the lever or pedal depressed, open the bleeder valve just enough to allow some fluid to leave the valve. Watch for air bubbles to exit the submerged end of the tube. When the fluid flow slows, close the valve again and have your assistant release the lever or pedal. If it's released before the valve is closed, air can be drawn back into the system. **Note:** *Avoid pulling the lever all the way back to the hand grip (front brake); this will cause overtravel of the piston and subsequent fluid leakage. This can be prevented by placing a 3/4-inch (20 mm) spacer between the hand grip and lever. Be careful when bleeding the rear brake also.*

9 Repeat Steps 7 and 8 until no air bubbles are seen in the fluid leaving the tube. Be sure to check the fluid level in the master cylinder frequently.

10 Completely tighten the bleeder valve.

11 Don't reuse old brake fluid. Some types attract moisture which will deteriorate the brake system components and can allow the fluid to boil, rendering the brakes inoperative. Also, don't mix different types of brake fluid.

12 Refill the master cylinder with fluid at the end of the operation.

Fluid replacement

13 Since brake fluid becomes contaminated while in use, it should be drained and the system filled with new fluid at the recommended intervals.

14 Before draining the old fluid, be sure to cover the painted surfaces of the gas tank and other components to prevent damage in the event that brake fluid is spilled on them.

15 Remove the reservoir cover (leave the rubber diaphragm in place) and attach a piece of clear plastic, rubber or vinyl tubing to the bleeder valve. Place the other end of the tubing in a plastic container.

16 Loosen the bleeder valve and slowly pump the lever or pedal to drain the fluid from the system. In order to completely drain the brake caliper, it must be removed and turned upside-down (refer to Chapter 6). If you're working on the front brake system, repeat the procedure on the other caliper of a two caliper system.

17 After the system has been drained, close the bleeder valve(s) and add new, clean brake fluid of the recommended type to the reservoir. Slowly pump the lever or pedal with the bleeder valve(s) open (tubing attached) to fill the system. Add more fluid to the reservoir as the level drops.

18 Once the system is full, it must be bled to remove all air as described in Steps 3 through 10 above.

19 Make sure the fluid level in the reservoir is correct, then install and tighten the reservoir cover. **Warning:** *Don't ride the motorcycle unless you're absolutely sure the brake system is capable of performing as it's designed! If the lever or pedal is spongy, recheck everything and perform the bleeding procedure again.*

33 Drive belt – check and adjustment

Check

Refer to illustration 33.3

1 Check the drive belt along its entire length for signs of cracking, damage and wear. Also check the condition of the front and rear sprockets.

2 Check for correct tension with the machine cold, resting on its sidestand and with the transmission in neutral. No rider should be seated and no luggage strapped to the motorcycle.

3 Firstly, find the tightest point on the belt by rotating the rear wheel and checking deflection in various positions along the belt's length. With this point positioned midway between the two sprockets on the lower run of the belt (at the point of the inspection window) apply 10lb (4.5 kg) of force upwards on the belt and measure the free play; if not as specified, adjust the belt as follows **(see illustration)**.

Adjustment

4 Remove the cotter pin from the rear axle nut and slacken the nut. Turn each adjuster nut on the swingarm ends by an equal amount to adjust the belt tension; turning clockwise tensions the belt and counterclockwise decreases tension.

5 It is important that correct alignment is maintained throughout adjustment. To check, hold a straightedge against the outside edge of the rear wheel sprocket, near the bottom of the sprocket. With the straightedge running parallel with the belt, make sure the belt is an equal distance from the straightedge for the entire length of the straightedge. Turn the adjuster nuts as required to align the rear wheel and recheck belt tension.

6 Tighten the rear axle nut to the specified torque and fit a new cotter pin, bending its ends to secure the nut.

Chapter 2 Engine, clutch and transmission

Contents

General information	1
Repair operations possible with the engine in the frame	2
Major engine repair – general information	3
Engine removal	4
Engine disassembly and reassembly – general information	5
Cylinder heads, barrels and tappets – removal	6
Starter motor and solenoid – removal	7
Clutch, primary drive sprocket and primary chain – removal	8
Ignition components – removal	9
Final drive sprocket – removal	10
Timing cover and camshafts – removal	11
Oil pump – removal, inspection and installation	12
Transmission components – removal and disassembly	13
Splitting the crankcases	14
Inspection and repair – general information	15
Crankshaft and connecting rod bearings – inspection	16
Cylinder barrels – inspection	17
Pistons – inspection	18
Valves, valve seats and valve gudes – servicing	19
Cylinder head and valves – disassembly, inspection and reassembly	20
Rocker arms/shafts and pushrods – inspection	21
Camshafts, timing gears, tappets and timing case bearings – inspection	22
Crankcases – inspection	23
Transmission components – inspection	24
Transmission gear cluster – disassembly and reassembly	25
Clutch components – inspection	26
Primary drive components – inspection	27
Kickstart mechanism – inspection	28
Starter drive assembly – inspection	29
Oil seals – inspection and replacement	30
Engine reassembly – general information	31
Crankcases – reassembly	32
Crankcase breather – timing (1976 and earlier models only)	33
Camshafts – installation	34
Ignition components – installation	35
Tappets – installation	36
Transmission gear cluster, primary drive and clutch components – installation	37
Starter motor – installation	38
Final drive sprocket – installation	39
Piston rings – installation	40
Pistons and cylinder barrels – installation	41
Cylinder heads – installation	42
Rocker boxes (pre-Evolution engine) – installation	43
Engine installation	44
Generator (early models) and primary chain cover – installation . .	45
Completing engine reassembly	46
Starting and running the rebuilt engine	47
Recommended break-in procedure	48

Specifications

General

Engine type	45-degree V-twin, four-stroke
Bore	
1970 and 1971	3.00 in (76.2 mm)
1972 through 1985	3.188 in (81.0 mm)
1986-on	
883 cc engine	3.00 in (76.2 mm)
1100 cc engine	3.350 in (85.09 mm)
1200 cc engine	3.498 in (88.85 mm)

Chapter 2 Engine, clutch and transmission

General (continued)

Stroke	3.812 in (96.8 mm)
Displacement	
1970 and 1971	883 cc (53.9 cu-in)
1972 through 1985	997.5 cc (60.9 cu-in)
1986 and 1987	883 cc (53.9 cu-in)/1100 cc (67.2 cu-in)
1988-on	883 cc (53.9 cu-in)/1200 cc (73.3 cu-in)
Oil pressure	
Pre-Evolution engine	
Under normal riding conditions	4 to 15 psi (0.27 to 1.04 Bars)
At idle	4 to 7 psi (0.27 to 0.48 Bars)
Evolution engine (1986 through 1990)	
Measured at adaptor between tappet guide blocks	
At 2500 rpm	5 to 30 psi (0.35 to 2.07 Bars)
At idle	1 to 7 psi (0.07 to 0.48 Bars)
Evolution engine (1991-on)	
Measured at oil pressure switch union	
At 2500 rpm	10 to 17 psi (0.69 to 1.17 Bars)
At 1000 rpm	7 to 12 psi (0.48 to 0.82 Bars)

Pre-Evolution engine

Valves and related components

Valve-to-guide clearance	
Standard	
Intake	0.0015 to 0.0035 in (0.038 to 0.088 mm)
Exhaust	0.0025 to 0.0045 in (0.063 to 0.114 mm)
Service limit	
Intake	0.006 in (0.152 mm)
Exhaust	0.007 in (0.178 mm)
Valve spring free length	
Standard	
1970 through early 1983	
Inner	1-23/64 in (34.54 mm)
Outer	1-1/2 in (38.1 mm)
Late 1983-on	
Inner	1-11/32 in (34.13 mm)
Outer	1-9/16 in (39.69 mm)
Service limit	
1970 through early 1983	
Inner	1-19/64 in (32.94 mm)
Outer	1-7/16 in (36.51 mm)
Late 1983-on	
Inner	1-5/16 in (33.34mm)
Outer	1-1/2 in (38.1 mm)
Rocker arm shaft-to-rocker box bushing clearance	
Standard	0.001 to 0.0025 in (0.025 to 0.063 mm)
Service limit	0.0035 in (0.089 mm)

Valve timing (identical for both cylinders)	**1970 through 1979**	**1980-on**
Intake valve opens	35.4 $\pm$ 3-degrees BTDC	7.5 $\pm$ 3-degrees BTDC
Intake valve closes	41.2 $\pm$ 3-degrees ABDC	42.5 $\pm$ 3-degrees ABDC
Exhaust valve opens	44.3 $\pm$ 4-degrees BBDC	36.0 $\pm$ 4-degrees BBDC
Exhaust valve closes	20.2 $\pm$ 4-degrees ATDC	8.0 $\pm$ 4-degrees ATDC

Tappets	
Fit in cam follower guide	
Standard	0.0005 to 0.001 in (0.013 to 0.025 mm)
Service limit	0.002 in (0.05 mm)
Roller fit	
Standard	0.0005 to 0.001 in (0.013 to 0.025 mm)
Service limit	0.0012 in (0.03 mm)
Roller end clearance	
Standard	0.008 to 0.010 in (0.23 to 0.25 mm)
Service limit	0.012 in (0.03 mm)
Valve clearance (COLD engine)	No lash (pushrods just free to rotate)

Pistons

Piston-to-cylinder clearance	
Standard	
Late 1985 only	0.0025 to 0.0035 in (0.06 to 0.09 mm)
All others	0.003 to 0.004 in (0.076 to 0.101 mm)

Pistons (continued)
Service limit
    Late 1985 only .... 0.0055 in (0.14 mm)
    All others .... 0.006 in (0.15 mm)
Piston ring side clearance
    1970 through 1982 .... 0.0035 to 0.006 in (0.088 to 0.15 mm)
    1983-on .... 0.004 to 0.006 in (0.10 to 0.15 mm)
Piston ring end gap
  Standard
    1970 through 1978 .... 0.015 to 0.025 in (0.38 to 0.63 mm)
    1979 through 1982 .... 0.010 in (0.254 mm)
    1983-on
      Compression rings .... 0.008 in (0.20 mm)
      Oil control ring .... 0.015 in (0.38 mm)
  Service limit
    1970 through 1982
      Compression rings .... 0.031 in (0.78 mm)
      Oil control ring .... 0.050 in (1.27 mm)
    1983-on
      Top compression ring .... 0.022 in (0.56 mm)
      Second compression ring .... 0.030 in (0.76 mm)
      Oil control ring .... 0.055 in (1.40 mm)

Crankshaft and connecting rods
Maximum flywheel runout .... 0.006 in (0.15 mm) at rim
Maximum shaft runout .... 0.002 in (0.051 mm)
Connecting rod end play
  1970 through 1978 .... 0.005 to 0.015 in (0.127 to 0.381 mm)
  1979-on .... 0.005 to 0.030 in (0.127 to 0.76 mm)

Camshafts
Idler gear shaft-to-bushing clearance .... 0.0005 to 0.003 in (0.013 to 0.076 mm)
Camshaft-to-bushing clearance .... 0.0005 to 0.003 in (0.013 to 0.076 mm)
Camshaft bearing clearance .... 0.0005 to 0.003 in (0.013 to 0.076 mm)
Camshaft end play
  1970 through 1976 .... 0.001 to 0.005 in (0.025 to 0.127 mm)
  1977 and 1978 (except rear intake) .... 0.0005 to 0.012 in (0.127 to 0.305 mm)
  1979-on (except rear intake) .... 0.005 to 0.025 in (0.127 to 0.635 mm)
  Rear intake camshaft .... 0.004 to 0.010 in (0.012 to 0.254 mm)
Camshaft gear backlash .... 0.0000 to 0.0005 in (0.00 to 0.0127 mm)

Torque specifications

	Ft-lbs (unless otherwise indicated)	Nm
Gear shaft nut	100 to 120	136 to 163
Sprocket shaft nut	100 to 120	136 to 163
Crankpin nuts	150 to 175	203 to 237
Crankshaft gear nut (right end of crankshaft)		
Through 1978	50	68
1979-on	35 to 45	47 to 61
Cylinder head bolts	55 to 65	75 to 88
Cylinder base nuts	25 to 35	34 to 47
Rocker arm cover bolts	14 to 19	19 to 26
Tappet adjusting locknut	6 to 11	8 to 15
Rear engine mount bolt/nut	16 to 24	22 to 33
Clutch center nut	110 to 150	149 to 203
Compensating sprocket threaded collar	Tighten securely	
Primary drive (engine) sprocket nut	150 to 165	203 to 224
Final drive sprocket nut	35 to 65	47 to 88
Transmission access cover bolts	13 to 15	18 to 20
Primary chaincase drain plug	14 to 21	19 to 28
Primary chain tensioner stud nut	8 to 12	11 to 16
Primary chaincase cover screws	80 to 110 in-lbs	9 to 12
Alternator stator mounting screws		
12-point socket-head/slotted	20 to 35 in-lbs	2.3 to 4
Torx-head	30 to 40 in-lbs	3.4 to 4.5

Evolution engine

Valves and related components
Valve seat width
  Standard .... 0.040 to 0.062 in (1.02 to 1.57 mm)
  Service limit .... 0.090 in (2.29 mm)

Chapter 2 Engine, clutch and transmission

Valve stem protrusion (measured from spring pocket in head)
  Standard .... 1.975 to 2.011 in (50.17 to 51.08 mm)
  Service limit .... 2.031 in (51.59 mm)
Valve-to-guide clearance
  Standard
    Intake .... 0.0008 to 0.0026 in (0.020 to 0.066 mm)
    Exhaust .... 0.0015 to 0.0033 in (0.038 to 0.084 mm)
  Service limit
    Intake .... 0.0035 in (0.09 mm)
    Exhaust .... 0.0040 in (0.10 mm)
Valve spring free length
  Standard
    Outer spring .... 2.105 to 2.177 in (53.47 to 55.30 mm)
    Inner spring .... 1.926 to 1.996 in (48.92 to 50.70 mm)
  Service limit
    Outer spring .... 2.105 in (53.47 mm)
    Inner spring .... 1.926 in (48.92 mm)
Rocker arm shaft-to-rocker arm bushing clearance
  Standard .... 0.0005 to 0.002 in (0.0127 to 0.050 mm)
  Service limit .... 0.0035 in (0.089 mm)
Rocker arm end play
  Standard .... 0.003 to 0.013 in (0.076 to 0.33 mm)
  Service limit .... 0.025 in (0.635 mm)
Rocker arm shaft-to-cover clearance
  Standard .... 0.0007 to 0.0022 in (0.018 to 0.056 mm)
  Service limit .... 0.0035 in (0.09 mm)
Tappets
  Fit in guide
    Standard .... 0.0008 to 0.0023 in (0.020 to 0.058 mm)
    Service limit .... 0.003 in (0.076 mm)
  Roller fit .... 0.0006 to 0.0013 in (0.015 to 0.033 mm)
  Roller end clearance .... 0.010 to 0.014 in (0.25 to 0.36 mm)

Cylinder head warpage limit .... 0.006 in (0.152 mm)

Cylinders
Bore
  Diameter (standard − ± 0.0002 in)
    883 .... 3.0005 in (76.213 mm)
    1100 .... 3.3505 in (85.103 mm)
    1200 .... 3.4978 in (88.844 mm)
  Taper limit .... 0.002 in (0.05 mm)
  Out-of-round limit .... 0.003 in (0.076 mm)
Gasket surface warpage limit
  Top (cylinder head) .... 0.006 in (0.15 mm)
  Base .... 0.008 in (0.20 mm)

Pistons
Piston ring end gap
  Standard
    Compression rings
      883 and 1100 .... 0.010 to 0.023 in (0.254 to 0.58 mm)
      1200 .... 0.007 to 0.020 in (0.178 to 0.508 mm)
    Oil control ring
      883 and 1100 .... 0.010 to 0.053 in (0.25 to 1.35 mm)
      1200 .... 0.009 to 0.052 in (0.15 to 1.32 mm)
  Service limit
    Compression rings .... 0.032 in (0.81 mm)
    Oil control ring .... 0.065 in (1.65 mm)
Piston ring side clearance
  Standard
    Compression rings
      883 and 1100 .... 0.002 to 0.0045 in (0.05 to 0.11 mm)
      1200 (top ring) .... 0.002 to 0.0045 in (0.05 to 0.11 mm)
      1200 (second ring) .... 0.0016 to 0.0041 in (0.04 to 0.10 mm)
    Oil control ring
      883 and 1100 .... 0.0014 to 0.0074 in (0.036 to 0.188 mm)
      1200 .... 0.0016 to 0.0076 in (0.04 to 0.193 mm)

Chapter 2 Engine, clutch and transmission

Pistons (continued)
Service limit
  Compression rings .... 0.0065 in (0.165 mm)
  Oil control ring .... 0.0094 in (0.24 mm)

Crankshaft and connecting rods
Maximum flywheel runout .... 0.010 in (0.25 mm) at rim
Maximum shaft runout .... 0.002 in (0.051 mm)
Connecting rod end play .... 0.030 in (0.127 mm) maximum
Crankshaft end play .... 0.001 to 0.005 in (0.025 to 0.127 mm)

Camshafts
Camshaft-to-bushing clearance .... 0.0007 to 0.003 in (0.018 to 0.076 mm)
Camshaft end play service limits
  Rear intake only .... 0.040 in (1.02 mm) maximum
  All others .... 0.025 in (0.64 mm) maximum

Primary drive side bearing
Outer race-to-crankcase .... 0.0004 to 0.0024 in (0.010 to 0.061 mm)
Inner race-to-shaft .... 0.0002 to 0.0015 in (0.005 to 0.038 mm)

Torque specifications
	Ft-lbs (unless otherwise stated)	Nm
Crankpin nuts	150 to 185	203 to 250
Crankshaft timing gear nut	35 to 45	47 to 61
Cylinder head bolts		
Step 1	7 to 9	9 to 12
Step 2	12 to 14	16 to 19
Step 3	Tighten an additional 90-degrees (1/4 turn)	
Rocker arm cover bolts (through 1990)		
1/4-inch	10 to 13	14 to 18
5/16-inch	15 to 18	20 to 24
Rocker arm cover bolts (1991-on)		
1/4-inch Allen head	90 to 120 in-lbs	10 to 14
1/4-inch hex head	10 to 13	14 to 18
5/16-inch	15 to 18	20 to 24
Tappet lockpin cover screw (1991-on)	80 to 110 in-lbs	9 to 12
Pushrod retaining plate screw (1991-on)	15 to 18	20 to 24
Final drive sprocket nut and lockscrew(s)	See Section 39	
Transmission access cover bolts		
Four-speed	14 to 19	19 to 26
Five-speed	13 to 17	18 to 23
Shifter shaft nuts	90 to 110 in-lbs	10 to 12
Primary chaincase oil drain plug	14 to 21	19 to 28
Primary chain tensioner locknut	20 to 25	27 to 34
Primary chaincase main cover screws	80 to 110 in-lbs	9 to 12
Primary chaincase outer cover screws (1995-on)	30 to 40 in-lbs	4.5 to 6.8
Alternator stator screws	30 to 40 in-lbs	3.3 to 4.5
Primary drive (engine sprocket) nut	150 to 165	203 to 224
Clutch nut (1991-on)	70 to 80	95 to 108

Clutch
Type
  1970 .... Dry, multi-plate
  1971-on .... Wet, multi-plate
Number of plates
  1970 .... 7 plain, 7 friction
  1971 through early 1984 .... 8 plain, 8 friction
  Late 1984 through 1990 .... 6 plain, 7 friction
  1991-on .... 7 plain, 8 friction
Number of springs
  1970 .... Six
  1971 through early 1984 .... Two
  Late 1984-on .... One (diaphragm spring)
Spring adjustment
  1970 .... 3/16 in (5 mm) from inner surface of spring tension adjusting plate-to-outer surface of spring cup flange
  1971 through 1973 .... 11/32 in (9 mm) from outer surface of outer drive plate-to-outer surface of releasing disc
  1974 through early 1984 .... Fixed spacers; 1.53 in (39 mm)

Chapter 2 Engine, clutch and transmission

Pushrod free play (1970 only)	0.095 to 0.115 in (2.42 to 2.92 mm)
Clutch plate thickness (late 1984 through 1990)	
Friction plate	
Standard	0.150 in (3.81 mm)
Minimum	0.130 in (3.3 mm)
Plain (steel) plate	
Standard	0.0629 in (1.60 mm)
Minimum	0.060 in (1.52 mm)
Warpage limit	0.010 in (0.254 mm)
Clutch plate thickness (1991-on)	
Friction plate	
Standard	0.0866 ± 0.0031 in (2.20 ± 0.079 mm)
Service limit (stacked together)	0.661 in (16.79 mm)
Plain (steel) plate	
Standard	0.0629 ± 0.0020 in (1.598 ± 0.051 mm)
Warpage limit	0.006 in (0.152 mm)

Transmission

Type	
1970 through 1990	Four-speed constant mesh
1991-on	Five-speed constant mesh
Sprocket teeth	
Engine (primary drive)	34 (4-speed), 35 (5-speed)
Clutch (primary drive)	59 (4-speed), 56 (5-speed)
Transmission (final drive)	
1970 and 1971 (XLCH only)	19
1970 through 1972	20
1973 through 1992 (chain drive)	21
1991-on (belt drive 883)	27
1991-on (belt drive 1200)	29
Rear wheel (final drive)	
1970 through 1981	51
1982 through 1992 (chain drive)	48
1991-on (belt drive)	61
Transmission oil capacity	See Chapter 1
Primary chain	
Type	3/8-inch triple row chain
Adjustment	See Chapter 1
Mainshaft clearances and wear limits (1970 through 1985)	
Clutch gear bearing-to-gear	0.0009 in (0.023 mm) (tight)
Clutch gear bearing-to-cover	0.001 to 0.0012 in (0.025 to 0.030 mm) (loose)
Clutch gear bushing-to-mainshaft	0.001 to 0.002 in (0.025 to 0.051 mm) (loose)
Roller bearing assembly (right side)	
1970 through 1978	0.0006 to 0.0014 in (0.015 to 0.036 mm) (loose)
1979-on	0.001 to 0.0034 in (0.025 to 0.086 mm) (loose)
End play	
Without axial play (minimum)	0.003 in (0.076 mm) min
With axial play (1979 through early 1984 only)	0.020 in (0.51 mm)
Late 1984-on	
Minimum	0.009 in (0.229 mm)
Maximum	0.015 in (0.381 mm)
Third gear end play	0.012 to 0.030 in (0.25 to 0.76 mm)
Third gear-to-shaft	0.002 to 0.003 in (0.051 to 0.076 mm) (loose)
Countershaft clearances and wear limits (1970 through 1985)	
Bearing clearance (shaft ends)	0.0005 to 0.003 in (0.013 to 0.076 mm) (loose)
Second gear-to-shaft	0.001 to 0.0025 in (0.025 to 0.064 mm) (loose)
First gear-to-shaft	0.0005 to 0.0016 in (0.013 to 0.041 mm) (loose)
First gear end play	0.004 to 0.009 in (0.102 to 0.229 mm)
Drive gear-to-shaft	0.0005 to 0.003 in (0.013 to 0.076 mm) (loose)
Drive gear end play	0.004 to 0.009 in (0.102 to 0.229 mm)
Countershaft end play	0.004 to 0.015 in (0.102 to 0.381 mm)
Clearance between gear faces (1970 through 1978)	
Countershaft First and Third gear	0.038 to 0.058 in (0.965 to 1.473 mm)
Countershaft Second and Third gear	0.038 to 0.058 in (0.965 to 1.473 mm)
Mainshaft Second and Third gear	0.043 to 0.083 in (1.092 to 2.108 mm)
Mainshaft Second and Fourth gear	0.043 to 0.083 in (1.092 to 2.108 mm)

Transmission (continued)

Clearance between gear faces (1979 through 1985)
    Countershaft First and Third gear .... 0.028 to 0.058 in (0.71 to 1.473 mm)
    Countershaft Second and Third gear .... 0.028 to 0.058 in (0.71 to 1.473 mm)
    Mainshaft Second and Third gear .... 0.028 to 0.058 in (0.71 to 1.473 mm)
    Mainshaft Second and Fourth gear .... 0.028 to 0.058 in (0.71 to 1.473 mm)
Shifter shaft end play (1970 through 1976) .... 0.010 to 0.030 in (0.254 to 0.762 mm)
Mainshaft clearances and wear limits (1986 through 1990)
    Clutch gear bearing-to-gear .... 0.0009 in (0.023 mm) (tight)
    Clutch gear bearing-to-cover .... 0.001 to 0.0012 in (0.025 to 0.030 mm) (loose)
    Roller bearing assembly (right side) .... 0.0004 to 0.0035 in (0.010 to 0.089 mm) (loose)
    End play
        Minimum .... 0.006 in (0.152 mm)
        Maximum .... 0.020 in (0.011 mm)
    Third gear end play .... 0.006 to 0.020 in (0.152 to 0.011 mm)
    Third gear-to-shaft .... 0.0027 to 0.004 in (0.069 to 0.102 mm) (loose)
Countershaft clearances and wear limits (1986 through 1990)
    Needle bearing journal diameter
        Access door end .... 0.7490 in (19.02 mm) minimum
        Right crankcase end .... 0.6865 in (17.44 mm) minimum
    Second gear-to-shaft .... 0.001 to 0.0025 in (0.025 to 0.064 mm) (loose)
    First gear needle bearing journal diameter .... 0.6865 in (17.44 mm) minimum
    Countershaft end play .... 0.004 to 0.015 in (0.102 to 0.381 mm)
Clearance between gear faces (1986 through 1990)
    Countershaft First and Third gear .... 0.040 to 0.080 in (1.02 to 2.03 mm)
    Countershaft Second and Third gear .... 0.040 to 0.080 in (1.02 to 2.03 mm)
    Mainshaft Second and Third gear .... 0.040 to 0.080 in (1.02 to 2.03 mm)
    Mainshaft Second and Fourth gear .... 0.040 to 0.080 in (1.02 to 2.03 mm)

1 General information

The engine used in the Harley-Davidson Sportster is a large capacity, 45-degree V-twin, mounted inline with the frame. In this Chapter, the early engine (through 1985) with cast-iron heads and cylinders is referred to as the pre-Evolution engine, while the later engine (1986-on) with aluminum heads and cylinders, as well as redesigned rocker boxes, is called the Evolution engine – the term used by the factory. The Evolution engine was built in three displacements: An 883 cc engine (all years), an 1100 cc engine (1986 and 1987 only) and a 1200 cc engine (1988-on).

Aluminum alloy castings are used throughout the engine, with the above-mentioned exception of the cylinder barrels and heads on pre-Evolution engines, which were cast-iron. The engine has overhead valves actuated by tappets and pushrods from the camshafts mounted in the gearcase at the right-hand side of the engine. The tappets on pre-Evolution engines are two-piece. The Evolution engine has one-piece hydraulic tappets and hollow, one-piece pushrods that feed oil to the rocker arm shafts, eliminating the need for external oil lines. A gerotor-type oil pump, driven off the crankshaft, provides pressurized oil to the main engine components. The transmission and primary drive share a common oil source, which is separate from the engine.

An electric starter is standard on most models. Early models were equipped with a kickstarter. The generator was mounted on the front of the engine through early 1984. Beginning in late 1984, the generator was replaced with an alternator mounted behind the clutch assembly on models through 1990, and behind the primary drive gear on 1991-on models.

A four-speed transmission was fitted on all models through 1990, and was uprated to five-speed from 1991-on. A multi-plate, dry clutch with coil springs was used on 1970 models, while a wet, multi-plate clutch is used on all others. The clutch from late 1984-on is a slightly different design with a large diaphragm spring instead of coil springs. Access to the transmission is available after the primary drive components and clutch have been removed.

Chain final drive to the rear wheel was used through 1992, with belt drive making an appearance on certain models in 1991; by 1993 all models were fitted with belt drive.

2 Repair operations possible with the engine in the frame

It's not necessary to remove the engine from the frame unless the crankcases have to be separated to gain access to the crankshaft, connecting rods or bearings. If only the cylinder heads and related valvetrain components (including the camshafts and tappets), or the pistons, rings or barrels require attention, the work can be done with the engine in the frame, provided the various disassembly procedures outlined in this Chapter are modified slightly. However, if major work or a complete overhaul is necessary, remove the engine from the frame.

Operations that can be completed with the engine in the frame include removal and installation of the ...

Generator or alternator
Starter motor
Transmission and related components
Clutch and primary drive components
Cylinder heads and cylinder barrels
Camshafts and timing gear
Oil pump and related components
Tappets and pushrods
Rocker arms and shafts
Valves, springs, seals and guides
Rocker box gaskets
Cylinder head and base gaskets

This does not, however, apply to the main crankcase assembly, which is split vertically and cannot be separated until the engine is out of the frame. Conversely, the crankcase assembly must be bolted together before it can be reinstalled in the frame.

3 Major engine repair – general information

1   It's not always easy to determine when or if an engine should be completely overhauled, as a number of factors must be considered.

Chapter 2 Engine, clutch and transmission

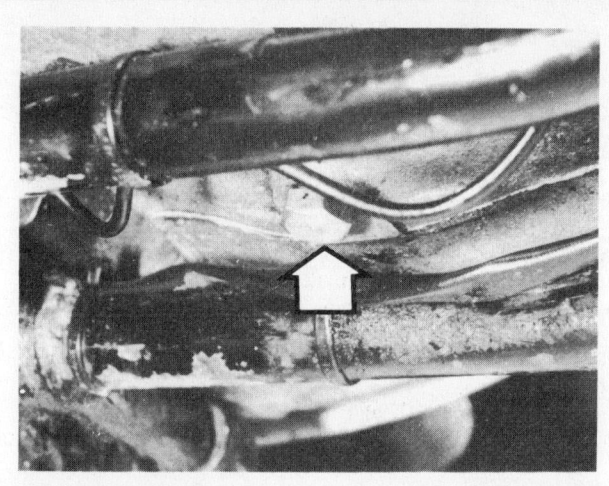

4.4 Location of the crankcase oil drain plug on early models

4.7 The front fuel tank bolt also secures the ignition coil assembly on early models

2 High mileage isn't necessarily an indication an overhaul is needed, while low mileage, on the other hand, doesn't preclude the need for an overhaul. Frequency of servicing is probably the single most important consideration. An engine that has regular and frequent oil and filter changes, as well as other required maintenance, will most likely give many miles of reliable service. Conversely, a neglected engine, or one that hasn't been broken in properly, may require an overhaul very early in its life.

3 Exhaust smoke and excessive oil consumption are both indications that piston rings and/or valve guides need attention. Make sure oil leaks aren't responsible before deciding the rings or guides are bad. Refer to Chapter 1 and perform a cylinder compression check to determine for certain the nature and extent of the work required.

4 If the engine is making obvious knocking or rumbling noises, the engine bearings are probably at fault. Main bearings are generally more durable than rod bearings or bushings, so begin with the latter when trying to make a diagnosis. Certain knocking noises may be caused by piston slap as well as bearings. Usually, valves, piston rings, piston pins, bushings and bearings will all need attention at the same time.

5 Loss of power, rough running, excessive valve train noise and high fuel consumption rates may also point to the need for an overhaul, especially if they're all present at the same time. If a complete tune-up doesn't remedy the situation, major mechanical work is the only solution. To diagnose and repair noisy tappets and valvetrain components, first check the oil pressure with the engine running at normal operating temperature. If it's not as specified, the oil may not be reaching the tappets. If oil is reaching the tappets, they may be dirty or defective. Inspect the pushrods and tappet blocks to make sure everything is installed correctly and look for unusual wear patterns. Check the camshaft lobes for unusual or excessive wear. Make sure the camshafts are correctly matched and not worn or damaged. Inspect the rocker arms and shafts for wear and binding. Check the valves to make sure they fit correctly in the guides and look for scuffing and galling on the stems. The valve seats must be tight in the head and the valves must seat properly.

6 An engine overhaul generally involves restoring the internal parts to the specifications of a new engine. During an overhaul the piston rings are replaced and the cylinders are bored and/or honed. If a rebore is done, then new pistons are also required. All major engine bearings are generally replaced with new ones and, if necessary, the crankshaft is also replaced. Generally the valves are serviced as well, since they're usually in less than perfect condition at this point. While the engine is being overhauled, other components such as the carburetor and the starter motor can be rebuilt also. The end result should be a like-new engine that will give as many trouble-free miles as the original.

7 Before beginning the engine overhaul, read through all of the related procedures to familiarize yourself with the scope and requirements of the job. Overhauling an engine isn't all that difficult, but it is time consuming.

Plan on the motorcycle being tied up for a minimum of four or five weeks. Check on the availability of parts and make sure any necessary special tools, equipment and supplies are obtained in advance.

8 Most work can be done with typical shop hand tools, although a number of precision measuring tools are required for inspecting parts to determine if they must be replaced. Often a dealer service department or motorcycle repair shop will handle the inspection of parts and offer advice concerning reconditioning and replacement. If special tools or equipment are needed for a particular job, let a dealer or repair shop handle it. As a general rule, time is the primary cost of an overhaul so it doesn't pay to install worn or substandard parts.

9 As a final note, to ensure maximum life and minimum trouble from a rebuilt engine, everything must be assembled with care in a spotlessly clean environment.

4 Engine removal

Refer to illustrations 4.4, 4.7, 4.8, 4.9, 4.10a, 4.10b, 4.17a, 4.17b, 4.17c, 4.18a, 4.18b, 4.19a, 4.19b, 4.20a, 4.20b, 4.20c, 4.20d, 4.21a, 4.21b, 4.22a, 4.22b, 4.26 and 4.27

1 Since the Sportster isn't equipped with a centerstand, the initial disassembly should be done with the machine resting on the sidestand, unless some form of rigid stand is available (like the ones used in many repair shops). Because the lower tubes of the cradle frame are quite close together, the machine won't be too stable if it's supported on the tubes.

2 If the motorcycle is on its side stand, be sure it's standing firmly on level ground and the wheels are chocked so it can't roll forward. Make sure the fuel valve is in the OFF position.

3 Place a drain pan beneath the transmission drain plug, remove the plug and drain the oil (see Chapter 1, Section 23).

4 Drain the engine oil as described in Chapter 1, Section 17. Note that early models also have a crankcase drain plug; remove this plug and allow any residue in the crankcases to drain **(see illustration)**.

5 When all the oil has drained, replace and tighten all the drain plugs, after making sure their sealing washers are clean and in good condition. Note the primary chaincase drain plug is magnetic; any metal particles that have collected should be cleaned off. Tighten the drain plugs now, otherwise this point may be overlooked during reassembly.

6 Disconnect the fuel hose from the carburetor at the fuel valve. The hose is retained by hose clamps.

7 Remove the fuel tank (see Chapter 3). Note that on some models the front bolt also holds the ignition coil assembly in position above the front cylinder head **(see illustration)**. Don't lose the two short spacers for the front fuel tank mount. Lift the fuel tank off and store it in a safe place. Disconnect the spark plug wires and the primary wires from the ignition coil.

Chapter 2 Engine, clutch and transmission

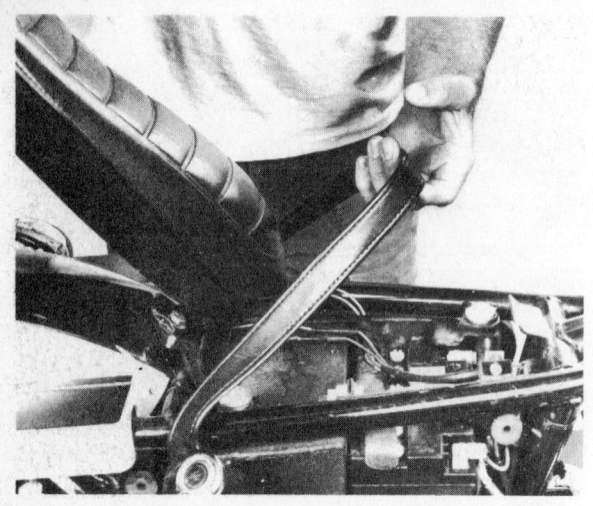

4.8 Remove the seat on later models to gain access to the battery

4.9 Remove the screws securing the air cleaner to the carburetor

4.10a The exhaust pipes and mufflers are retained by clamps on early models (arrow)

4.10b Note how the muffler clamp is attached to the frame

Label the wires so they're reinstalled on the proper terminals on the coil. Remove the coil from the frame. On 1993-on 883 models, disconnect its wires and remove the horn.

8 On early models, release the battery retaining strap and take off the top cover of the battery box. Tilt the side cover to clear the frame and release it. On later models, remove the seat to gain access to the hold down bracket and the battery terminals **(see illustration)**. Disconnect the two battery cables (negative cable first) and remove the battery. It should be placed in a safe place, as it's easily damaged and will spread a corrosive liquid, if spilled. Note how the vent pipe is routed.

9 Remove the air cleaner from the carburetor intake by unscrewing the bolts through the outer cover. Take off the outer cover of the air cleaner assembly and lift out the element. Access is now available to the screws/bolts that retain the remainder of the assembly to the carburetor and brackets or cylinder head **(see illustration)**.

10 Remove the exhaust system, either separately on early models, or as a complete unit on later models with the crossover pipe. The cylinder head joint will be either a screw clamp type **(see illustration)** or a nut and stud arrangement, depending on the year of manufacture. The mufflers are retained to brackets at various points, although note that on later models the rear brake master cylinder linkage may well obscure the front cylinder pipe nut; disconnect the linkage at the clevis for access **(see illustration)**.

11 Remove the feed and return hoses from the oil tank. The clamps securing the hoses from the factory must be cut off and cannot be reused. Remove the oil tank mounting bolts and nuts and lift the tank out of the frame.

12 In order to remove the battery carrier, which is rubber mounted to offset the effect of engine vibration, you'll have to detach the wires to the regulator and remove the regulator (not all models). Make a note of the connections to aid reassembly. When the regulator is detached, access is available to the rear battery carrier mount.

13 Disconnect the starter motor cable from the starter solenoid.

14 Where a rear disc brake is fitted, detach the master cylinder from the sprocket cover. Release any hydraulic pipe/hose ties and secure the unit to the frame to prevent strain or distortion of the hydraulic line.

15 Remove the sprocket cover from the right-hand side, noting that if a kickstart is fitted the lever must first be removed, and return spring tension released slowly using a screwdriver. Also, on 1970 models release the clutch cable from its mechanism. Where fitted, remove the speedometer drive cable. When removing the sprocket cover bolts take note of their

Chapter 2 Engine, clutch and transmission

4.17a Unscrew the tachometer cable from the timing cover

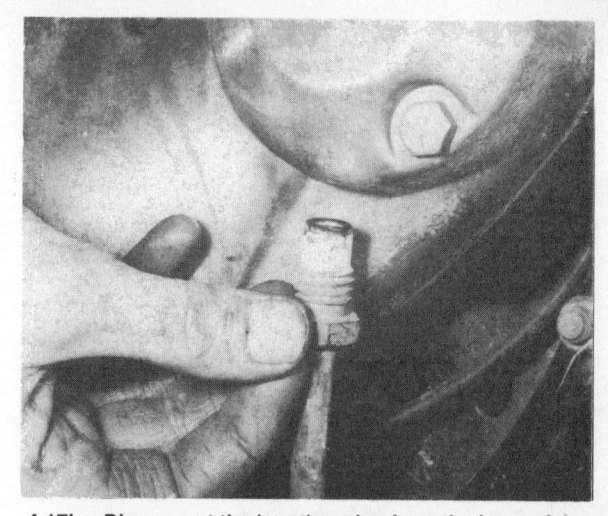

4.17b Disconnect the breather pipe from the base of the circular housing

4.17c Disconnect the wire (arrow) from the oil pressure switch – early model location

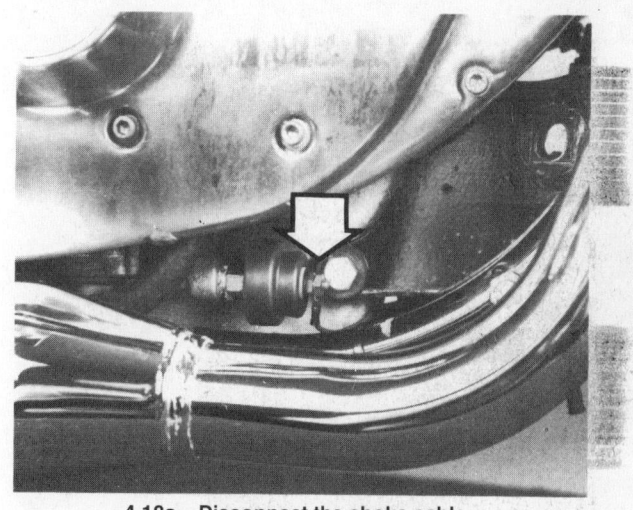

4.18a Disconnect the choke cable . . .

4.18b . . . and the throttle cable from the carburetor

length to ensure they are installed in their original positions. If dowels are fitted remove these for safekeeping. On chain drive models, disconnect the chain at its master link and lift it off the sprocket. On belt drive models, slacken the rear axle nut and chain adjusters to allow the rear wheel to be moved forward, then slip the belt off its front sprocket. Disconnect the wire from the neutral switch.

16 On later models, the right-hand footpeg and brake pedal will have been detached with the sprocket cover, whereas on early models they must be removed separately. **Note:** *On 1970 through 1974 models the gear shift lever is on the right-hand side.*

17 Where a cable-driven tachometer is fitted, disconnect its drive cable from the top of the timing cover **(see illustration)**. On early models, unscrew the breather pipe union from the timing cover **(see illustration)**. On all models, disconnect the wire from the oil pressure switch, located at the base of the timing cover on early models, and on the filter housing on later models **(see illustration)**.

18 Disconnect the control cables from the carburetor by loosening the set screws **(see illustrations)**. Note they have specially protected ends to prevent the cable from spreading or fraying when pressure is applied. On later models the cables must be disconnected at the throttle first. Disconnect the choke cable.

19 Remove the left-hand footpeg and either the rear brake pedal, on

Chapter 2 Engine, clutch and transmission

4.19a Remove the left footpeg and the shift lever on 1975 and later models

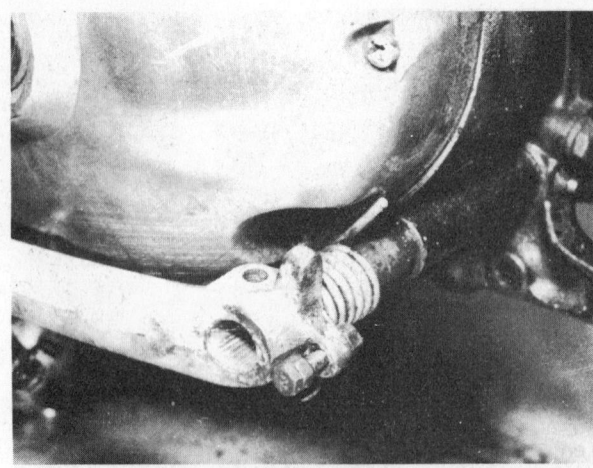

4.19b Release the return spring tension before removing the brake lever on 1970 through 1974 models

4.20a The primary chaincase cover can be removed without disconnecting the clutch cable

4.20b On pre-1981 electric start models, the thrust washer on the starter motor shaft extension is loose and easily lost

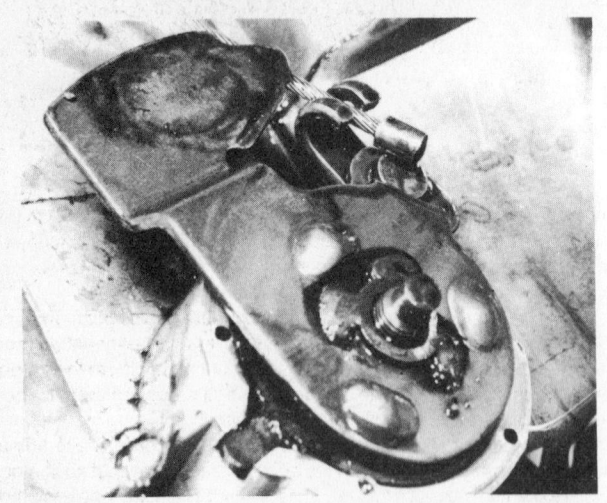

4.20c Separate the clutch cable from the operating arm inside the chaincase cover – early type can be maneuvered out of bracket . . .

4.20d . . . but later type requires dismantling of mechanism to allow cable disconnection

1970 through 1974 models, or the shift lever on 1975 and later models **(see illustrations)**. Ease off and release the tension on the brake pedal return spring before the lever is withdrawn.

20 Before removing the primary chaincase, slacken the primary chain adjuster locknut and back off the adjuster a few turns. On late-1984-on models, remove the small inspection cover to gain access to the clutch adjuster mechanism; lift out the spring and hex piece, then screw the adjuster all the way in so that the nut can be withdrawn. On all models remove all of the chaincase screws and lift the chaincase off the crankcase **(see illustration)**. The casing screws are of differing lengths, so keep them in a

Chapter 2 Engine, clutch and transmission

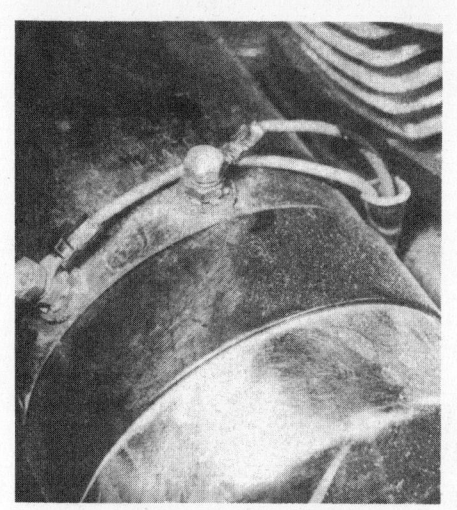

4.21a Label the wires and terminals, then disconnect the wires from the generator

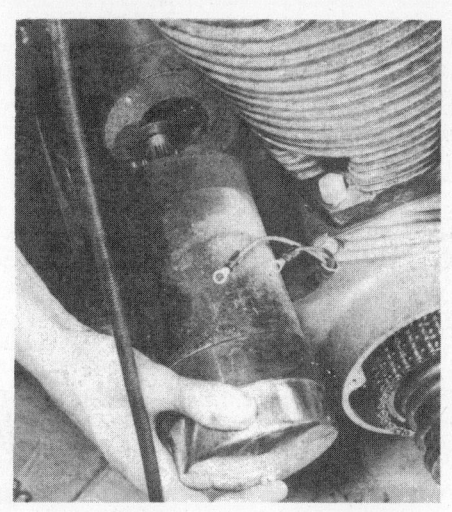

4.21b Tilt the generator while removing it, to give the oil thrower clearance

4.22a Note how the bracket is bolted to the cylinder head – early model shown

4.22b Note the arrangement of the washers at the frame connections – early model shown

4.26 Small plates form the front engine mounts

cardboard template of the cover (the old gasket will do) to ensure they are returned to their original locations. Note the thrust washer on the starter shaft of pre-1981 models, and place it with the chaincase for safekeeping **(see illustration)**. To remove the casing completely you'll have to detach the clutch cable. On early models maneuver the cable trunnion out of its bracket, but on later models dismantle the release mechanism to free it **(see illustration)**.
21 On models through early 1984, disconnect its wires and withdraw the generator from the front of the engine by removing the two bolts on the right side of the motorcycle **(see illustration)**. Tilt the generator as it is withdrawn so that its oil thrower clears the idler gear **(see illustration)**.
22 Disconnect the horn (except 1993-on models), ignition/lighting switch, vacuum-operated electric switch (later models) and choke cable from the upper engine steady bracket. Taking note of the exact position of any washers and ground connections, remove the bracket from the cylinder heads **(see illustrations)**.
23 On later models, unbolt and remove the front mounting bracket securing the front cylinder to the frame tubes; again take note of the washer positions.
24 Remove the carburetor and intake manifold (see Chapter 3).
25 Release the ignition module wiring and on 1979 models, release the cable ties securing the module wiring to the frame tubes. On late 1984 (alternator equipped) models, separate the regulator wiring at the connector along the frame tubes. On earlier (generator-equipped) models, disconnect the individual wires from the regulator if not already done.
26 Keeping a note of washer and ground connection positions, remove the bolts and brackets which form the front lower engine mounting **(see illustration)**. Similarly disconnect the rear mounting pad from the frame. Many of the mounting bolts are of differing lengths – return them to their relative positions once the engine has been removed.
27 The engine is now ready to be lifted out of the frame. Make a final check that all wiring connections have been detached and that no components are likely to prevent removal. Enlist the aid of at least one assistant to help take the weight of the engine as it is removed, and if the operation is being undertaken on the side stand it is a good idea to have someone else hold the frame steady. Special care is necessary as the engine has only minimal clearance in the frame at certain points. On 1970 through 1985 models the engine must be removed from the left-hand side, while on 1986-on Evolution models it must be removed from the right **(see illustration)**. Tip the top of the engine slightly off center to ease removal. **Caution:** *Do not lay an Evolution engine down on its primary side otherwise the clutch cable adjusting screw or cable end fitting will be damaged.*

4.27 Lift the engine out from the left side of the frame on pre–Evolution models

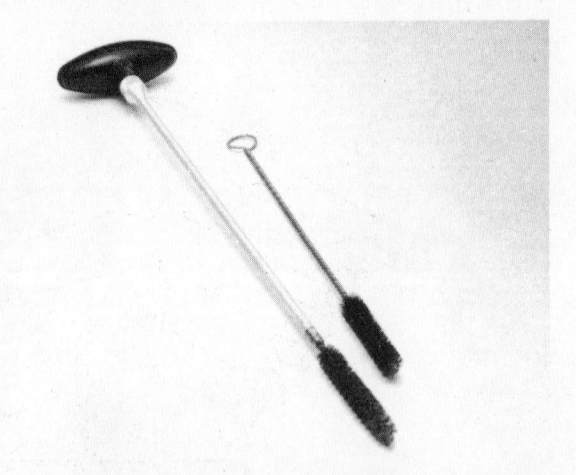

5.2 A selection of brushes is required for cleaning holes and passages in the engine components

5 Engine disassembly and reassembly – general information

Refer to illustration 5.2

1 Before beginning work on the engine, the external surfaces should be cleaned thoroughly. A motorcycle engine has very little protection from road grit and other foreign material, which sooner or later will find its way into the dismantled engine if this simple precaution isn't followed.

2 In addition to the precision measuring tools mentioned earlier, you will need a torque wrench, a valve spring compressor, oil line brushes **(see illustration)**, a motorcycle piston ring removal and installation tool and a piston ring compressor (low-profile type). Some new, clean engine oil, some engine assembly lube or moly-base grease and several types of Loctite will also be required.

3 An aerosol engine degreaser can be used effectively, especially if it's allowed to penetrate the film of grease and oil before it's washed away. When rinsing it off, make sure the water can't get into the internal areas of the engine – plug or cover all openings.

4 If the engine hasn't been disassembled before, either purchase or borrow an impact screwdriver. This will prevent damage to the screws used for engine assembly, most of which will be very tight. If an impact screwdriver isn't available, you can use a screwdriver bit attached to an extension and a sliding T-handle as a substitute.

5 Disassemble the engine on a clean surface and have a supply of clean, lint-free rags available.

6 Never use force to remove any stubborn part unless specific mention is made of it in the text. There's usually a good reason why a part is difficult to remove, often because the disassembly operation has been tackled in the wrong sequence.

7 Disassembly will be easier if a simple engine stand is constructed that will accept the engine mounting bolts. This arrangement will permit the engine to be clamped rigidly to the workbench, leaving both hands free for disassembly.

8 When disassembling the engine, keep "mated" parts together (including gears, cylinders, pistons, etc. that have been in contact with each other during engine operation). These "mated" parts must be reused or replaced as an assembly.

6 Cylinder heads, barrels and tappets – removal

1 Begin by removing the intake manifold (if it's still in place). On pre-Evolution engines, it's secured by two clamps. Loosen both clamp screws and pull the intake manifold out of the head fittings (note that there's an O-ring inside each port, to ensure an airtight joint). On Evolution engines, the manifold is attached to each head with one Allen-head screw and one nut.

Pre-Evolution engine

Refer to illustrations 6.2a, 6.2b, 6.3a, 6.3b, 6.4a, 6.4b, 6.5, 6.6a and 6.6b

2 Remove the rocker boxes. Begin by unscrewing the fittings and removing the small diameter external oil lines, which are easily damaged **(see illustrations)**.

6.2a Disconnect the two oil feed lines (arrows) to the rocker boxes (pre-Evolution engine)

6.2b Note the rubber seal inside the union nut

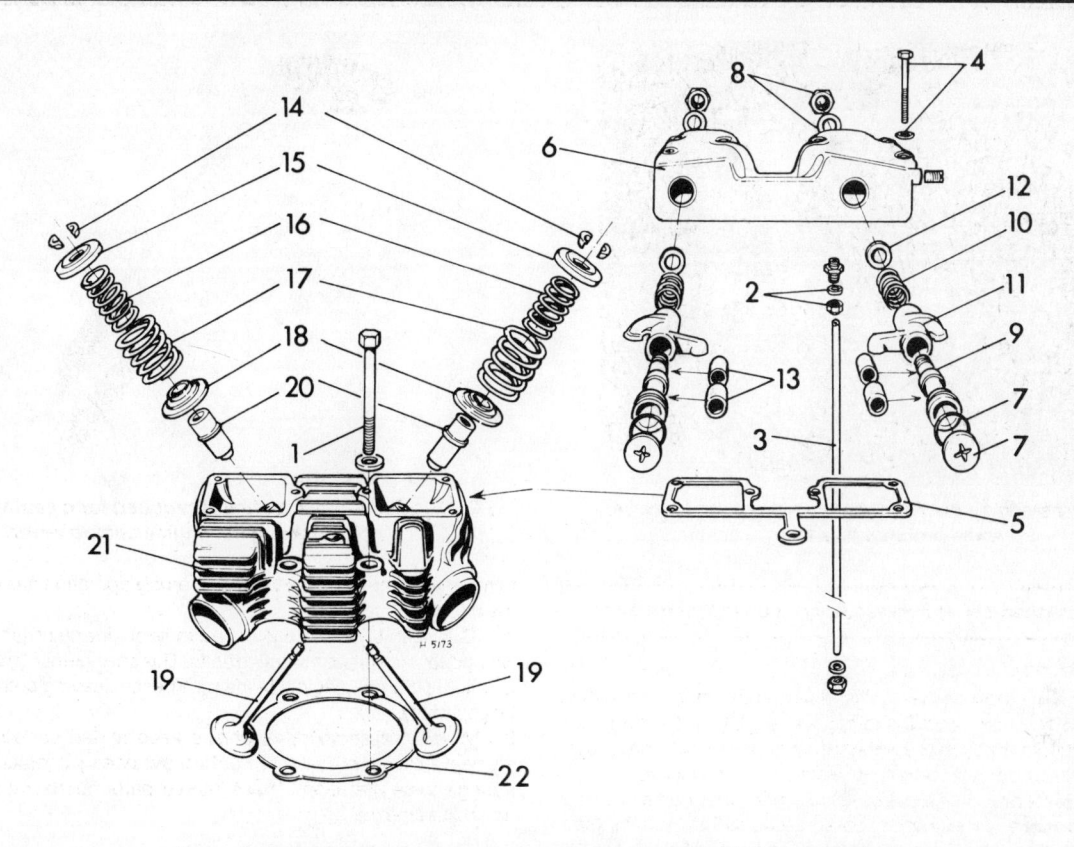

Chapter 2 Engine, clutch and transmission

6.3a Cylinder head components – exploded view (pre-Evolution engine)

1 Cylinder head bolt and washer
2 Oil line union nut and rubber seal
3 Rocker box cover-to-crankcase oil line
4 Rocker box cover bolt and washer
5 Rocker box cover gasket
6 Rocker box cover
7 Rocker arm shaft end cap and O-ring
8 Rocker arm shaft acorn nut and washer
9 Rocker arm shaft
10 Rocker arm spring
11 Rocker arm
12 Rocker arm spacer
13 Rocker arm bushing
14 Valve keepers
15 Valve retainer
16 Inner valve spring
17 Outer valve spring
18 Valve spring seat
19 Intake and exhaust valve
20 Intake and exhaust valve guides
21 Cylinder head
22 Cylinder head gasket

6.3b After the bolts are removed, detach the rocker boxes (if they're stuck, dislodge them with a soft-face hammer)

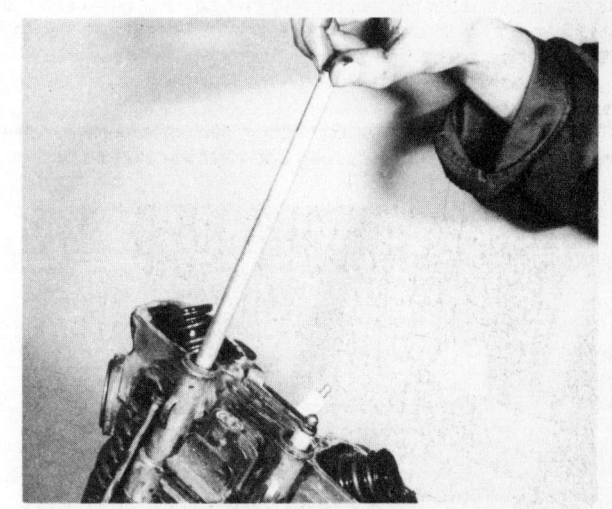

6.4a Mark the pushrods during removal – they must be reinstalled in their original locations

3 Each rocker box is secured by seven bolts (**see illustration**), which should be loosened in 1/4-turn increments. Make sure the valves of the cylinder involved are closed, to avoid stress on the casting from an open valve. Mark the rocker boxes so they can't be interchanged and remove them (**see illustration**).

4 The pushrods can be lifted out of the tubes and clearly marked to make sure they're eventually replaced in their original locations (**see illustration**). The pushrod tubes can then be detached, each as a complete assembly (**see illustration**).

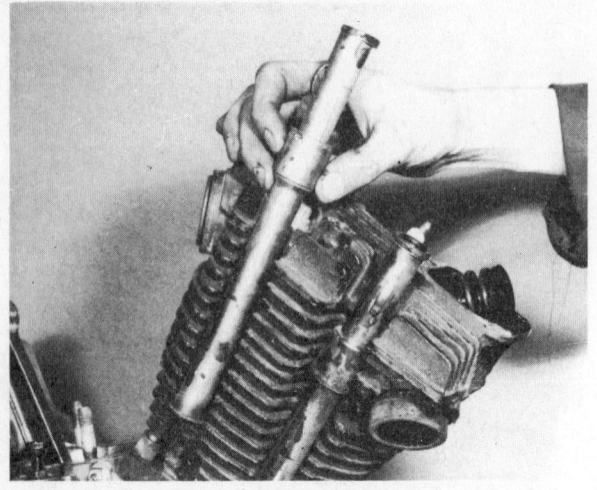

64   Chapter 2   Engine, clutch and transmission

6.4b   Remove the pushrod tubes as an assembly

6.5   DO NOT pry the cylinder head off the barrel – damage could result

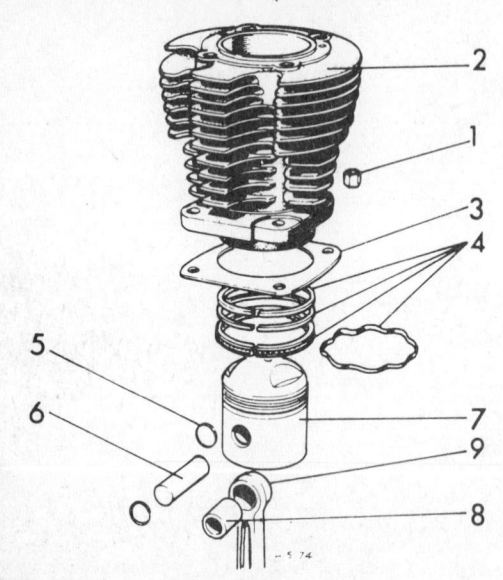

6.6a   Cylinder barrel, piston and related components – exploded view (pre-Evolution engine)

1   Cylinder base nut
2   Cylinder barrel
3   Cylinder base gasket
4   Piston ring set
5   Circlips
6   Piston pin
7   Piston
8   Connecting rod bushing
9   Connecting rod

6.6b   Carefully lift the barrels off after removing the base nuts – don't let the piston strike the crankcase

Evolution engine

Refer to illustrations 6.7, 6.8, 6.9, 6.10 and 6.14

7   On models through 1990, push down on the front pushrod tube spring retainer **(see illustration)**, then remove the keeper and pull the upper tube out of the recess in the underside of the cylinder head (it's a good idea to start with the front head). Repeat the procedure for the remaining pushrod tube.

8   On all models, remove the Allen-head screws and washers from the upper rocker arm cover, then detach the cover **(see illustration)**. If the cover is stuck, tap it gently with a soft-face hammer to break the gasket seal.

9   Remove the middle rocker arm cover **(see illustration)**. Remove and discard all rocker cover gaskets and use new ones during reassembly.

10   Turn the crankshaft until both valves in the affected head are closed, then remove the large bolts retaining the lower rocker arm cover **(see illustration)**.

11   On models through 1990, carefully tap the rocker arm shafts out of the lower rocker arm cover (from the left side), then lift out the rocker arms and remove the pushrods and tubes (**Note:** *the pushrods must not be interchanged or turned end-for-end, so make sure they're marked!*). Make sure the parts are marked or stored in marked containers – they must be returned to their original locations during reassembly.

12   On all models, remove the lower rocker arm cover-to-cylinder head bolts. There are two Allen-head bolts and three hex-head bolts.

13   Detach the lower rocker arm cover and gaskets. Discard the gaskets and use new ones during reassembly.

5   Each cylinder head is retained by four bolts, which will be very tight **(see illustration 6.3a)**. Loosen them in 1/4-turn increments, following a criss-cross pattern, until they can be removed by hand. If the cylinder head isn't free after the bolts have been removed, tap it lightly with a soft-face hammer in an attempt to break the seal. Don't attempt to pry it off, or fins will be broken. There's no point in marking the cylinder head, since they are different and can't be interchanged **(see illustration)**.

6   Each cylinder barrel is retained by four nuts around the base flange. When the nuts are removed, the cylinder barrel can be detached **(see illustrations)**. Be sure to catch the piston as it emerges from the bore. If it's allowed to fall free, either the piston or the piston rings may be damaged. If only a top-end overhaul is required, pad the mouth of the crankcase with a clean rag as each barrel is removed. This will prevent particles of broken piston ring or other foreign matter from dropping into the crankcase as each piston emerges from its bore. Otherwise you would have to strip the engine completely in order to retrieve the broken parts.

Chapter 2 Engine, clutch and transmission

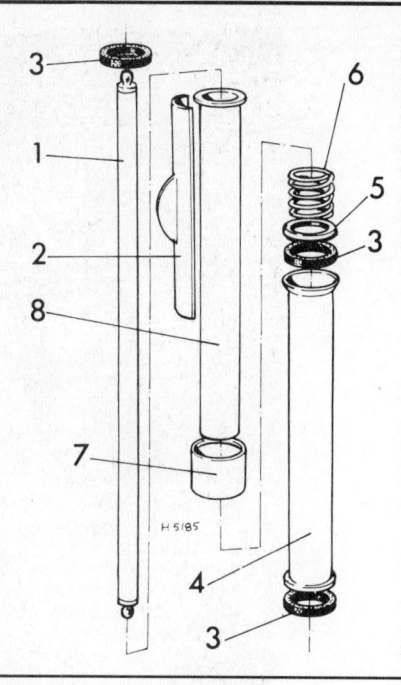

6.7 Pushrod tube assembly – exploded view

1. Pushrod
2. Keeper
3. Sealing washer
4. Lower tube
5. Washer
6. Spring
7. Spring retainer
8. Upper tube

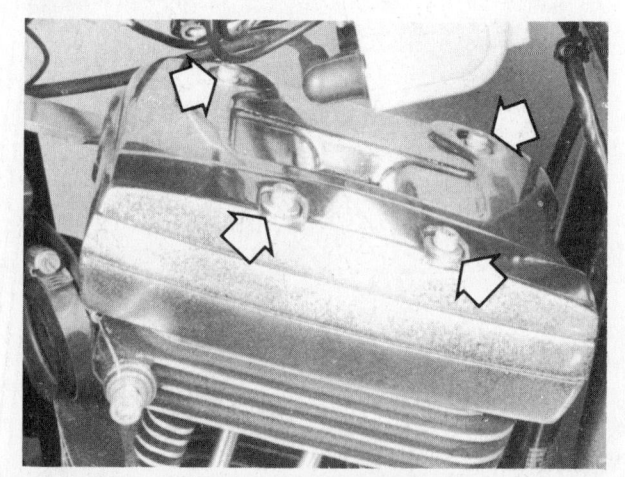

6.8 Remove the four Allen-head screws (arrows) and detach the upper rocker arm cover, . . .

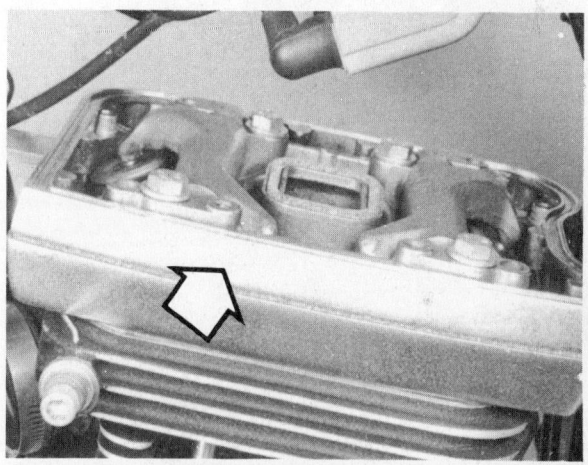

6.9 . . . then lift off the middle rocker arm cover and gasket (arrow) (Evolution engine)

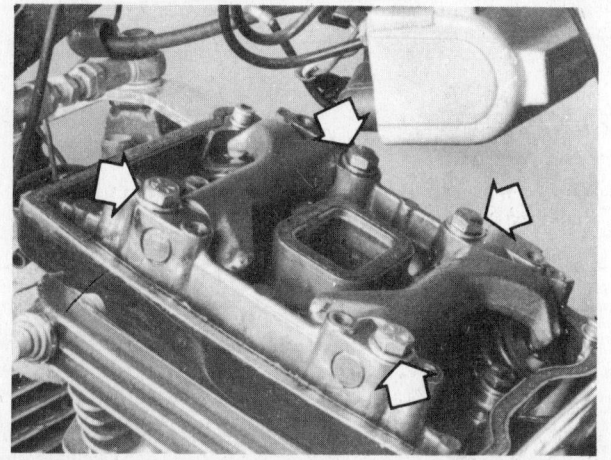

6.10 Remove the four large bolts (arrows) (the two on the right side retain the rocker arm shafts), then make sure the valves are completely closed and drive out the shafts from the left side, which will release the rocker arms

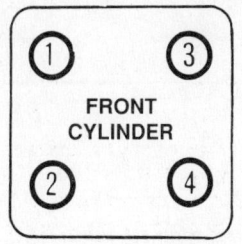

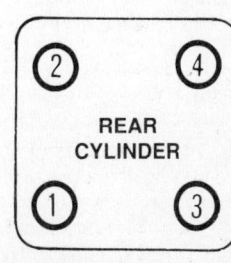

FRONT CYLINDER

1 3
2 4

REAR CYLINDER

2 4
1 3

6.14 Cylinder head bolt loosening/tightening sequence

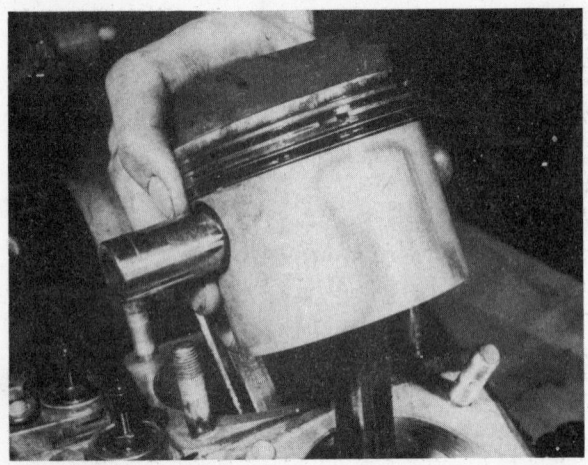

Chapter 2 Engine, clutch and transmission

6.21 Carefully remove one circlip and press the piston pin out

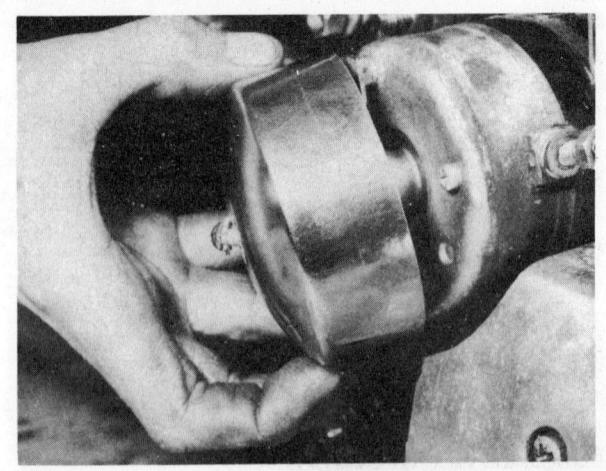

7.2a Remove the end cover from pre-1981 model starters (it's held by a single screw) . . .

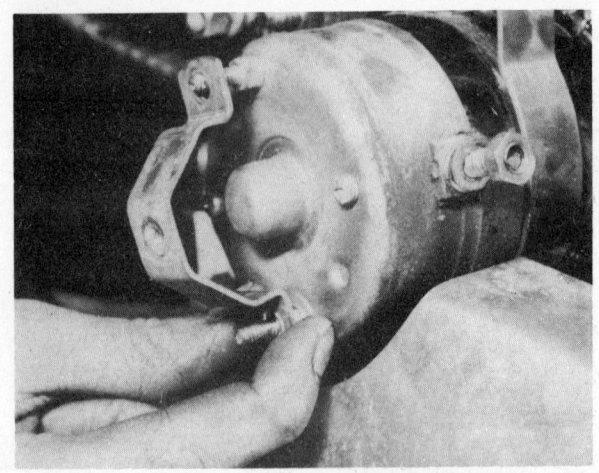

7.2b . . . and take off the mounting bracket, then . . .

7.2c . . . unscrew the long through bolts

14 Loosen the cylinder head bolts in 1/8-turn increments, following the recommended sequence **(see illustration)**. If they aren't loosened gradually, the head, cylinder or crankcase studs may be distorted. Remove the bolts and any washers installed under them (on 1988 and later models, washers aren't used).

15 Carefully lift off the cylinder head, then remove the gasket and O-rings around the dowel sleeves. Discard the gasket and O-rings and use new ones during installation.

16 Mark the cylinder (front or rear), then raise it enough to place a clean shop rag under the piston. This will prevent debris (such as broken piston rings) from falling into the crankcase.

17 Remove the cylinder. Be very careful not to scratch or bend the crankcase studs and don't allow the piston to fall and be damaged. Once the cylinder is off, slip six-inch sections of rubber hose over the studs to protect them and the piston. If the studs are damaged in any way, they could fail during engine operation.

18 On 1991-on models, the rocker shafts can be removed from the lower rocker cover at this stage. Tap them out from the left side, then lift out the rocker arms.

19 On 1991-on models the pushrods and tappets can now be removed. Dealing with each pushrod assembly separately, remove the Allen-head screw and washer to allow the retainer plate to be slid up and off the pushrod tube. Pull the tube out of the crankcase and lift out the pushrod. **Note:** Label the pushrod to identify its bore and top orientation.

Both engines
Refer to illustration 6.21

20 Wear safety glasses when removing the piston pin circlips. Remove one circlip from each piston and discard it – circlips should never be re-used. To prevent the circlip from flying out, place your thumb over it as it's removed from the groove with an awl.

21 Press the pin out and remove the piston **(see illustration)**. If the piston pin is tight, warm the piston crown to expand the metal. To do this, place a clean rag in boiling water, wring it out and wrap it around the piston crown. Mark each piston inside the skirt, to ensure it's installed facing the right direction in the correct cylinder during reassembly. **Note:** *The pistons used in the Evolution engine are very hard and should be handled with care. If they get scratched or gouged, they could score the cylinder(s) during engine operation.*

22 To release the tappets on all models through 1990, release the single bolt and withdraw the guide block. Lift out the tappet. Label both tappet and guide block (if no manufacturer's marks exist) to ensure they are returned to their original positions on installation.

23 To release the tappets on 1991-on models, remove the triangular cover at the centre of the pushrod bores and pull out the tappet retaining pin, using pliers if necessary; lift the tappet out of its bore and label it for identification.

7 Starter motor and solenoid – removal

Caution: *Ensure that the battery negative cable is disconnected before working on the starter motor.*

1970 through 1980
Refer to illustrations 7.2a, 7.2b, 7.2c, 7.2d, 7.3a and 7.3b

1 To remove the starter motor, first detach the large wire. If the cam fol-

Chapter 2 Engine, clutch and transmission

7.2d The starter motor can now be detached

7.3a Press in the spring loaded plunger and withdraw the pin

7.3b The solenoid can be unscrewed from the rear of the chaincase

7.7 Starter motor wires – 1981-on (arrow)

lowers have been removed, the clamp around the starter motor body will already be loose, as it's retained by the rear cam follower housing bolt.

2 Remove the starter motor end cover and the mounting bracket. This will permit access to the two long through bolts that retain the starter motor to the rear of the primary chaincase **(see illustrations)**. When these two bolts have been loosened, the starter motor can be removed as a complete unit **(see illustrations)**.

3 Moving to the left-hand side of the machine, remove the primary chaincase (see Section 4, Step 20) and press in the spring loaded plunger of the starter solenoid so the pin that normally seats in the retainer cap can be removed **(see illustration)**. Lift off the cup and spring. The solenoid itself can now be removed from the back of the chaincase by disconnecting the wire and removing the two mounting bolts **(see illustration)**.

4 Remove the single countersunk screw found in the top of the chaincase casting, immediately above the clutch. This will release the solenoid reaction lever which, when removed, will permit the starter drive assembly to be lifted out, together with the casting bolted to the back of the primary chaincase. Don't lose the bronze washer on the end of the shaft.

1981-on

Refer to illustrations 7.7 and 7.9

5 Remove the left footpeg, footpeg bracket and the shift lever.

6 Remove the primary chaincase cover on the left side of the motorcycle as described in Section 4, Step 20. Without disconnecting the clutch cable, pivot the chaincase cover out of the way so the starter mounting bolts are accessible.

7 Disconnect the wires from the starter motor and the solenoid **(see illustration)**.

8 Remove the exhaust system to gain clearance to remove the starter motor. On some models it's possible to remove the rear exhaust pipe only.

9 Remove the starter motor mounting bolts from inside the primary chaincase and detach the starter motor from the engine **(see illustration)**.

7.9 Starter motor mounting bolts – 1981-on

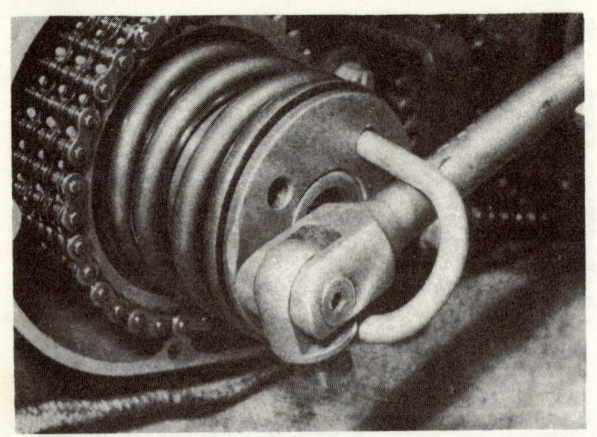

Chapter 2 Engine, clutch and transmission

8.2a A large U-bolt can be used to unscrew the compensating sprocket threaded collar

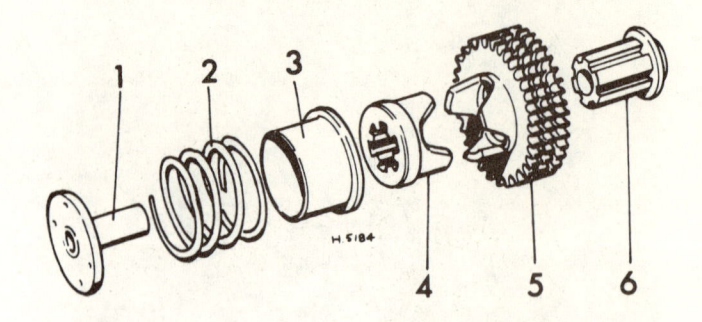

8.2b Compensating sprocket components – exploded view

| 1 | Threaded collar | 3 | Outer collar | 5 | Engine sprocket |
| 2 | Compensating spring | 4 | Sliding cam | 6 | Splined collar |

8 Clutch, primary drive sprocket and primary chain – removal

Primary drive (engine) sprocket

Refer to illustrations 8.2a, 8.2b, 8.2c and 8.3

1 Because a triple row endless chain is employed for the primary drive, the engine and clutch sprockets have to be lifted off in unison with the chain. This makes it necessary to remove the clutch and the primary drive sprocket (compensating sprocket on models produced before 1976) first. In order to do this, the transmission must be shifted into gear (temporarily install the shift lever) and the drive sprocket must be held with a chain wrench or other tool to keep it from turning as the nut/threaded collar is loosened. If the engine is in the frame, shift the transmission into gear and apply the rear brake with the rear tire in firm contact with the ground. Another way to lock the transmission, if the cylinders and heads are still in place, is to remove the spark plugs and rotate the engine to top dead center. Back the engine up about 1/8-revolution and fill the combustion chamber with nylon cord inserted through the spark plug hole until no more cord can be fed in. Be sure the end of the cord is still outside of the engine. On certain models from 1992-on, a lockplate may be fitted over the sprocket nut; remove its lock screw and lift the plate off the sprocket nut before attempting nut removal.

2 On early models (with compensating sprocket), if the Harley-Davidson service tool is not available note that the threaded collar can be loosened with a large U-bolt and screwdriver or pry bar **(see illustration)**. Unscrew the collar and remove the spring, outer collar and the splined sliding cam **(see illustrations)**.

3 On later models, unscrew the sprocket nut fully and remove it from the crankshaft end **(see illustration)**. **Warning:** *This nut will be very tight – ensure that you have adequate means of locking the engine to prevent rotation.*

4 On all models the engine sprocket is now free to be removed from the shaft, but as stressed in Step 1, it must be withdrawn as a unit with the clutch and chain. Refer to the appropriate sub-Section which follows and remove the clutch from the mainshaft.

8.2c The compensating spring pressure is released before the end of the collar threads

8.3 Primary drive sprocket is retained by a flanged nut on later models

8.5a Heavy springs require a great deal of caution when removing the release disc

Chapter 2 Engine, clutch and transmission

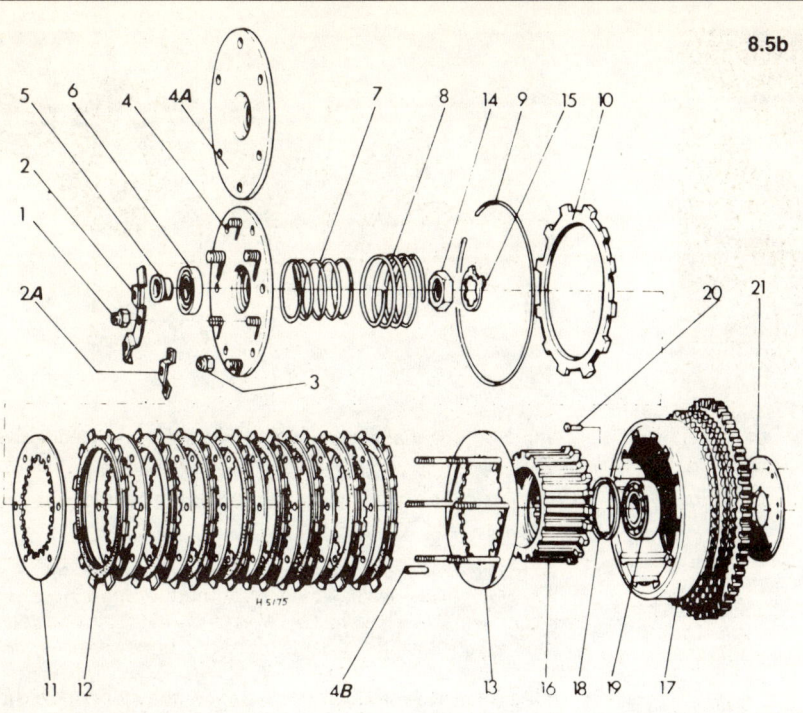

8.5b Wet-type clutch (1971 through early 1984 models) – exploded view

1. Retainer nut
2. Retainer (early 1971)
2A. Retainer (late 1971-on)
3. Pressure plate nut
4. Release disc (thru early 1974)
4A. Release disc (late 1974-on)
4B. Stud spacer (late 1974-on)
5. Release disc collar
6. Release disc bearing
7. Inner spring
8. Outer spring
9. Retaining circlip – outer drive plate
10. Outer drive plate
11. Plain plate
12. Friction plate
13. Pressure plate
14. Clutch center nut
15. Lock washer
16. Clutch center
17. Clutch drum/sprocket assembly
18. Retaining circlip – clutch drum bearing
19. Clutch drum bearing
20. Rivet
21. Kickstarter ratchet

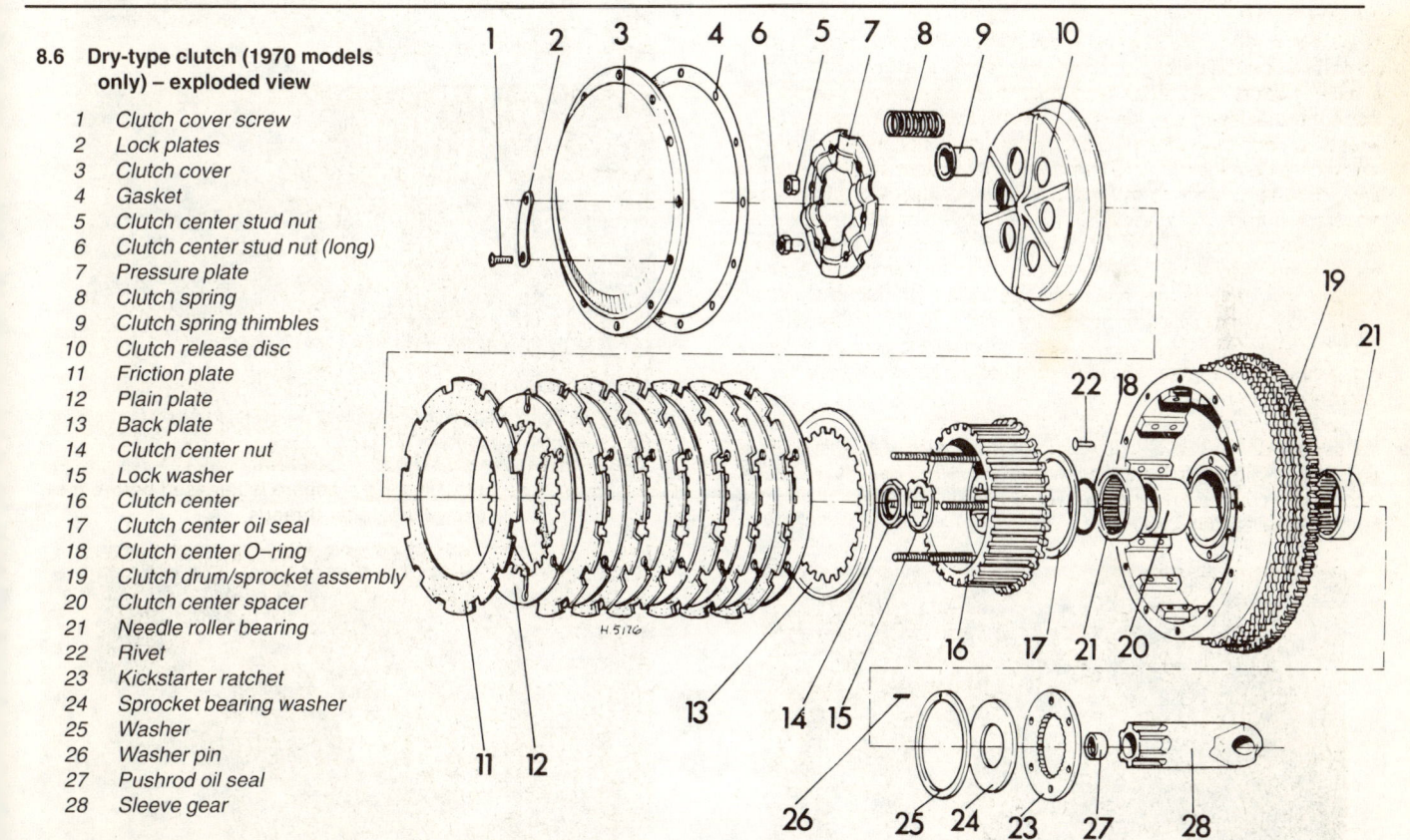

8.6 Dry-type clutch (1970 models only) – exploded view

1. Clutch cover screw
2. Lock plates
3. Clutch cover
4. Gasket
5. Clutch center stud nut
6. Clutch center stud nut (long)
7. Pressure plate
8. Clutch spring
9. Clutch spring thimbles
10. Clutch release disc
11. Friction plate
12. Plain plate
13. Back plate
14. Clutch center nut
15. Lock washer
16. Clutch center
17. Clutch center oil seal
18. Clutch center O-ring
19. Clutch drum/sprocket assembly
20. Clutch center spacer
21. Needle roller bearing
22. Rivet
23. Kickstarter ratchet
24. Sprocket bearing washer
25. Washer
26. Washer pin
27. Pushrod oil seal
28. Sleeve gear

Clutch components

1970 through early 1984

Refer to illustrations 8.5a, 8.5b, 8.6, 8.7a, 8.7b, 8.8 and 8.9

5   Special care is needed when removing the clutch, as a very strong inner and outer spring applies pressure to the release disc **(see illustration)**. Never loosen the six nuts around the edge of the release disc without first taking adequate precautions. If the proper Harley-Davidson service tool isn't available, use a three-jaw gear puller to keep the release disc from moving while the six nuts are removed. The puller can then be loosened slowly, to relieve the spring pressure in a controlled fashion. Note that even when the nuts are removed, the spring still exerts considerable pressure when the release disc has passed beyond the limit of the studs **(see illustration)**.

6   The 1970 clutch, which is designed to run dry, has a somewhat different arrangement **(see illustration)**. In this instance you have to remove

Chapter 2 Engine, clutch and transmission

8.7a Remove the inner circlip to release the clutch plates

8.7b The clutch plates will pull out from the center of the clutch

the domed clutch cover, which acts as an oil excluder, by removing the twelve screws around its edge. Note that they're linked in pairs by a lock plate. The cover will then lift off, exposing the gasket. This type of clutch has six separate springs, which seat within the thimbles inserted into the release disc. Removal of the six retaining nuts (three long, three short) will permit the release disc to be removed without need for special precautions or service tools.

7 When the release disc has been removed, along with the internal circlip, the clutch plates and release disc can be pulled out of the clutch drum **(see illustrations)**. On 1970 models, take out the shouldered clutch release rod.

8 The clutch center is retained by a large nut, which is very tight. Bend back the tab washer, then use a tight-fitting socket on the nut, with a long extension, to jar the nut loose. Hold the clutch center to keep it from turning as the nut is loosened **(see illustration)**.

9 The clutch center will now come off. It may be necessary to use a puller. Before removing the engine sprocket, clutch sprocket and primary chain in unison, detach the chain tensioner **(see illustration)**. They will then come off easily. Slide the clutch bushing off the end of the gear shaft sleeve.

Late 1984-on

Refer to illustrations 8.10a and 8.10b

Note: *The clutch can be removed from the mainshaft as a complete unit as described below. Disassembly of the clutch, however, requires a Harley-Davidson service tool to safely compress its diaphragm spring while the large circlip is removed – no other means of dismantling the clutch should*

8.8 A metal "sprag" can be used to lock the center of the clutch while the nut is loosened

be used. It is recommended that the clutch be disassembled by a Harley-Davidson service department.

10 Remove the circlip from the clutch boss, and withdraw the adjuster assembly (comprising a small circlip, guide/release plate, bearing and adjuster screw) from the pressure plate **(see illustrations)**.

8.9 Remove the chain tensioner (arrow) to increase the slack in the chain

8.10a Remove the circlip shown (not the large diaphragm spring circlip) . . .

8.10b . . . and lift out the adjuster assembly – 1991-on type shown

Chapter 2 Engine, clutch and transmission

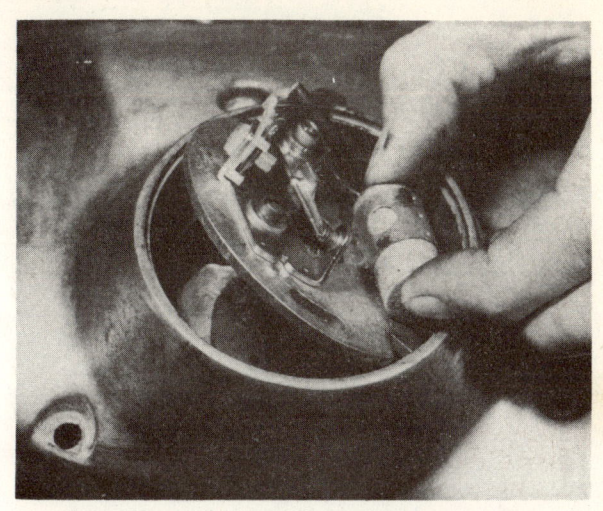

9.5 Lift out the contact breaker point assembly

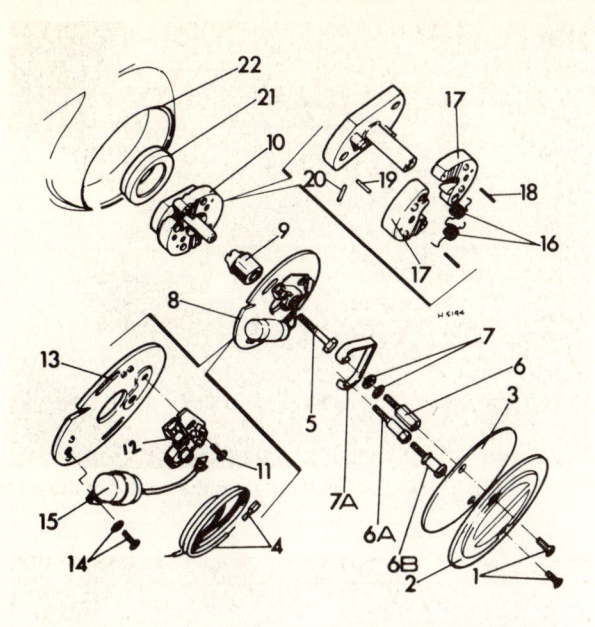

9.4 Ignition components – exploded view (1971 through 1978 models)

1   Screws
2   Cover
3   Gasket
4   Primary wire/terminal
5   Contact breaker cam retaining bolt
6   Contact breaker baseplate screw (early 1971)
6A Contact breaker baseplate screw (late 1971 and 1972)
6B Contact breaker baseplate screw (1973-on)
7   Lock washer/washer (early 1971)
7A Retainer (late 1971 and 1972)
8   Contact breaker baseplate assembly
9   Contact breaker cam
10 Mechanical advance assembly
11 Point adjusting screw (thru early 1972)
12 Contact breaker point assembly
13 Contact breaker baseplate
14 Condenser, screw and lock washer (late 1972-on)
15 Condenser
16 Counterweight spring
17 Counterweight
18 Pin
19 Cam stop pin
20 Register pin
21 Camshaft seal
22 Timing cover

11 On models through 1990, remove the external circlip and spacer from the mainshaft. On 1991-on models remove the mainshaft nut (noting that it has left-hand threads and is slackened clockwise) and remove the conical washer. On all models, remove the clutch, engine sprocket and chain from the crankshaft and mainshaft as a single unit, noting that there will be some magnetic resistance as the alternator rotor and stator separate. On 1991-on models slide the spacer off the crankshaft end if it is loose.

12 If the alternator rotor requires removal from the rear of the clutch sprocket (models through 1990) or engine sprocket (1991-on), refer to Chapter 8.

Kickstart mechanism (where fitted)

13 After the primary drive has been removed, access is available to the kickstart mechanism (if so equipped). Rotate the kickstart crank gear so the kickstart ratchet gear can be lifted off the transmission sleeve gear assembly **(see illustration 37.13)**. Remove the nut from the end of the kickstart shaft (after bending back the tab washer) and tap the end of the shaft so it's driven through the center of the kickstart crank gear. The kickstart shaft can now be withdrawn, together with the oil seal, shims and thrust plate. Note the number of shims used – all of them must be replaced during reassembly.

9 Ignition components – removal

1 Lay the engine on its left side, with the timing cover facing up. Make sure the surface the engine is resting on is clean, otherwise there's a risk of damaging the primary chaincase mating surface, resulting in oil leaks.

2 Remove the circular ignition cover from the timing cover. The circular ignition cover is held in place with screws on 1971 through 1979 models and is riveted in place on 1980 and later models. 1970 models have a separate distributor, which is dealt with later in this Section.

1970

3 The contact breaker assembly is contained within the distributor, mounted in the top of the timing cover. The distributor is retained by a clamp, which is released by removing the two screws. When the clamp is detached, the distributor can be lifted out of the timing cover (see Chapter 4).

1971 through 1978

Refer to illustrations 9.4, 9.5 and 9.6

4 Remove the center retaining bolt that secures the contact breaker cam and the two screws that secure the contact breaker baseplate assembly **(see illustration)**.

5 Scribe a line across the baseplate and the gearcase, then pull out the baseplate **(see illustration)**. The scribed line is to assist in reassembly so it won't be necessary to retime the ignition.

6 Pull off the contact breaker cam and the mechanical advance unit **(see illustration)**.

1979

Refer to illustration 9.7

7 Lift the ignition module out of the case after the outer cover is removed **(see illustration)**.

8 Remove the screws securing the timer plate, then remove the bolt from the center of the trigger rotor.

9 Disconnect the sensor from the timer plate and the wire from the module to the timer plate.

10 Lift the timer plate out of position and pull the trigger rotor off the advance assembly. The advance assembly can now be removed from the gearcase cover.

72   Chapter 2 Engine, clutch and transmission

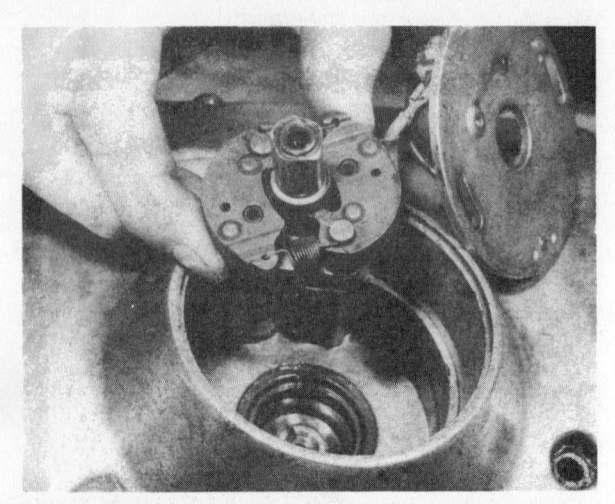

9.6 Pull off the mechanical advance unit under the contact breaker point assembly

1980-on

Refer to illustration 9.7

11 These systems can be disassembled in basically the same way, the only difference being the addition of a vacuum operated electric switch (VOES) on 1983 and later models.

12 Use a 3/8-inch drill bit to remove the rivets securing the outer cover and detach the cover **(see illustration 9.7).** Remove the screws holding the inner cover in place. Remove the inner cover and the gasket.

13 Unplug the electrical connector between the sensor plate and the ignition module. Mark the wires and their locations in the connector, then remove the wires from the connector on the sensor plate side.

14 Remove the screws securing the sensor plate to the gearcase and lift it out of position. Pull the wires through the opening in the gearcase one at a time.

15 Remove the screw and washer from the center of the rotor and detach the rotor.

16 If necessary, the camshaft oil seal can be pried out carefully and replaced with a new one.

17 The VOES on 1983 and later models doesn't have to be removed or disconnected to remove the ignition components from the gearcase.

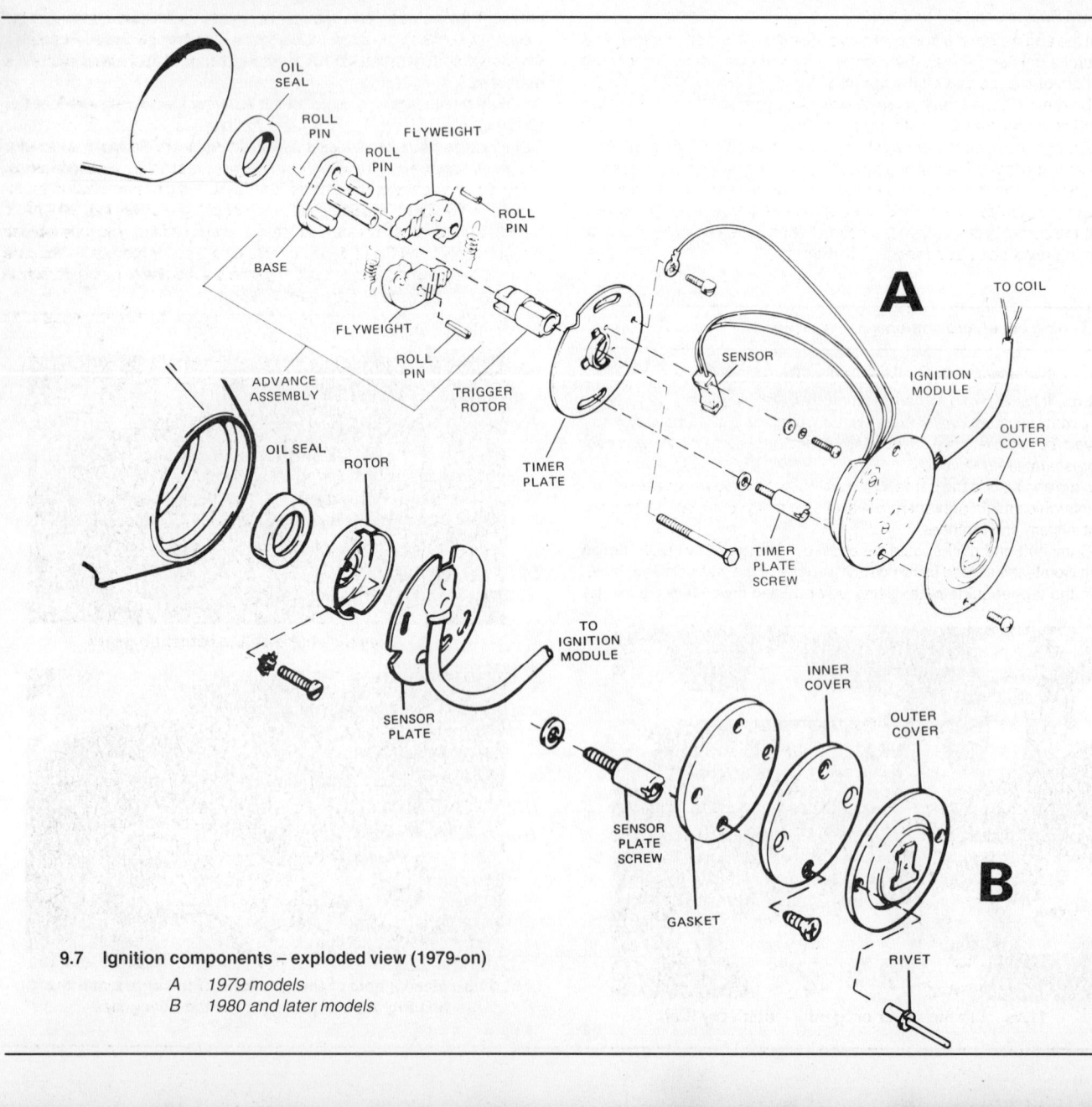

9.7 Ignition components – exploded view (1979-on)
  A   1979 models
  B   1980 and later models

Chapter 2 Engine, clutch and transmission

10.3 Lock the engine and unscrew the sprocket retaining nut

10.4 Use a puller if the sprocket is a tight fit on the splines

10 Final drive sprocket – removal

Refer to illustrations 10.3 and 10.4

1 If the engine is in the frame, remove the exhaust system, footpeg and rear brake master cylinder, then remove the sprocket cover. Disconnect the chain or slip the belt off the sprocket.
2 On pre-1971 models, remove the clutch pushrods (3) from inside the transmission mainshaft, if not already removed.
3 Either bend back the tab washer, remove the lock screw, or remove the screw and lock plate according to the model being worked on **(see illustration)**. Lock the transmission through the engine and remove the sprocket nut. **Note:** *The nut has a left-hand thread on 1991-on models.*
4 If the sprocket is stuck on the splined shaft, use a sprocket puller to free it to prevent damage **(see illustration)**.

11 Timing cover and camshafts – removal

Refer to illustrations 11.4a, 11.4b, 11.5, 11.6a, 11.6b and 11.7
Caution: *If the timing cover is not being removed as part of a general engine overhaul procedure, valve train pressure on the camshafts must be relieved (by removing the pushrods and tappets) before the timing cover or camshafts are disturbed.*
1 If the engine is in the frame, drain the engine oil and remove any components which might prevent removal of the timing cover, such as the exhaust system and footpegs.
2 Camshaft end play should be checked before removing the timing cover. Rotate the engine by hand until the lobe of the first camshaft is visible in the tappet guide hole. Using a flat-bladed screwdriver, push the camshaft towards the timing cover and using a feeler blade, measure the clearance between the crankcase and the camshaft shoulder. Repeat on the other three camshafts and make a note of any which have excessive end play.
3 Remove the ignition components as described in Section 9 of this Chapter.
4 Place an oil drain tray beneath the timing cover. Remove all screws and bolts from the cover and withdraw it from the crankcase **(see illustration)**. It may require a few taps from a soft-faced hammer around its periphery to release the gasket seal – don't pry the cover free otherwise damage will result. On models with a rear chain oil feed, you may have to disconnect or bend the oil line to enable timing cover removal. Take care as the cover is withdrawn to spot any shims which have come detached from their shafts **(see illustration)**.
5 Note the exact location of any shims on the shaft ends and inspect the

11.4b Note the shims on the camshaft gears

11.4a Lift the cover off gently – don't pry it off

11.5 Make a note of the locations of the shims and the bearing retainer plates under the cam gears

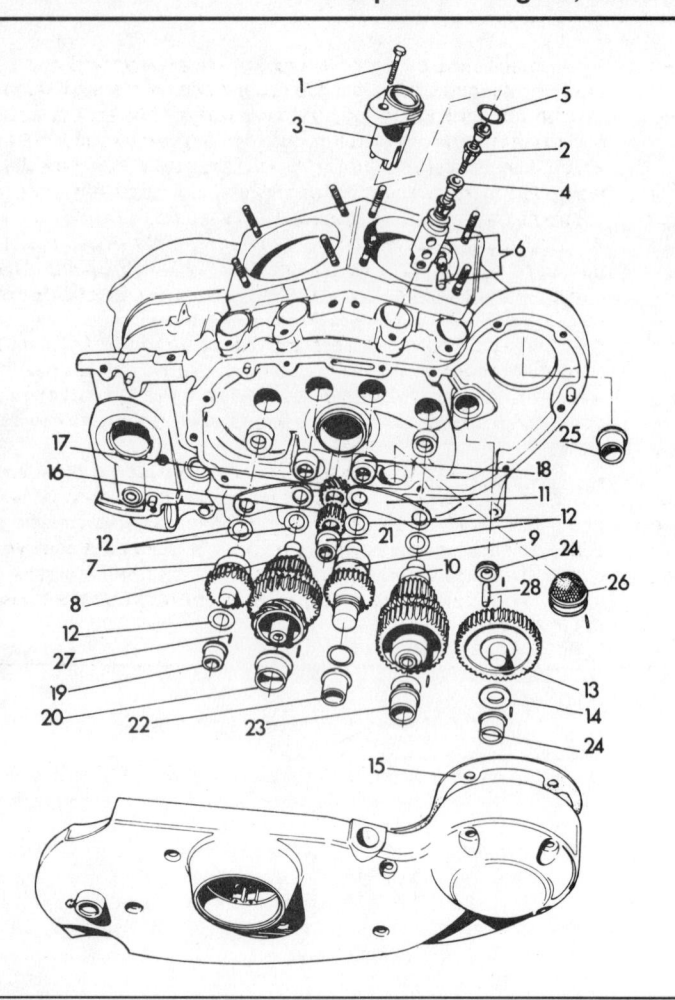

11.6a Gearcase (timing) cover and related components (early models) – exploded view

1. Tappet guide bolt
2. Tappet adjusting screw/locknut
3. Tappet guide
4. Tappet and roller
5. O-ring
6. Tappet roller kit
7. Rear cylinder exhaust camshaft
8. Rear cylinder intake camshaft
9. Front cylinder intake camshaft
10. Front cylinder exhaust camshaft
11. Bearing retaining plate
12. Camshaft shim (as required)
13. Idler gear
14. Idler gear shaft fiber washer
15. Timing cover gasket
16. Crankshaft gear
17. Oil pump drive gear
18. Camshaft needle roller bearing
19. Rear exhaust camshaft bushing (timing cover)
20. Rear intake camshaft bushing (timing cover)
21. Crankshaft pinion bushing (timing cover)
22. Front intake camshaft bushing (timing cover)
23. Front exhaust camshaft bushing (timing cover)
24. Idler gear bushing
25. Oil separator bushing
26. Crankcase oil strainer
27. Timing cover bushing pin
28. Idler gear shaft

camshaft gears and crankshaft gear for alignment marks **(see illustration 34.4b)**. If the marks are faint, highlight them with a felt marker. Also check the camshafts for identification marks denoting their orientation; again if none are found make your own. When this information has been recorded, withdraw the camshafts. On models through early 1984, free the generator and remove its idler gear from the crankcase. Note the location of any washers or bearing retainer plates on the crankcase side; keep them with their respective shafts or mark them to ensure correct reassembly **(see illustration)**.

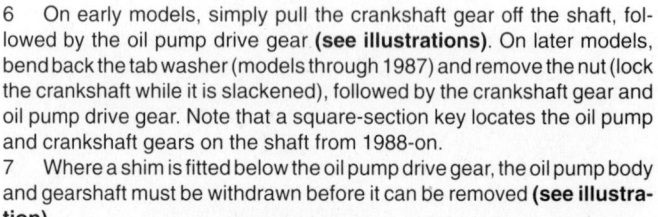

6 On early models, simply pull the crankshaft gear off the shaft, followed by the oil pump drive gear **(see illustrations)**. On later models, bend back the tab washer (models through 1987) and remove the nut (lock the crankshaft while it is slackened), followed by the crankshaft gear and oil pump drive gear. Note that a square-section key locates the oil pump and crankshaft gears on the shaft from 1988-on.

7 Where a shim is fitted below the oil pump drive gear, the oil pump body and gearshaft must be withdrawn before it can be removed **(see illustration)**.

11.6b The crankshaft gear can be pried off the shaft, along with the oil pump worm drive gear (early models only – on later models the gear is held on the shaft by a nut)

11.7 The large shim can't be removed until the oil pump is removed

Chapter 2 Engine, clutch and transmission

12 Oil pump – removal, inspection and installation

1970 through 1976

Refer to illustrations 12.1, 12.2a, 12.2b, 12.2c, 12.2d, 13.3a, 12.3b, 12.5 and 12.7

1 The oil pump should not be removed unless absolutely necessary. To remove it, drain the oil tank first and unscrew the crankcase drain plug. Detach the oil pressure switch by unscrewing it from the front of the oil pump housing, after the wire has been removed. Disconnect the main oil feed line and remove the five nuts and washers that retain the oil pump on the studs that project from the crankcase. The pump will probably separate into three sections as it's withdrawn, although the upper two sections cannot be separated without further disassembly **(see illustration)**.

2 The base plate will lift off, exposing the two scavenge pump gears **(see illustration)**. The idler gear will lift out; the drive gear is retained by two half rings or a circlip, which will have to be pried out of position first **(see illustration)**. It's keyed in position; don't lose the small Woodruff key **(see illustration)**. The second part of the body can now be detached and the oil pump pressure switch mount unscrewed to release the check valve spring and ball bearing **(see illustration)**. The feed pump idler gear will remain in the top half of the housing, together with its small detachable shaft.

3 The feed pump drive gear will probably remain attached to the breather valve gear and shaft, within the upper portion of the pump assembly **(see illustration)**. Before the breather valve gear and shaft can be lifted out, it will be necessary to lift off the feed pump drive gear and withdraw the small pin that transfers the drive from the shaft to the gear **(see illustration)**. Note there's a small wire screen around the breather valve, which should be cleaned prior to reassembly of the pump.

4 Damage is most likely to occur if some engine component has broken up and particles of metal have found their way into the pump. This will damage the oil pump gears, which will be obvious. Replacement gears will have to be obtained; the originals cannot be reused.

5 Unscrew the check valve assembly from the front of the oil pump body **(see illustration)**. Examine the check valve spring and ball bearing for wear. The free length of the spring should be 1-15/16 inch. If there's any doubt about the condition of either component, replace them with new ones.

6 Examine the ball valve seat. If only minor damage is evident, such as pit marks or dents, the seat can often be repaired by inserting the ball and giving it a sharp blow with a punch and hammer when it's resting on its seat. If damage is bad, a new pump body will be required. A badly seating check valve will cause oil to drain into the crankcase from the oil tank, making starting difficult and permitting the discharge of oil from the crankcase breather pipe.

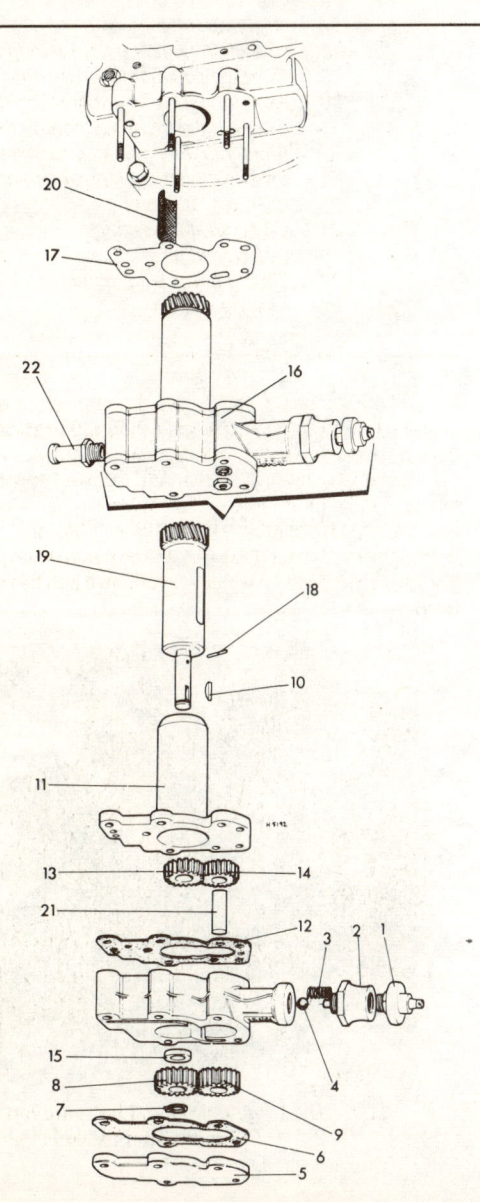

12.1 Oil pump (1970 through 1976 models) – exploded view

1. Oil pressure switch
2. Oil pump fitting
3. Check valve spring
4. Ball check valve
5. Base plate
6. Gasket
7. Retainer (half ring or circlip)
8. Scavenge pump drive gear
9. Scavenge pump idler gear
10. Breather valve shaft key
11. Oil pump cover
12. Body cover gasket
13. Feed pump drive gear
14. Feed pump idler gear
15. Oil pump seal
16. Oil pump body
17. Body gasket
18. Drive pin
19. Breather valve gear and shaft
20. Breather valve filter
21. Idler gear shaft
22. Feed line fitting

Chapter 2 Engine, clutch and transmission

12.2a Remove the base plate to expose the scavenge pump gears

12.2b Remove the circlip securing the oil pump drive gear

12.2c Note how the gear is keyed to the shaft

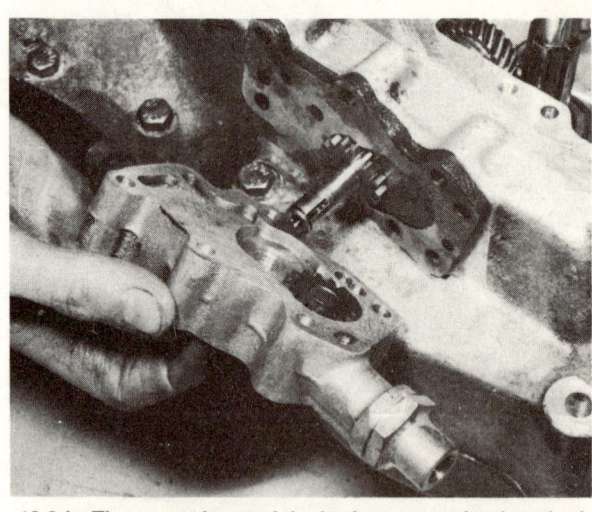

12.2d The second part of the body can now be detached

12.3a The feed pump idler gear will remain attached to the shaft

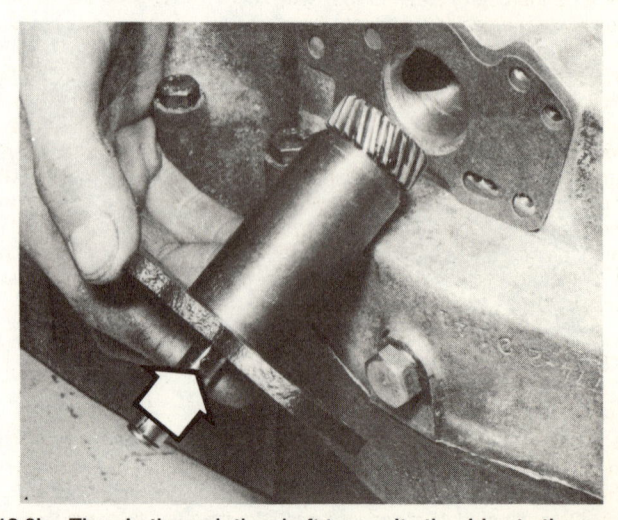

12.3b The pin through the shaft transmits the drive to the gear

Chapter 2 Engine, clutch and transmission

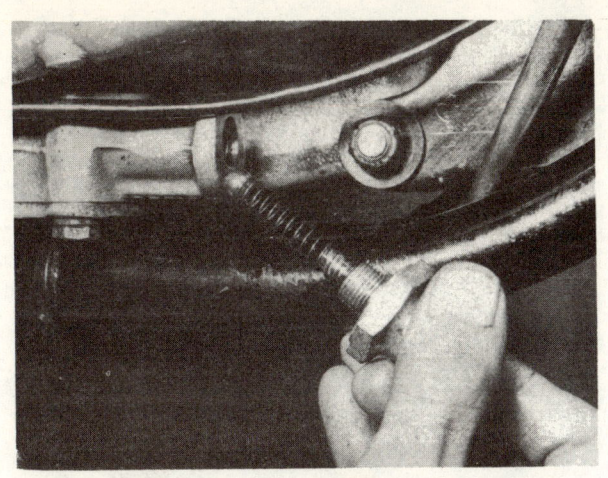

12.5 Examine the check valve spring, ball and seat

12.7 Note the small oil seal in this portion of the oil pump body

7   Reassemble the pump by reversing the disassembly procedure using new gaskets and a new oil seal **(see illustration)** and circlip. All parts must be thoroughly cleaned with solvent and lightly oiled prior to reassembly. It's particularly important that each gasket is checked to ensure it does not mask off any oilways or obstruct the free movement of the pump gears. Don't use gasket cement at any of the joints. Tighten the pump mounting nuts a little at a time during reassembly and make sure it revolves freely at each stage. Keep checking to make sure the pump revolves freely, as uneven tension may cause it to misalign. Even a badly cut gasket can cause the gears to bind by fouling them during rotation. Note that during reassembly you'll have to remove the timing cover so the breather valve can be timed correctly, as described in Section 33. This is important.

8   Reconnect the oil line and install a new clamp. Check the line first, to ensure it isn't split or kinked.

1977 through 1990

Refer to illustration 12.10

9   The oil pump can be removed from the engine with the engine still in the frame. Label the oil lines to assist in reassembly, then disconnect them.

10   On early models, disconnect the wire from the oil pressure switch **(see illustration)**; on later models the switch is mounted on the oil filter adaptor.

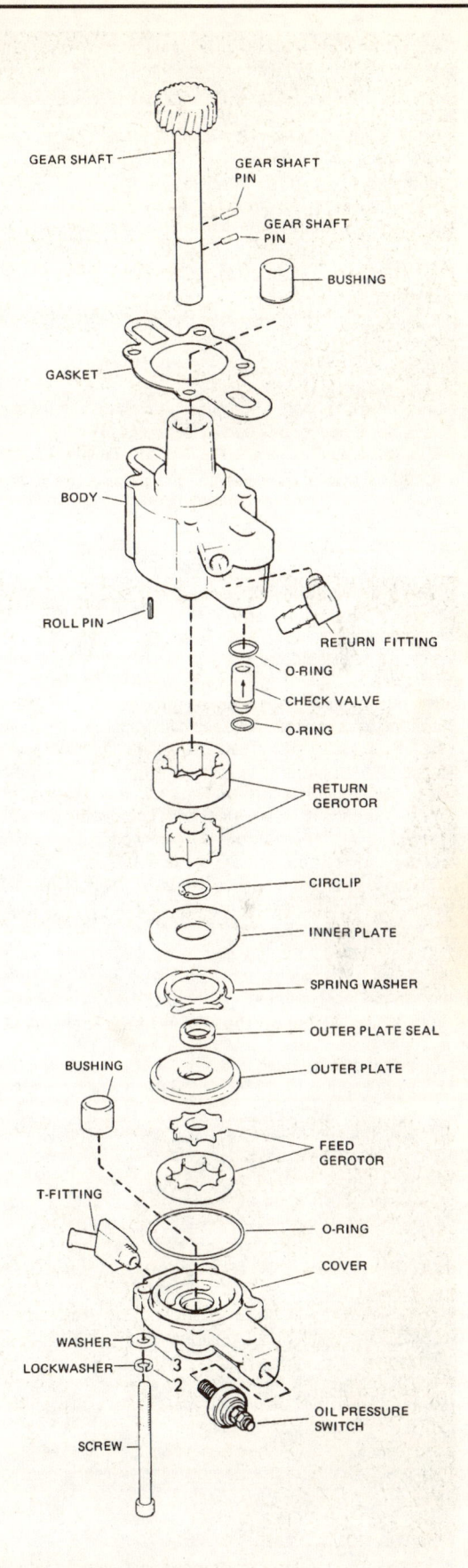

12.10 Oil pump (1977 and later models) – exploded view

11 Remove the screws that secure the oil pump to the crankcase. Pull the oil pump away from the engine and throw away the old gasket.
12 Separate the cover from the body of the oil pump and remove the O-ring.
13 Lift the lower gerotor set off the gear shaft and pull the pin out of the gear shaft with a needle-nose pliers.
14 Withdraw the outer plate, the plate seal, the spring washer, the inner plate and the outer gerotor (from the return set) from the gear shaft.
15 Remove the circlip from the gear shaft and lift the inner gerotor off the shaft.
16 Pull the remaining pin from the gear shaft and withdraw the shaft from the oil pump body.
17 On early models, press the check valve out the bottom of the pump housing by inserting a 5/16-inch pin punch through the oil pump outlet opening. Remove the O-rings from the check valve. On later models the check valve is installed in the oil filter.
18 Remove the oil seal from the outer plate and replace it with a new seal.
19 Check the spring loaded cup inside the check valve for binding and make sure it closes completely. Replace the valve if there's any doubt about it working properly.
20 Check the spring washer for damage such as broken or cracked fingers.
21 Clean all of the parts with solvent and dry them thoroughly. If available, use compressed air to blow out all of the passages.
22 Check the entire pump body and cover for cracks and evidence of wear. Look closely for ridge wear on the gerotors and where the gerotors ride on the pump housing.
23 Place the inner gerotors inside of the outer gerotors and check the clearances with a feeler gauge. There should be no more than 0.004-inch clearance between the inner and outer gerotors.
24 Measure the thickness of the feed (lower) gerotors with a micrometer. If the thicknesses vary they must be replaced as a set.
25 Install the feed gerotors inside the cover and place a straightedge over the components. Insert a feeler gauge between the straightedge and the gerotor components. If the inner and outer gerotors aren't the same height, the cover is warped and must be replaced. The gerotors should extend out of the cover between 0.001 and 0.011-inch. If they extend less than 0.001-inch, the cover must be sanded down until the required measurements are obtained. To do the cover modification, place a piece of 280-grit sandpaper on a flat surface, such as a pane of glass, and invert the cover on the piece of sandpaper. Move the cover in a circular motion until the desired dimension is obtained. Finish the ridge of the cover with 400-grit sandpaper. Thoroughly clean the cover with solvent to prevent sand from getting into the oil. If the gerotors extend out of the cover more than 0.011-inch, the cover must be replaced with a new one.
26 Measure the clearances between the gear shaft and the two gear shaft bushings. It must be 0.005-inch. If the bushings are scored, damaged or worn, replace them.
27 The gear shaft bushings must be pressed into position. The bushing in the pump housing must be installed 0.100-inch below the surface, while the bushing in the cover must be installed 0.120-inch below the surface.
28 Inspect the teeth on the gear shaft for wear or damage and replace the shaft if necessary.
29 Assembly of the oil pump is essentially the reverse of disassembly. Be sure to lubricate all moving parts and the seals and O-rings with clean oil during assembly.
30 Place the new seal in the outer plate with the lip of the seal facing the feed (lower) gerotors. Secure the seal in the outer plate with Loctite Lock N 'Seal Adhesive.
31 Install the spring plate with the fingers away from the inner plate.
32 When installing the outer plate, align the slot with the pin in the pump housing. Be sure the seal side faces in.
33 Install new O-rings on the check valve and the cover.
34 Place the new oil pump gasket in position with non-hardening gasket sealant.

1991-on models

Oil pump

35 The oil pump can be removed with the engine in the frame, but first drain the engine oil (see Chapter 1).
36 Disconnect the oil lines from the pump, labeling them FEED, RETURN and VENT to aid installation. Remove the two long Allen-head screws (those without washers) to free the complete pump from the crankcase. Peel off the gasket; a new one must be used on installation.
37 To disassemble the pump, remove the remaining two Allen-head screws (shorter) from its cover and lift the cover and its O-ring off the body. Slide the feed gerotor set (thinner), separator plate, and return gerotor set (thicker) off the gearshaft.
38 To remove the gearshaft from the pump body, remove the circlip and washer, and withdraw the gearshaft from the other side.
39 Clean all parts with solvent and dry thoroughly. If available, use compressed air to blow out all oil passages.
40 Inspect the gerotors for signs of excessive wear and scoring; replace if wear is excessive. Mesh each set of gerotors together and using a feeler blade, measure the clearance between their tips. If either set exceeds the service limit of 0.004 inches (0.10 mm) replace the rotors.
41 Using a micrometer, measure the thickness of the feed (thinner) gerotors; they must be exactly the same thickness.
42 The oil pump must be reassembled in clean conditions, using only new engine oil on the moving components. Slide the gearshaft into the body, install the washer and insert the circlip in its groove.
43 Lubricate and fit the inner return gerotor over the gearshaft, followed by its outer geroter. Install the separator plate, aligning its cutouts with those of the body. Finally, install the feed gerotor set over the gear shaft end.
44 Using a new O-ring install the pump cover and secure with the two shorter screws, not forgetting their washers; tighten to 125 to 150 in-lbs (14 to 17 Nm).
45 Using a new gasket install the pump on the crankcase, making sure that its gear engages correctly. Install the two long screws and tighten to 125 to 150 in-lbs (14 to 17 Nm). Reconnect the oil lines to their unions using new clamps.
46 Refill the oil tank and check for leaks. Check that oil pressure builds up to the correct level.

Check and regulator valves

47 The oil filter adaptor on the front portion of the crankcase houses the pressure regulator valve (1991 models only), check valve and oil pressure switch.
48 To gain access to the check valve, unscrew the oil filter and then unscrew the threaded boss from the center of the adaptor. Lift out the ball and spring behind it.
49 The regulator valve can only be accessed after the gearcase has been removed. Unscrew its end plug, then remove the spring and valve.

13.2a After the access cover bolts are removed, the transmission can be withdrawn from the case as an assembly

Chapter 2 Engine, clutch and transmission

13.2b Note how the uncaged needle bearings have collapsed

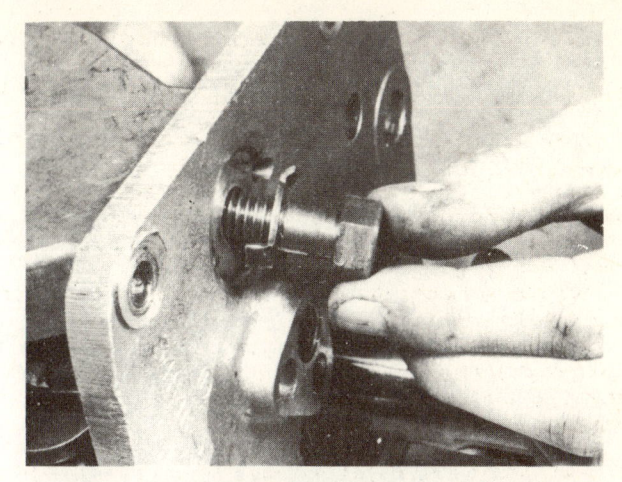

13.3a Remove the bolt and tab washer to free the selector mechanism

13.3b Note the arrangement and number of shims installed

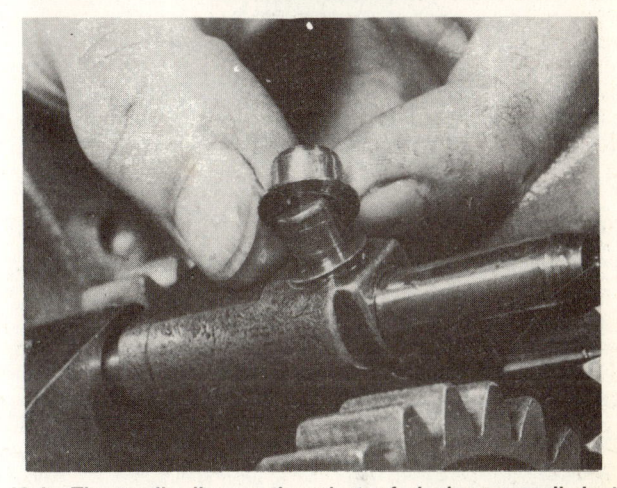

13.4 The small rollers on the selector fork pins are easily lost

13 Transmission components – removal and disassembly

1   As previously mentioned, there's no need to disassemble the engine if the transmission components are the only parts requiring attention. Removal is via the access cover in the left crankcase half. Remove the clutch and primary drive components (Section 8), the final drive sprocket (Section 10), and on late 1984 through 1990 models remove the alternator stator from the access cover (Chapter 7).

1970 through 1990 models (4-speed)

Refer to illustrations 13.2a, 13.2b, 13.3a, 13.3b, 13.4, 13.5a and 13.5b

2   To remove the transmission as a complete unit, remove the four bolts around the outside of the square shaped access cover in the primary chaincase. The cover can then be pulled out, bringing with it the complete gear cluster and selector mechanism **(see illustration)**. The right-hand needle roller bearing, through which the mainshaft passes, will collapse as the bearings are uncaged **(see illustration)**. Don't lose any of them.

3   To release the cam plate assembly, remove the single bolt that retains the assembly to the access plate **(see illustration)**. The assembly has two dowels to locate the pawl carrier support and is shimmed. Note how many shims are installed and set them aside in a safe place for reassembly **(see illustration)**.

4   It's also easy to lose the rollers that slip over the selector fork pins as they disengage from the cam plate tracks **(see illustration)**. They should be taken off and kept in a safe place until reassembly. The selector forks will lift off the gears as the gear cluster is disassembled.

5   There's no need to disassemble the gear cluster unless examination shows obvious defects such as chipped or broken teeth, rounded dogs or worn bushings. See Section 24 for further information **(see illustrations)**.

6   The foregoing information relates specifically to transmissions manufactured after 1971. Pre-1971 transmissions differ in certain minor respects, although the same basic disassembly and reassembly techniques apply.

13.5a The complete four-speed transmission gear cluster

Chapter 2 Engine, clutch and transmission

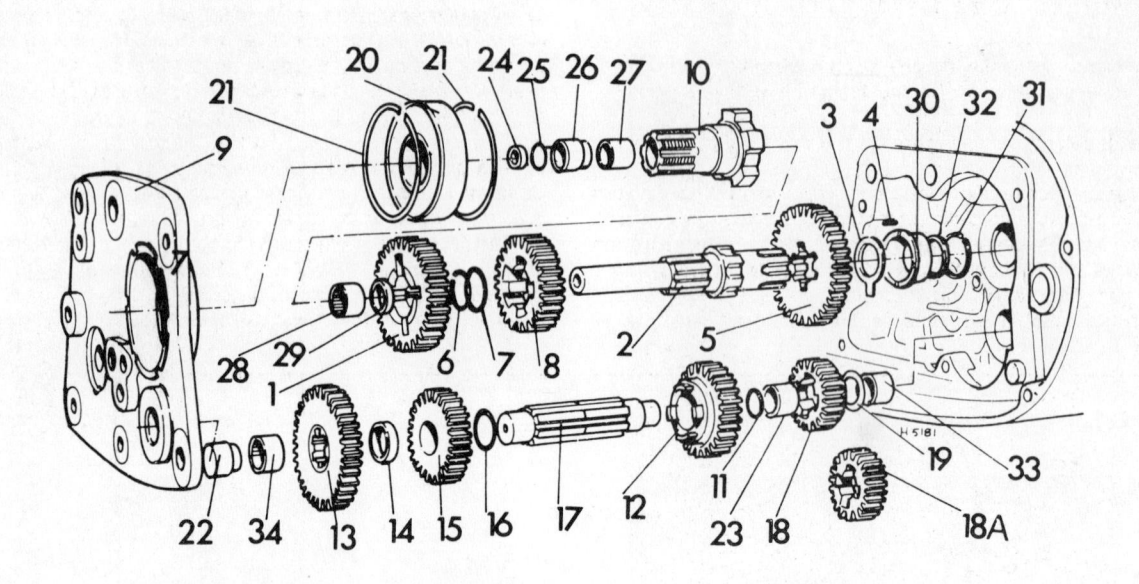

13.5b Transmission gear shafts (4-speed) – exploded view

1. Mainshaft second gear
2. Mainshaft
3. Mainshaft tabbed thrust washer
4. Mainshaft roller
5. Mainshaft first gear
6. Mainshaft third gear snap-ring (DO NOT reuse)
7. Mainshaft third gear washer
8. Mainshaft third gear
9. Access cover
10. Sleeve gear
11. Countershaft first gear washer
12. Countershaft third gear
13. Countershaft drive gear
14. Countershaft gear spacer
15. Countershaft second gear
16. Countershaft second gear washer
17. Countershaft
18. Countershaft first gear (1972 and earlier)
18A. Countershaft bottom gear (1973-on)
19. Countershaft gear washer (1973-on)
20. Mainshaft ball bearing
21. Mainshaft ball bearing snap-ring
22. Countershaft blanking plug
23. Countershaft first gear bushing
24. Pushrod oil seal (1970 only)
25. Sleeve gear O-ring (1970 only)
26. Sleeve gear oil seal extension (1970 only)
27. Sleeve gear bushing
28. Sleeve gear needle roller bearing
29. Mainshaft thrust washer
30. Mainshaft roller bearing
31. Mainshaft roller bearing retaining ring
32. Mainshaft roller bearing washer
33. Countershaft bearing (closed end)
34. Countershaft bearing (open end)

13.10 Five-speed gearbox – withdraw the gearshafts and shift drum as an assembly with the access cover

13.11 Special tools are required to extract the mainshaft 5th gear from the casing

Chapter 2 Engine, clutch and transmission

1991-on models (5-speed)

Refer to illustrations 13.10, 13.11, 13.12, 13.13a, 13.13b, 13.13c and 13.13d

7   If you intend to dismantle the gear shafts, take this opportunity to slacken the Torx-head screw in the countershaft end while the shaft is locked in gear.
8   Release the detent lever spring from its anchor pin, but do not remove the lever's pivot bolt.
9   The shifter shaft is retained to the crankcase by two locknuts. Before removal, verify the claw arm-to-pin setting (see Section 37, Step 11) as a guide to reassembly. Release the spring clip from the end of the shift drum and withdraw the cam. Remove the two locknuts and washers and pull the shifter shaft free of the casing and drum pins.
10 Remove the five bolts from the access cover, and very gently pry the cover off the crankcase complete with gear shafts and shift drum/forks **(see illustration)**.
11 The mainshaft 5th gear will remain in the right-hand crankcase **(see illustration)**. It is a tight fit in its bearing and will require a press to draw it out of the bearing. Harley-Davidson has a service tool for this purpose (Part No HD-35316A and HD-3531691) which is essentially a long draw-bolt that bears across the casing aperture and passes through the 5th gear, through a thick washer with a OD larger than the 5th gear boss, and then into a nut. As the nut is tightened the 5th gear is drawn into the casing.
Caution: *It is advised that the correct tool is used due to the strain alternatives might place on the casing.* Before attempting removal, first pry the blind seal out of the gear end and remove the spacer and sealing ring from the threaded portion. Oil seal removal is best done with the shaft removed.
12 Examine the gear shafts as described in Section 24, and if necessary dismantle them as described in Section 25 **(see illustration)**.

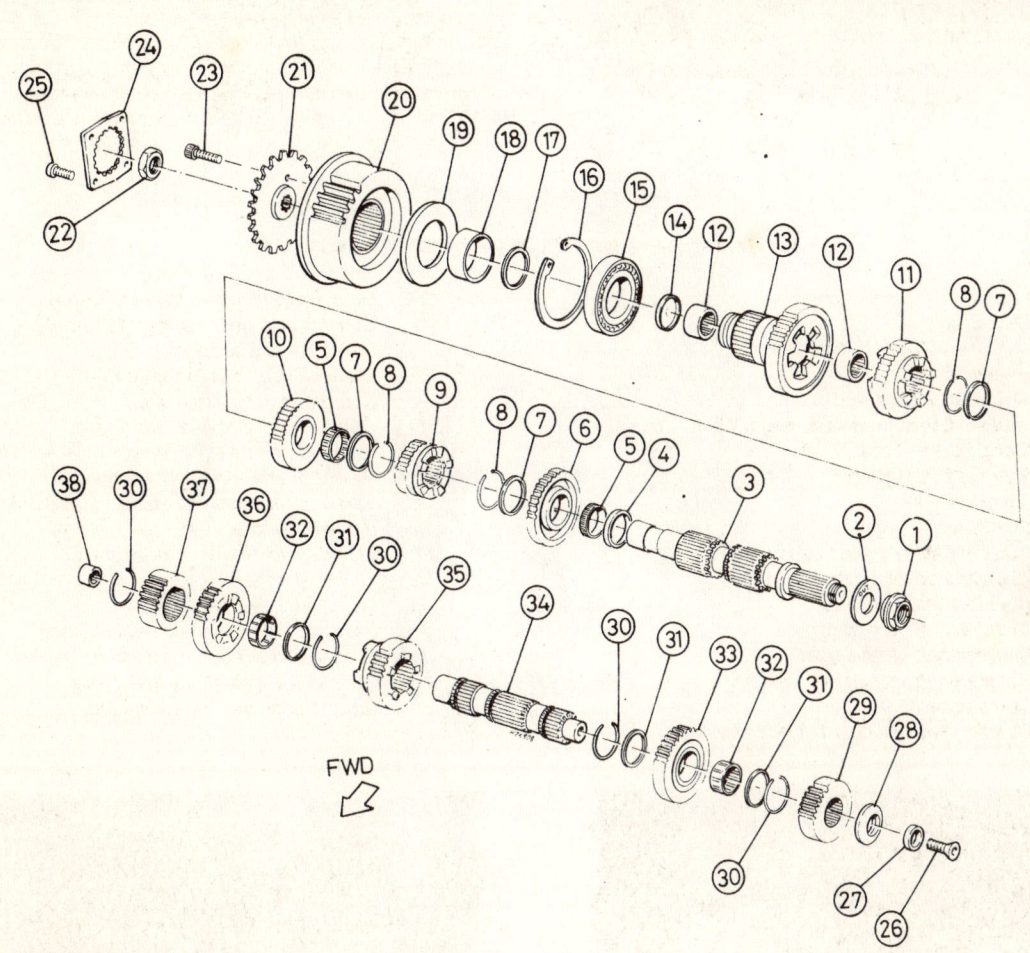

13.12 Transmission gear shafts (5-speed) – exploded view

1	Clutch nut	14	Blind plug	26	Torx screw
2	Conical washer	15	Bearing	27	Retaining plate
3	Mainshaft	16	Circlip	28	Angled spacer
4	Spacer	17	Seal	29	Countershaft 4th gear
5	Split bearing	18	Spacer	30	Circlip
6	Mainshaft 4th gear	19	Oil seal	31	Thrust washer
7	Thrust washer	20	Sprocket (belt drive)	32	Split bearing
8	Circlip	21	Sprocket (chain drive)	33	Countershaft 1st gear
9	Mainshaft 1st gear	22	Nut	34	Countershaft
10	Mainshaft 3rd gear	23	Lock bolt (chain drives and 1991 belt drive)	35	Countershaft 3rd gear
11	Mainshaft 2nd gear	24	Lock plate (1992-on belt drive)	36	Countershaft 2nd gear
12	Needle roller bearing	25	Lock bolt (1992-on belt drive)	37	Countershaft 5th gear
13	Mainshaft 5th gear			38	Bearing

Chapter 2 Engine, clutch and transmission

13 To remove the gear shift drum, first position the drum in neutral (roll pin in right-hand end at 12 o'clock). Remove the cotter pins from the forks and using a magnetic rod, withdraw the shift guide pins from the fork bores. Remove the nut from the detent lever pivot bolt and lift out the reinforcing and locking plates from the drum groove (you may have to pull the drum out of position slightly to clear the roll pin). Withdraw the shift drum from the forks and, having made note of their position, lift the forks off the gear shafts **(see illustrations)**.

14 Splitting the crankcases

Note: *Before the crankcases can be separated, the engine must be removed from the frame and all components (including the gearbox) detached as described in Sections 5 through 13.*
Refer to illustrations 14.1a, 14.1b, 14.4a, 14.4b and 14.6

1 On early models with a compensating sprocket attached to the crankshaft, remove the sprocket shaft extension. Use a screwdriver to pry it off the crankshaft while tapping lightly around the outside with a soft-face hammer **(see illustrations)**. Don't use excessive force or drive the screwdriver into the gap between the collar and the crankcase itself – it's too

13.13a Remove their cotter pins and then extract the guide pins from the forks

13.13b Remove the nut (arrow), followed by the reinforcing and locking plates . . .

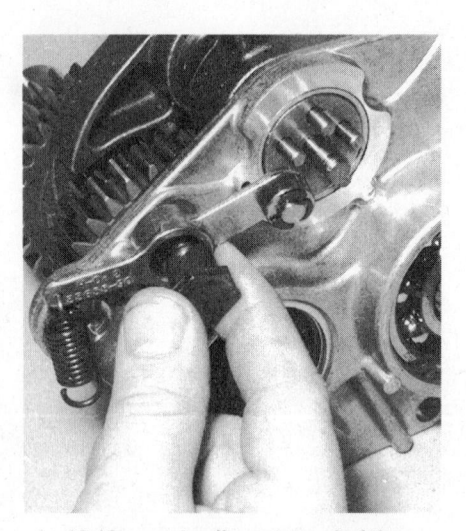

13.13c . . . to allow removal of the detent lever pivot bolt

13.13d Lift the shift drum and forks off the gearshafts

14.1a Tap the compensating sprocket collar with a hammer to loosen it, . . .

14.1b . . . then gently pry it off the end of the crankshaft

Chapter 2 Engine, clutch and transmission

easy to damage the bearing housing and oil seal located in front of the main bearing.

2 On models through 1990 remove the circlip (where fitted) from the crankshaft right-hand end. On 1991-on models detach the alternator stator from the left-hand crankcase (if not already done), referring to Chapter 7, Section 10 for details. Slackening them evenly to avoid distortion, remove all crankcase bolts and stud nuts. On early models there will be five bolts and three stud nuts on the right side, two bolts and one stud nut on the left side and the engine rear mount bolts. On later models there will be three bolts and three nuts on the left side, eight bolts on the right side and the engine rear mount bolts. Disconnect the tachometer drive unit (if equipped). In all cases, two of the fasteners mentioned above will also secure the engine front mounting bracket.

3 On 1991-on models, remove the four Allen-head bolts, and five hex-head bolts (two of these retain the engine front mounting bracket) from the crankcase left side.

4 With the engine resting on its left side (primary case), lift off the right-hand half to leave the crankshaft in the left-hand crankcase half **(see illustrations)**. If necessary, tap the crankcase gently with a soft-faced hammer to break the seal. **Note:** *As the crankcase leaves the right-hand main bearing on early models the bearing rollers will probably fall out of position; make sure that none are lost.*

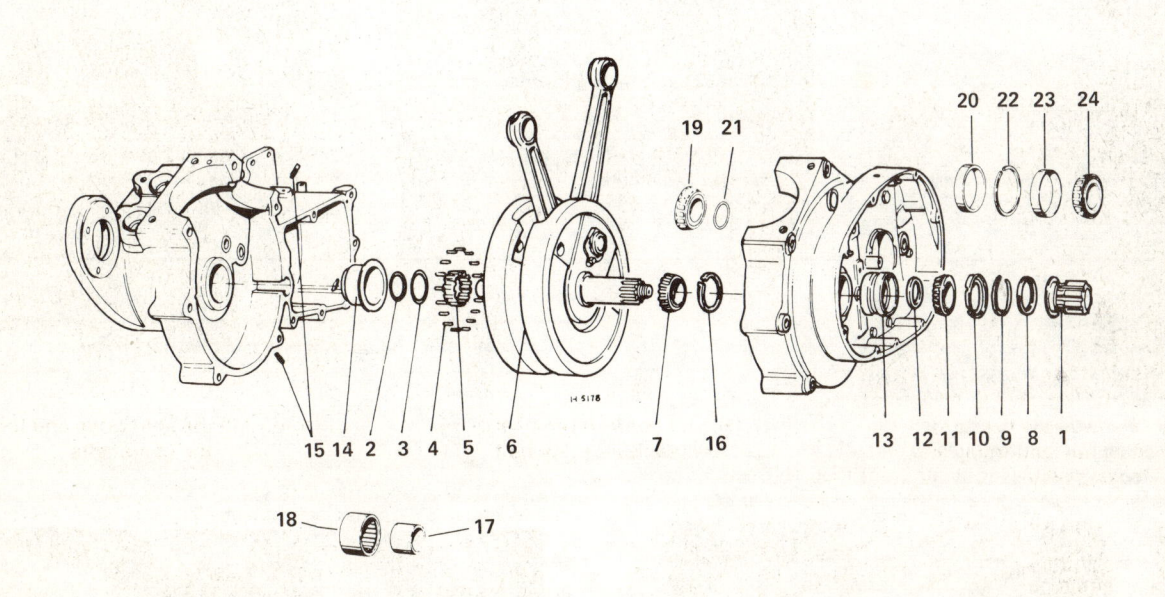

14.4a The right-hand crankcase will lift off the left-hand case (note the needle roller main bearing)

14.4b Crankcases, crankshaft/flywheel and main bearings (early models) – exploded view

1 Compensating sprocket splined collar
2 Snap-ring (pre-1977)
3 Thrust washer (pre-1977)
4 Needle roller main bearing (pre-1977)
5 Bearing cage (pre-1977)
6 Crankshaft/flywheel assembly
7 Right-hand tapered roller main bearing (pre-1977)
8 Crankshaft oil seal
9 Spring ring (pre-1977)
10 Compensating sprocket spacer (pre-1977)
11 Left-hand tapered roller main bearing (pre-1977)
12 Tapered roller main bearing spacer (pre-1977)
13 Tapered roller main bearing combined outer race (pre-1977)
14 Crankshaft main bearing bushing (pre-1977)
15 Bushing retaining screw
16 Snap-ring (pre-1977)
17 Crankshaft bearing inner race (post 1976)
18 Crankshaft bearing (post 1976)
19 Right main bearing (post 1976)
20 Right main bearing outer race (post 1976)
21 Spacer (post 1976)
22 Snap-ring (post 1976)
23 Left main bearing outer race (post 1976)
24 Left main bearing (post 1976)

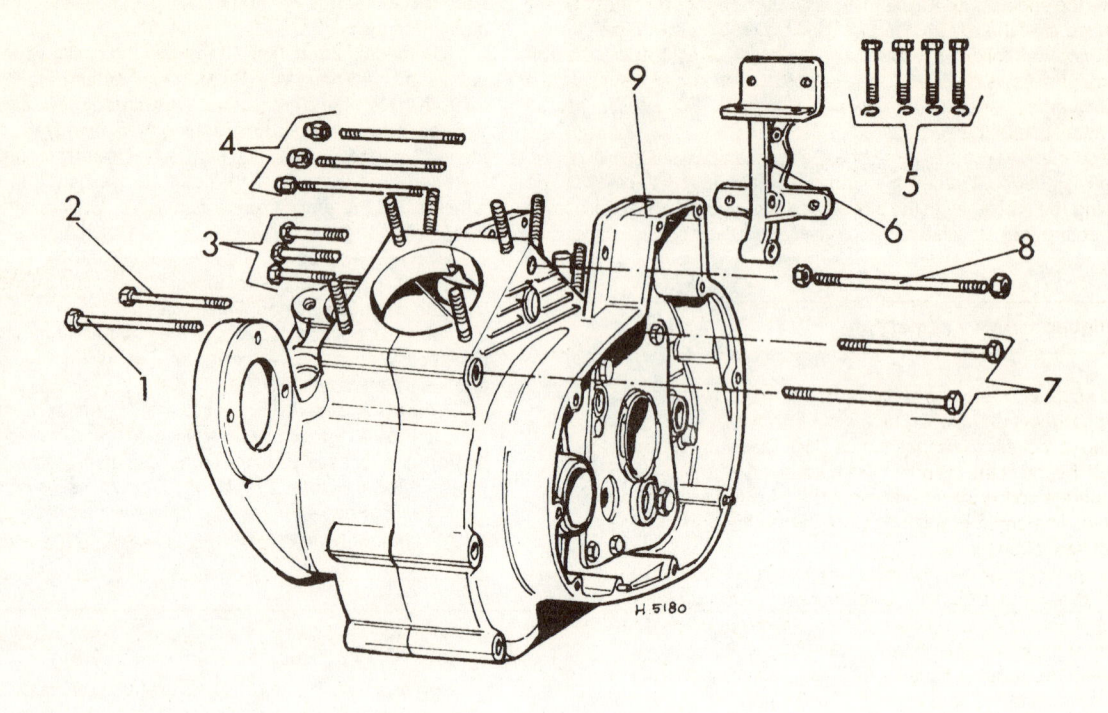

14.6 Crankcase fasteners (early models)

1 Engine bolt – 4-7/16 inch long
2 Engine bolt – 4-1/16 inch long
3 Engine bolt – 2-3/8 inch long
4 Rear mounting stud and locknut
5 Rear mounting bolt and lock washer
6 Rear engine mount
7 Crankcase bolt
8 Crankcase center stud and locknut
9 Crankcase castings

5   If you have to remove the complete flywheel/crankshaft assembly (to replace the big-end bearings for example) the whole assembly must be pressed out of the left-hand crankcase with a special press. Don't hammer the flywheel assembly out of the left-hand crankcase – it will upset the alignment of the flywheels.
6   When the crankcases are parted on early models, it's possible to remove the rear engine mount **(see illustration)**. It must be replaced before the crankcases are rejoined; it cannot be installed afterwards.

15 Inspection and repair – general information

1   Before examining the parts of the disassembled engine for wear, clean them thoroughly. Use solvent to remove all traces of old oil and sludge that accumulated in the engine.
2   Examine all castings for cracks and other damage, especially the crankcase castings. If a crack is discovered, it'll require professional repair or replacement of the part.
3   Carefully examine each part to determine the extent of wear, using the figures listed in the Specifications Section of this Chapter. If there's any question or doubt, be safe and replace the component. Notes included in the following text will indicate what type of wear can be expected and whether the part concerned can be reused.
4   Use a clean lint-free rag for cleaning and drying the various components. This will reduce the risk of small particles obstructing the internal oilways, causing the lubrication system to fail.
5   Above all, work in clean, well-lit surroundings so problems don't pass undetected. Failure to detect damage or signs of advanced wear may necessitate another complete teardown at a later date, due to the premature failure of the part concerned.

16 Crankshaft and connecting rod bearings – inspection

1   The crankshaft assembly consists of a pair of heavy flywheels joined together by a single crankpin and big-end assembly, which serves both cylinders. An ingenious overlapping arrangement permits the two connecting rods to share the common big-end assembly without fouling each other during rotation. The front connecting rod fits inside the forked rear rod.
2   It isn't possible to separate the flywheel assembly without access to a large press or to realign the components to a high enough standard of accuracy without a lathe. As a result, if repairs are required, take the flywheel assembly to a Harley-Davidson dealer.
3   Failure of the big-end bearing is usually accompanied by a pronounced knock from within the crankcase, which gets progressively worse. Vibration will also be experienced. There should be no vertical play whatsoever in either of the connecting rods after the old oil has been washed out of the big-end bearing. If even a small amount of play is evident, it must not be ignored; the bearing should be replaced. Don't run the engine with a worn big-end bearing – the connecting rods or crankpin will break and cause extensive damage.
4   Don't confuse big-end wear with side play in the bearing (a small amount of side play is acceptable). It should be maintained within the limits in the Specifications. The side play should be checked with a feeler gauge.
5   Worn main bearings are evidenced by an audible rumble from the lower end of the engine and vibration that's particularly noticeable through the footpegs. If the primary chaincase cover is removed, play can be detected by attempting to move the end of the crankshaft up-and-down. In the case of the big-end assembly, if any play is evident, the main bearings are due for replacement.

Chapter 2 Engine, clutch and transmission

6   While the engine is apart, examine the main bearings for signs of wear and damage after the oil has been washed out. The usual indications are chipped or broken rollers, scuff marks in the hardened surface of the rollers and roughness as the bearing is revolved. If a bearing fails soon after a complete overhaul, another teardown of the engine will be necessary, so don't reinstall questionable parts.

7   The connecting rod small end bearings are plain bushings. When replacement is needed, the new bushing can be pressed in to displace the old bushing (be sure to align the oil hole in the bushing with the oil hole in the connecting rod). It'll also have to be reamed to size.

17 Cylinder barrels – inspection

1   The usual indication of badly worn cylinder barrels is excessive oil consumption, accompanied by blue smoke from the exhaust and possibly piston slap (a metallic rattle that occurs when there's little or no load on the engine). If the top of the cylinder bore is examined carefully, you probably will find a ridge on the thrust side, denoting the limit of travel of the upper piston ring. The depth of this ridge will vary according to the amount of wear that has taken place.

2   If you're working on an Evolution engine, check both gasket surfaces on each cylinder barrel for warpage with a straightedge and feeler gauge(s). If either surface is warped beyond the specified limit, the barrel (and piston) must be replaced with new parts.

3   Using a telescoping gauge and outside micrometer, or an inside micrometer, measure the bore about 1/2-inch below the limit of top ring travel, first from back-to-front in the barrel, then from side-to-side. This will show whether the bore has worn out-of-round. Repeat the measurements just above the lower limit of oil ring travel and also approximately half-way between these two points. If the bore hasn't worn more than 0.002-inch, isn't out-of-round or badly scored, the cylinder can be reused.

4   If the bore wear exceeds 0.002-inch or the bore is out-of-round or damaged in any way, a rebore is required. Oversize pistons are available in various rebore sizes.

5   If the cylinder barrel is already at the maximum rebore size, you'll have to install a replacement cylinder barrel. It's not possible to go beyond the maximum bore limit without seriously weakening it.

6   Have the barrels honed at a dealer service department and install new piston rings regardless of what is done with the pistons.

18 Pistons – inspection

Refer to illustration 18.9

1   Before the inspection process can be carried out, the pistons must be cleaned and the old piston rings removed.

2   Using a piston ring installation tool, carefully remove the rings from the pistons. Don't nick or gouge the pistons in the process. Make a note of how the rings are installed so they'll be reinstalled properly.

3   Scrape all traces of carbon from the tops of the pistons. A hand-held wire brush or a piece of fine emery cloth can be used once the majority of the deposits have been scraped away. Do not, under any circumstances, use a wire brush mounted in a drill motor to remove deposits from the pistons; the piston material is soft and will be eroded away by the wire brush.

4   Use a piston ring groove cleaning tool to remove any carbon deposits from the ring grooves. If a tool isn't available, a piece broken off an old ring will do the job. Be very careful to remove only the carbon deposits. Don't remove any metal and don't nick or gouge the sides of the ring grooves.

5   Once the deposits have been removed, clean the pistons with solvent and dry them thoroughly. Make sure the oil return holes in the back sides of the oil ring grooves are clear.

6   If the pistons aren't damaged or worn excessively and if the cylinders aren't damaged, new pistons won't be necessary. Normal piston wear appears as even, vertical wear on the thrust surfaces of the piston and slight looseness of the top ring in its groove. New piston rings, on the other hand, should always be used when an engine is rebuilt.

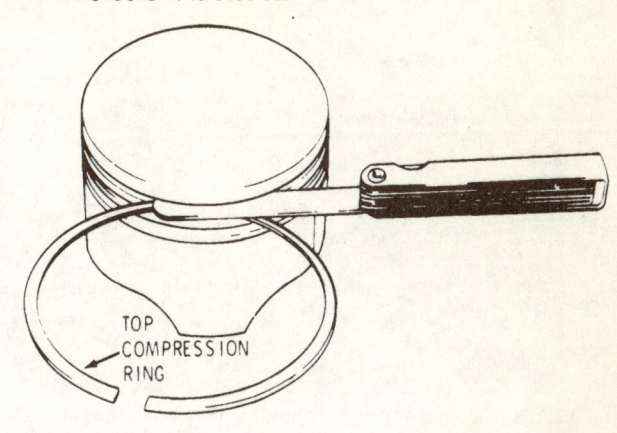

18.9 Measuring piston ring side clearance with a feeler gauge

7   Carefully inspect each piston for cracks around the skirt, at the pin bosses and at the ring lands.

8   Look for scoring and scuffing on the thrust faces of the skirt, holes in the piston crown and burned areas at the edge of the crown. If the skirt is scored or scuffed, the engine may have been suffering from overheating and/or abnormal combustion, which caused excessively high operating temperatures. The oil pump should be checked thoroughly. A hole in the piston crown, an extreme to be sure, is an indication that abnormal combustion (pre-ignition) was occurring. Burned areas at the edge of the piston crown are usually evidence of spark knock (detonation). If any of the above problems exist, the causes must be corrected or the damage will occur again.

9   Measure the piston ring side clearance by laying a new piston ring in the ring groove and slipping a feeler gauge in beside it **(see illustration)**. Check the clearance at three or four locations around the groove. Be sure to use the correct ring for each groove; they are different. If the clearance is greater than specified, new pistons will have to be used when the engine is reassembled.

10   On pre-Evolution engines, check the piston-to-bore clearance by measuring the bore and the piston diameter. Make sure the pistons and cylinders are correctly matched. Measure the piston across the skirt on the thrust faces at a 90-degree angle to the piston pin, about 3/8-inch up from the bottom of the skirt. Subtract the piston diameter from the bore diameter to obtain the clearance. If it's greater than specified, the cylinders will have to be bored and new oversize pistons and rings installed.

11   If the appropriate precision measuring tools aren't available, the piston-to-cylinder clearances can be obtained, though not quite as accurately, using feeler gauge stock. Feeler gauge stock comes in 12-inch lengths and various thicknesses and is generally available at auto parts stores.

12   To check the clearance, select a 0.004-inch feeler gauge and slip it into the cylinder along with the appropriate piston. The cylinder should be upside-down and the piston must be positioned exactly as it normally would be. Place the feeler gauge between the piston and cylinder on one of the thrust faces (90-degrees to the piston pin bore).

13   The piston should slip through the cylinder (with the feeler gauge in place) with moderate pressure (don't try to force it). If it falls through, or slides through easily, the clearance is excessive and a new piston will be required. If it won't fit into the cylinder, the clearance is less than 0.004-inch and progressively thinner feeler gauges will have to be used to determine the clearance. If the piston binds at the lower end of the cylinder and is loose toward the top, the cylinder is tapered, and if tight spots are encountered as the piston/feeler gauge is rotated in the cylinder, the cylinder is out-of-round.

14   Repeat the procedure for the remaining piston and cylinder. Be sure to have the cylinders and pistons checked by a Harley-Davidson dealer

Chapter 2 Engine, clutch and transmission

20.6a A valve spring compressor is needed to remove the keepers (arrows)

20.6b Remove the valve springs, . . .

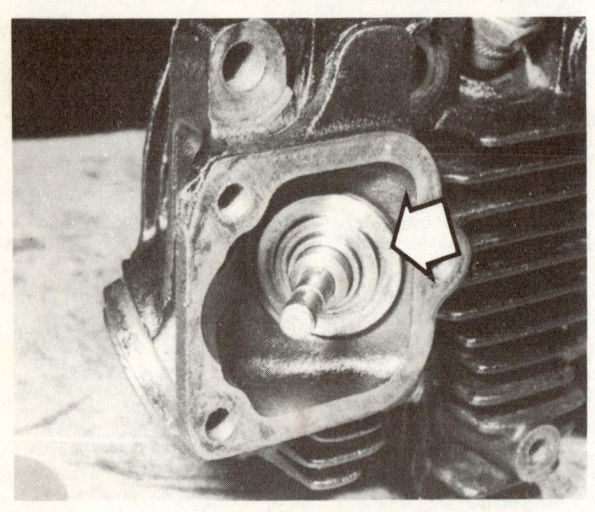

20.6c . . . the spring seat . . .

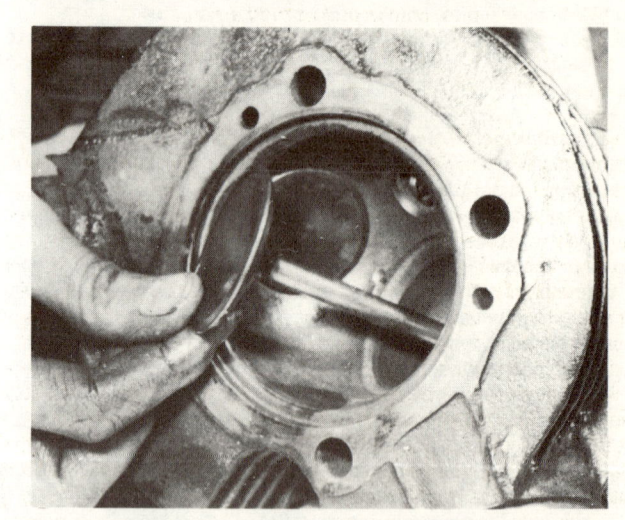

20.6d . . . and the valve from the head

service department or a motorcycle repair shop to confirm your findings before purchasing new parts. **Note:** *The pistons and cylinders for Evolution engines must be checked and matched by a dealer service department.*

19 Valves, valve seats and valve guides – servicing

1   Because of the complex nature of this job and the special tools and equipment required, servicing of the valves, the valve seats and the valve guides (commonly known as a valve job) is best left to a professional.
2   The home mechanic can, however, remove and disassemble the heads, do the initial cleaning and inspection, then reassemble and deliver the heads to a dealer service department, motorcycle repair shop or even an automotive machine shop for the actual valve servicing.
3   The dealer service department will remove the valves and springs, recondition or replace the valves and valve seats, replace the valve guides, check and replace the valve springs, spring retainers and keepers (as necessary), replace the valve seals (if used) with new ones and reassemble the valve components. **Note:** *If the valve seats are reconditioned, the valve stem protrusion must be checked **(see illustration 20.14c)**. If it's not as specified, the seat(s) must be replaced.*
4   After the valve job has been performed, the heads will be in like-new condition. When the heads are returned, be sure to clean them again very thoroughly before installation on the engine to remove any metal particles or abrasive grit that may still be present from the valve service operations. Use compressed air, if available, to blow out all the holes and passages.

20 Cylinder head and valves – disassembly, inspection and reassembly

1   As was mentioned in Section 19, valve servicing and valve guide replacement should be left to a dealer service department or motorcycle repair shop. However, disassembly, cleaning and inspection of the valves and related components can be done (if the necessary special tools are available) by the home mechanic. This way no expense is incurred if the inspection reveals that service work isn't required at this time.
2   To properly disassemble the valve components without the risk of damaging them, a valve spring compressor is absolutely necessary. If the special tool isn't available, have a dealer service department or motorcycle repair shop handle the entire process of disassembly, inspection, service or repair (if required) and reassembly of the valves

Disassembly

Refer to illustrations 20.6a, 20.6b, 20.6c and 20.6d

3   Before the valves are removed, scrape away any traces of gasket material from the head gasket sealing surface. If you're working on an Evolu-

Chapter 2 Engine, clutch and transmission

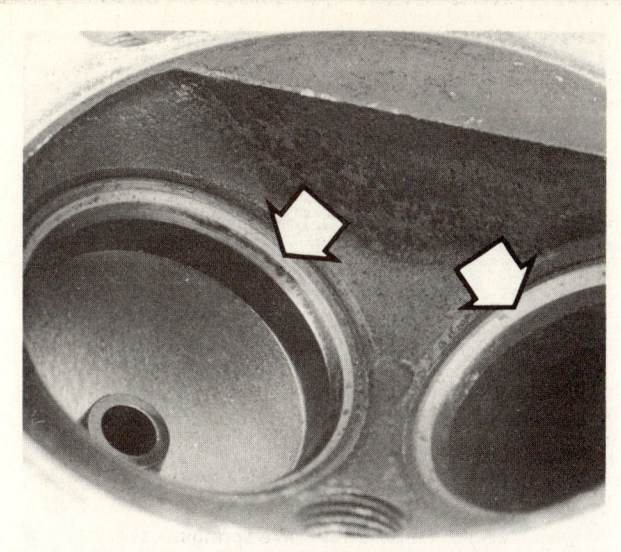

20.14a Check the valve seats (arrows) in each head – look for pits, cracks and burned areas

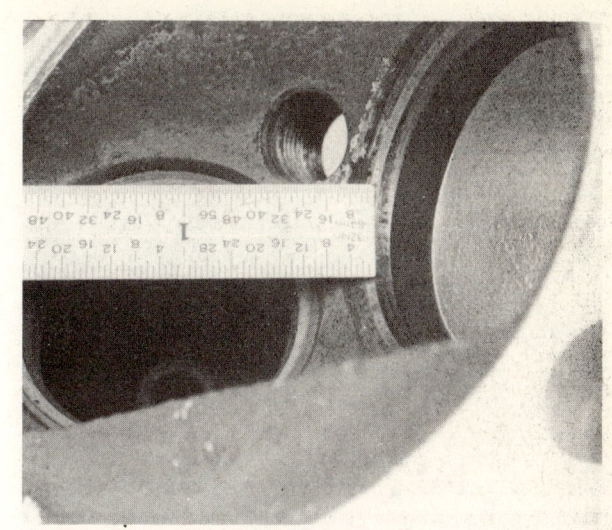

20.14b Use a ruler to measure the width of each valve seat

tion engine, be careful – don't nick or gouge the soft aluminum head. Gasket removing solvents, which work very well, are available at most motorcycle shops and accessory stores.

4 Carefully scrape all carbon deposits out of the combustion chamber. A hand held wire brush or a piece of fine emery cloth can be used once the majority of deposits have been scraped away.

5 Before proceeding, arrange to label and store the valves along with their related components so they can be kept separate and reinstalled in the same valve guides they are removed from.

6 Compress the valve springs on the first valve with a spring compressor, then remove the keepers **(see illustration)** and the retainer from the valve assembly. Don't compress the springs any more than absolutely necessary. Carefully release the valve spring compressor and remove the valve spring retainer, the springs, the spring seat and the valve from the head **(see illustrations)**. If the valve binds in the guide (won't pull through) push it back into the head and deburr the area around the keeper groove with a very fine file or whetstone.

7 Repeat the procedure for the remaining valves. Remember to keep the parts for each valve together so they can be reinstalled in the same location.

8 Once the valves have been removed and labeled, pull off the valve stem seals (if used) with pliers and discard them. The old seals should never be reused.

9 Next, clean the cylinder heads with solvent and dry them thoroughly. Compressed air will speed the drying process and ensure that all holes and recessed areas are clean.

10 Clean all of the valve springs, keepers, retainers and spring seats with solvent and dry them thoroughly. Do the parts from one valve at a time so no mixing of parts between valves occurs.

11 Scrape off any deposits that may have formed on the valves, then use a motorized wire brush to remove deposits from the valve heads and stems. Again, make sure the valves don't get mixed up.

Inspection

Refer to illustrations 20.14a, 20.14b, 20.14c, 20.15a, 20.15b, 20.16a, 20.16b, 20.17, 20.18a and 20.18b

12 Inspect the heads very carefully for cracks and other damage. If cracks are found a new head is in order.

13 Using a straightedge and feeler gauge, check the head gasket mating surface for warpage (Evolution engine only). Lay the straightedge diagonally (corner-to-corner), intersecting the head bolt holes, and try to slip a 0.006-inch feeler gauge under it at each location. If the feeler gauge can

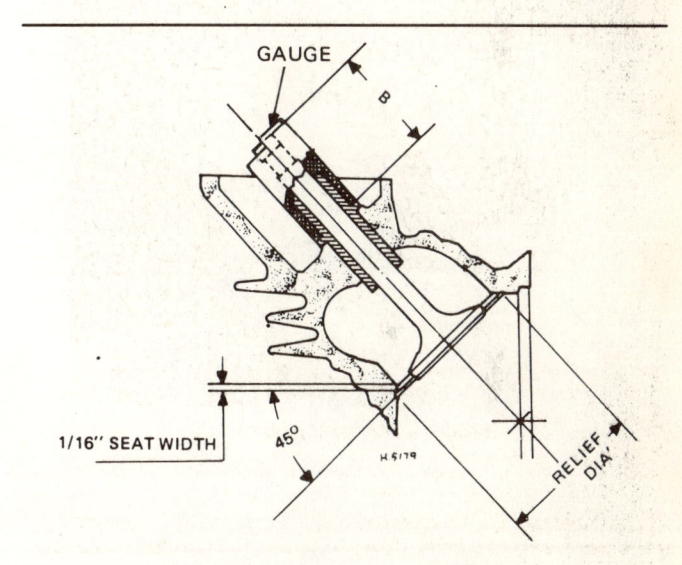

20.14c Critical valve/seat dimensions (pre-Evolution engine)

Intake valve relief diameter = 2.120-inch
Exhaust valve relief diameter = 1.620-inch
B minimum value = 1.375-inch
B maximum value = 1.420-inch

be inserted between the head and the straightedge, the head is warped and must be replaced with a new one.

14 Examine the valve seats in each of the combustion chambers **(see illustration)**. If they're pitted, cracked or burned, the heads will require valve service that's beyond the scope of the home mechanic. Measure the valve seat width and compare it to the Specifications **(see illustration)**. If it's not within the specified range, or if it varies around its circumference, valve seat service is required. **Note:** *If the valve seats are reconditioned, the valve stem protrusion must be checked **(see illustration)**. If it's not as specified, the seat(s) must be replaced.*

15 Clean the valve guides to remove any carbon buildup, then measure the inside diameters of the guides (at both ends and the center of the

Chapter 2 Engine, clutch and transmission

20.15a A small hole gauge can be used to determine the inside diameter of each valve guide

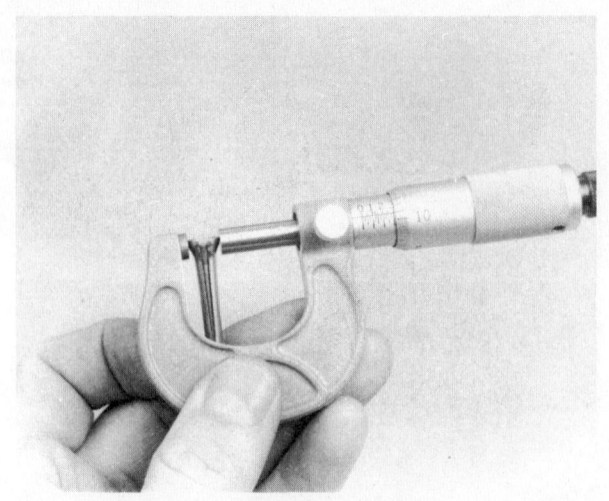

20.15b Measure the small hole gauge with a micrometer to obtain the actual size of the guide

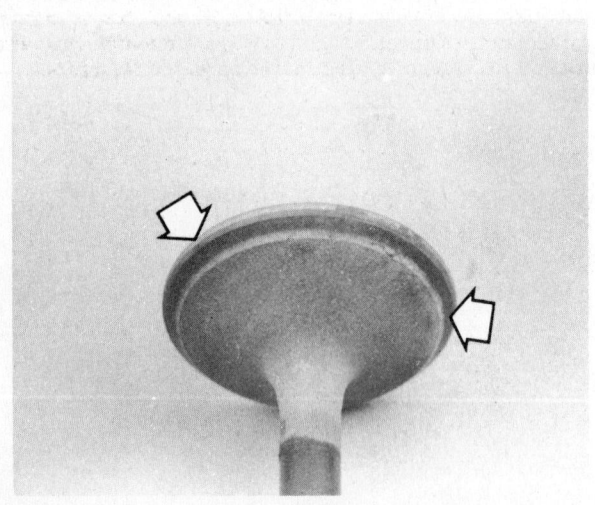

20.16a Check the valve face and margin for wear and cracks

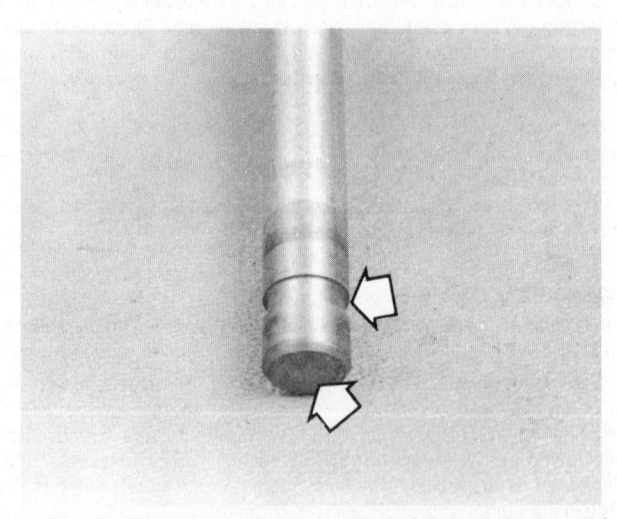

20.16b Look for wear on the very end of the valve stem and make sure the keeper groove isn't distorted

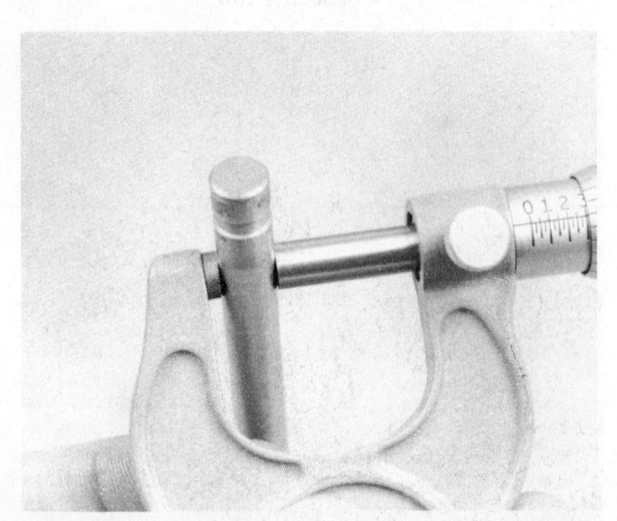

20.17 Measure the valve stem diameter with a micrometer

guide) with a small hole gauge and a 0-to-1 inch micrometer **(see illustrations)**. Record the measurements for future reference. These measurements, along with the valve stem diameter measurements, will enable you to compute the valve-to-guide clearance. This clearance, when compared to the Specifications, will be one factor that will determine the extent of valve service work required. The guides are measured at the ends and at the center to determine if they're worn in a bell-mouth pattern (more wear at the ends). If they are, guide replacement is an absolute must.

16 Carefully inspect each valve face for cracks, pits and burned spots **(see illustration)**. Check the valve stem and the keeper groove area for cracks **(see illustration)**. Rotate the valve and check for any obvious indication that it's bent. Check the end of the stem for pitting and excessive wear. The presence of any of the above conditions indicates the need for valve servicing.

17 Measure the valve stem diameter **(see illustration)**. By subtracting the stem diameter from the valve guide diameter, the valve-to-guide clearance is obtained. If the valve-to-guide clearance is greater than specified, the guides will have to be replaced and new valves may have to be installed, depending on the condition of the old ones. Insert each valve into its guide and hold it tightly against the seat, then measure the distance from the cylinder head spring seat surface to the very end of the valve stem. If it's greater than specified, the valve and/or seat must be replaced.

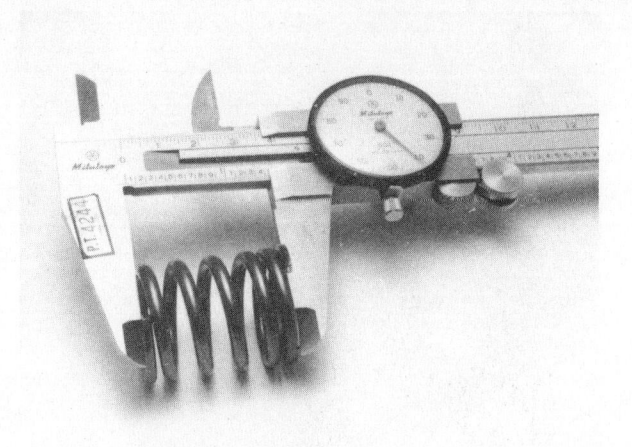

Chapter 2 Engine, clutch and transmission

20.18a Measure the valve spring free length with a dial or vernier caliper

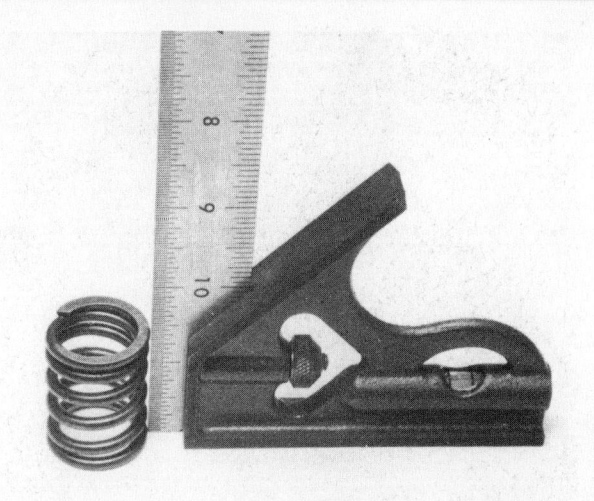

20.18b Check each valve spring for squareness with an accurate machinist's square

18 Check the end of each valve spring for wear and pitting. Measure the free length and compare it to the Specifications **(see illustration)**. Any springs that are shorter than specified have sagged and shouldn't be reused. Stand the spring on a flat surface and check it for squareness **(see illustration)**.

19 Check the spring retainers and keepers for obvious wear and cracks. Questionable parts shouldn't be reused – extensive damage will occur in the event of failure during engine operation.

20 If the inspection indicates that no service work is required, the valve components can be reinstalled in the heads.

Reassembly

Refer to illustrations 20.22, 20.23a, 20.23b, 20.24a, 20.24b, 20.27, 20.28, 20.30, 20.31a and 20.31b

21 Before installing the valves in the head, they should be lapped to ensure a positive seal between the valves and seats. This procedure requires fine valve lapping compound (available at auto parts stores) and a valve lapping tool. If a lapping tool isn't available, a piece of rubber or plastic hose can be slipped over the valve stem (after the valve has been installed in the guide) and used to turn the valve.

22 Apply a small amount of fine lapping compound to the valve face **(see illustration)**, then slip the valve into the guide. **Note:** *Make sure the valve is installed in the correct guide and be careful not to get any lapping compound on the valve stem.*

23 Attach the lapping tool (or hose) to the valve and rotate the tool between the palms of your hands. Use a back-and-forth motion rather than a circular motion **(see illustration)**. Lift the valve off the seat at regular intervals to distribute the lapping compound properly **(see illustration)**.

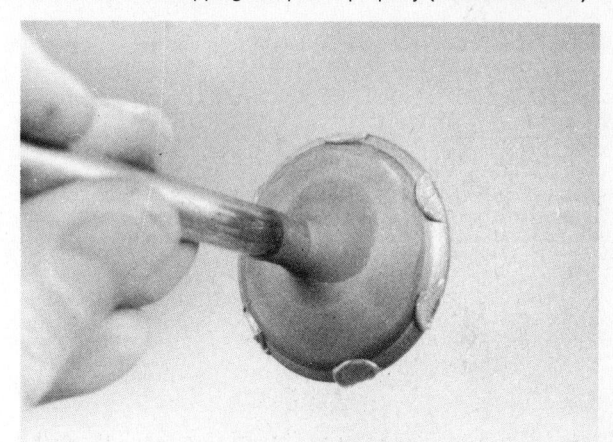

20.22 Apply the lapping compound very sparingly, in small dabs, to the valve face only

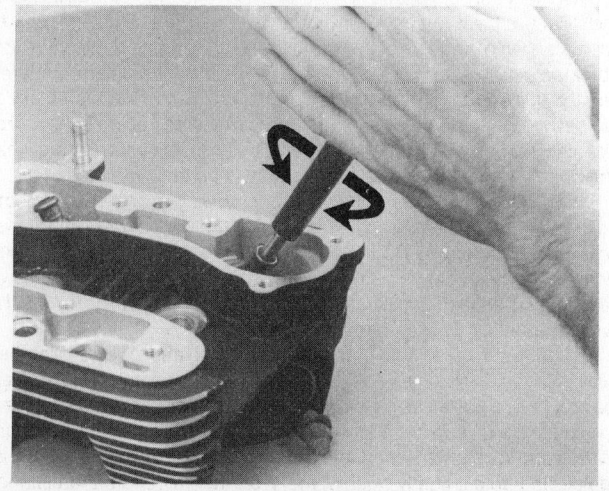

20.23a Rotate the lapping tool or hose back-and-forth between the palms of your hands

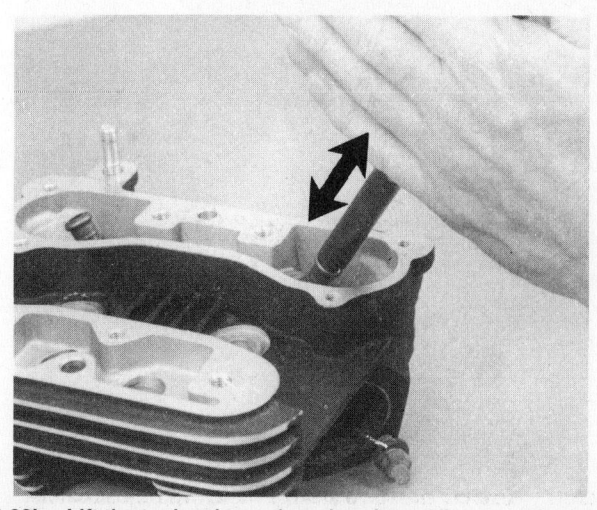

20.23b Lift the tool and turn the valve about 1/3-turn periodically to redistribute the lapping compound on the valve face and seat

Chapter 2 Engine, clutch and transmission

20.24a After lapping, the valve face should exhibit a uniform, unbroken contact pattern (arrow) . . .

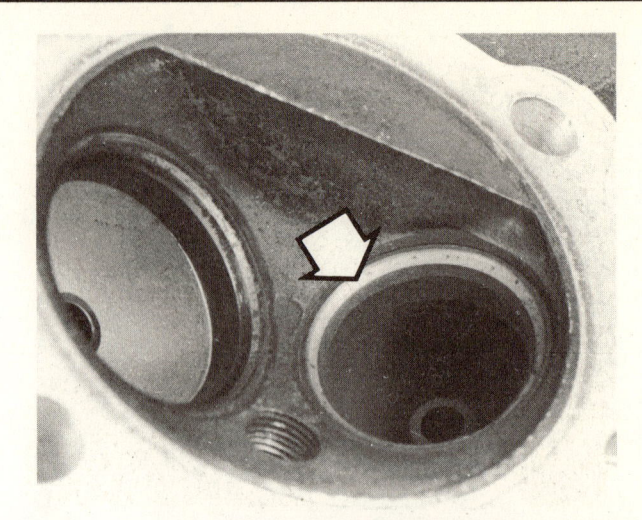

20.24b . . . and the seat should be the specified width (arrow), with a smooth, unbroken appearance

24 Continue the lapping procedure until the valve face and seat contact area is of uniform width and unbroken around the entire circumference of the valve face and seat **(see illustrations)**.

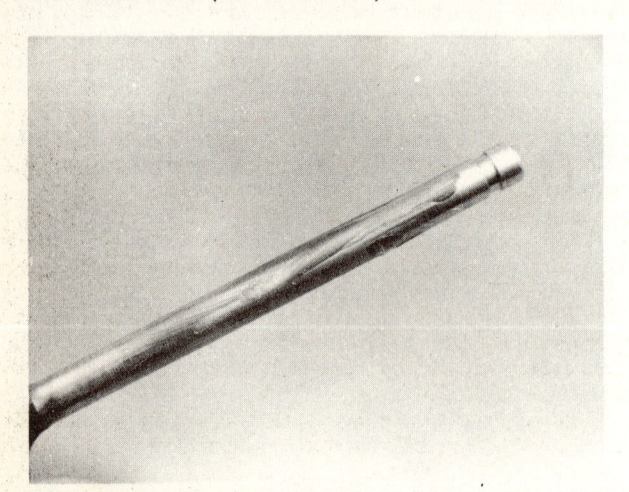

25 Carefully remove the valve from the guide and wipe off all traces of lapping compound. Use solvent to clean the valve and wipe the seat area thoroughly with a solvent-soaked cloth. Repeat the procedure for the remaining valves.

26 Once all of the valves have been lapped, check for proper valve sealing by pouring a small amount of solvent into each of the head ports with the valves in place and held tightly against the seats. If the solvent leaks past the valve(s) into the combustion chamber area, repeat the lapping procedure, then reinstall the valve(s) and repeat the check. Repeat the procedure until a satisfactory seal is obtained.

27 Coat the first valve stem with grease (preferably moly-based) then slide it into the guide **(see illustration)**.

28 Lay the spring seat in place in the cylinder head **(see illustration)**.

29 On models equipped with valve stem seals, place a protective sleeve over the valve stem so the keeper area of the stem is covered (if the seal is installed without the sleeve, the seal will be damaged).

30 Slide a new valve seal into position over the valve stem. Use a hammer and deep socket to install the seal **(see illustration)**. You'll be able to feel when the seal seats completely on the guide (don't continue hammering on it after it seats – you may damage it). Don't twist or cock it or it won't seal properly and DO NOT remove the valve from the guide once the seal is in place.

20.27 Lubricate the valve stem with moly-base grease before installing the valve in the cylinder head

20.28 Lay the spring seat in the head, over the guide, . . .

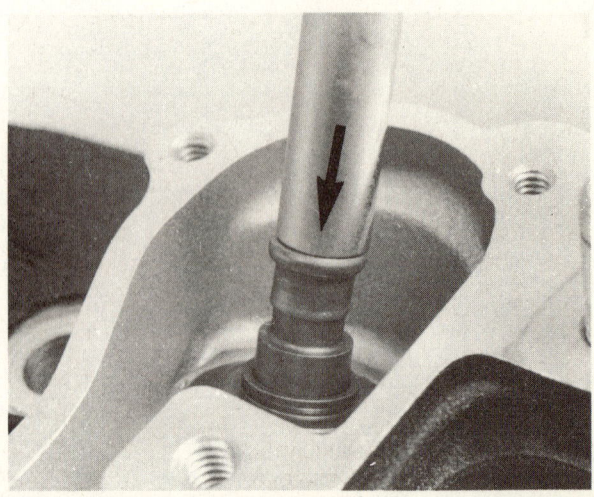

20.30 . . . THEN install the valve guide seal and make sure it's seated on the guide

Chapter 2 Engine, clutch and transmission

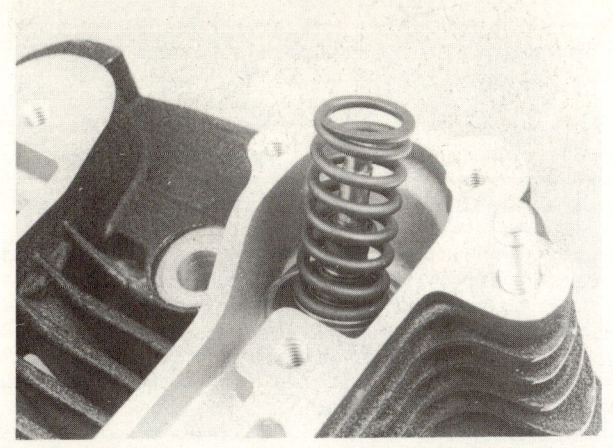

20.31a Install the inner and outer valve springs, then position the retainer and compress the springs so the keepers can be installed

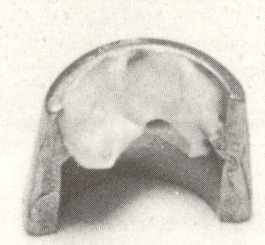

20.31b A small dab of grease will help hold the keepers in place on the valve while the spring is released

31 Position the valve springs and retainer over the valve stem and attach the valve spring compressor **(see illustration)**. Depress the valve springs only as far as absolutely necessary to slip the keepers into place. Make certain the keepers are securely locked in their retaining grooves. Apply a small amount of grease to the keepers to hold them in place as the pressure is released from the springs **(see illustration)**.
32 Repeat the procedure for the remaining valve in the first head and both valves in the other head.
33 Support the cylinder head on blocks so the valves can't contact the workbench top and gently tap each of the valve stems with a soft-face hammer. This will help seat the keepers in the grooves.

21 Rocker arms/shafts and pushrods – inspection

Refer to illustrations 21.2a, 21.2b and 21.5

1 It's unlikely that excessive wear will occur in either the rocker arms or the shafts unless lubrication failure has occured or the machine has covered a lot of miles. A clicking noise from the rocker area is the usual symptom of wear in the rocker arms, and shouldn't be confused with a somewhat similar noise caused by excessive valve clearances.
2 Check the rocker arm bushings and the shafts for evidence of excessive wear and damage **(see illustrations)**.
3 Measure the outside diameter of the shaft (where it rides in the bushings) and the inside diameter of the rocker arm bushings. Subtract the shaft diameter from the bushing diameter to obtain the clearance. If it's excessive, install new parts or have the bushings in the rocker arm(s) replaced by a dealer service department.
4 Check each rocker arm where it contacts the pushrod and valve stem. If cracks, scuffing or breakthrough in the case hardened surface are evident, install a new rocker arm.
5 Check the ends of the pushrods where they ride in the tappets and the rocker arms **(see illustration)**. Check for distortion by rolling them on a flat surface such as a pane of glass. If wear or distortion is evident, new pushrods should be installed.
6 To disassemble a rocker box on pre-Evolution engines, unscrew the rocker arm shaft end caps found on the right-hand side of each rocker box

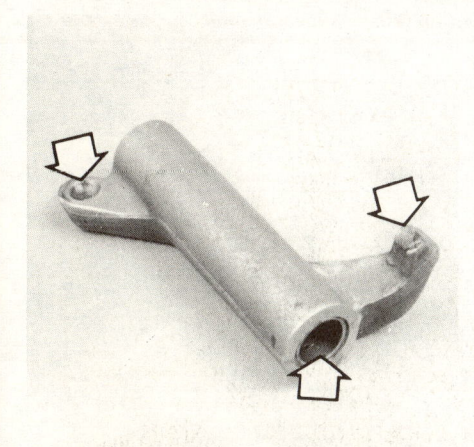

21.2a Check the rocker arm bushings, the pushrod recesses and the valve stem pads for wear and damage (arrows)

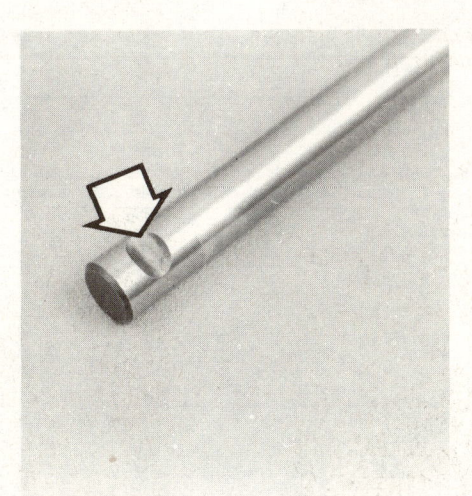

21.2b Check the rocker arm shafts for wear at the areas where the rocker arm bushings ride – the Evolution engine rocker arm shafts have a cutout in one end (arrow) for the bolt to pass through

21.5 Check the pushrod ends for wear and make sure the oil holes (Evolution engine) are clear

Chapter 2 Engine, clutch and transmission

(see illustration 6.3a). Early models have caps with a screwdriver slot, later models a hex-shaped socket to accept an Allen wrench. Remove both caps and the O-ring behind each one (if equipped). Discard the O-ring.

7 Unscrew the acorn nuts from the other end of the rocker box and tap the shaft out of position from the acorn nut end.

8 Remove the end play control spacer and the rocker arm. This may take a certain amount of manipulation. The disassembled components will include the rocker arm, a spacer and the shaft itself. Mark all the parts before tapping out the second shaft so they're eventually replaced in their original locations.

9 Replacement of the rocker arm bushings should be done by a dealer service department, since the new ones may have to be reamed to size to fit the shafts.

22 Camshafts, timing gears, tappets and timing case bearings – inspection

Refer to illustrations 22.3 and 22.7

Note: *On all models, it is important when selecting a replacement timing cover, bushings, camshafts or timing gear, that the advice of a Harley-Davidson dealer is sought. Unless the replacement gears installed are suitably matched, and certain gears replaced in matched pairs, excessive gear noise and a pronounced wear rate will result.*

1 The timing gears are unlikely to require attention unless the engine has very high mileage or if there has been a lubrication failure. Wear will be evident in the form of excessive backlash between the individual gears, with a characteristic "clacking" noise.

2 The cam lobes should be checked for wear and damage in the form of pit marks, scuffing or flaking of the case hardened surfaces. Wear will be particularly evident on the flanks of the lobes, at the point where they begin to lift the tappet. If there's any doubt about the condition of the cam lobes, the camshafts should be replaced as a precaution.

3 If the cams are worn or damaged, more than likely the tappets will require attention too. Referring to the information given in the Specifications, check the tappets and their guides for wear **(see illustration)**. Inspect the roller on the tappet end. Excessive wear between the tappet and guide will cause the tappet to tilt and create additional mechanical noise. In the case of the hydraulic tappets fitted to Evolution engines, soak the tappets in fresh engine oil to await installation.

4 Check the end of the tappet which bears on the pushrod; wear of its hardened surface will necessitate tappet replacement. Similarly check the pushrod ends as described in Section 21.

5 The needle roller bearings in the timing case should be replaced if there's any play evident or if there's any sign of roughness. This also applies to any plain bushings. Replacement of the bearings should be done by a dealer service department, especially in the case of bushings, which may have to be reamed to size. Don't forget the timing cover, which has bushings also. The bushings are pegged in position and require expert attention when replacement is necessary.

6 On pre-Evolution engines a figure is given for acceptable camshaft backlash (see Specifications). Backlash can be measured with a dial indicator. If there's excessive backlash between the gears, larger rear exhaust and front intake cam gears can be installed. If an increase of two or more color codes is required, a larger crankshaft gear should be installed also.

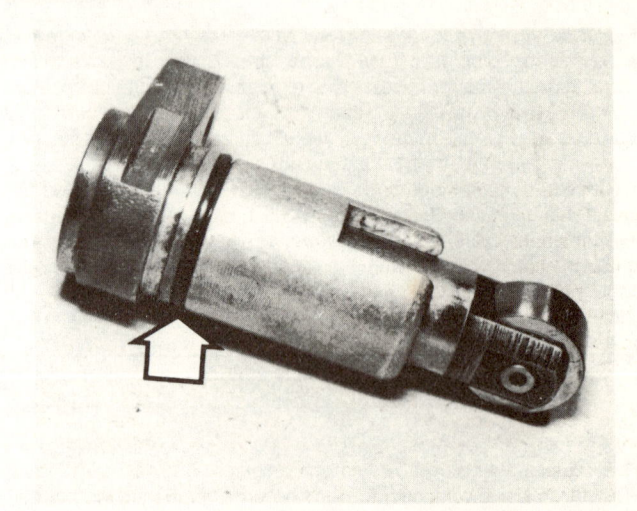

7 Also on pre-Evolution engines, cam gear diameter can be used to provide information on the correct replacement part when compared with the gear color code/size table **(see illustration)**. Each gear must be measured across opposite pins using a micrometer.

8 On Evolution engines, cam and timing gear replacement must be done by careful selection according to the color code of the individual gears or stamped letter, and the color code marking on the inside face of the timing cover, next to the shaft bushing locations. It is advised that the motorcycle or labeled component parts be taken to a Harley-Davidson dealer for inspection. If there's excessive backlash between the gears, larger rear exhaust and front intake cam gears can be installed. If an increase of two or more color codes is required, a larger crankshaft gear should be installed also.

22.3 Examine the tappets/guides carefully for wear – install a new O-ring (arrow) on each guide

Color Code	Rear Exhaust	Rear Intake Inner	Rear Intake Outer	Front Intake	Front Exhaust Inner	Front Exhaust Outer	Crank-shaft	Idler
Brown	1.8893 1.8903	1.8953 1.8943	2.3902 2.3912	1.8893 1.8903	1.8953 1.8943	2.3902 2.3912	1.2681 1.2671	3.0223 3.0213
Blue	1.8903 1.8913	1.8943 1.8933	2.3912 2.3922	1.8903 1.8913	1.8943 1.8933	2.3912 2.3922	1.2671 1.2661	3.0213 3.0203
Red	1.8913 1.8923	1.8933 1.8923	2.3922 2.3932	1.8913 1.8923	1.8933 1.8923	2.3922 2.3932	1.2661 1.2651	3.0203 3.0193
White	1.8923 1.8933	1.8923 1.8913	2.3932 2.3942	1.8923 1.8933	1.8923 1.8913	2.8932 2.3942	1.2651 1.2641	3.0193 3.0183
Green	1.8933 1.8943	1.8913 1.8903	2.3942 2.3952	1.8933 1.8943	1.8913 1.8903	2.3942 2.3952	1.2641 1.2631	3.0183 3.0173
Yellow	1.8943 1.8953	1.8903 1.8893	2.3952 2.3962	1.8943 1.8953	1.8903 1.8893	2.3952 2.3962	1.2631 1.2621	3.0173 3.0163
Black	1.8953 1.8963	1.8893 1.8883	2.3962 2.3972	1.8953 1.8963	1.8893 1.8883	2.3962 2.3972	1.2621 1.2611	3.0163 3.0153

22.7 Camshaft gear color codes/sizes (pre–Evolution engine – typical)

Chapter 2 Engine, clutch and transmission

23 Crankcases – inspection

1   After thorough cleaning, the crankcases should be examined for cracks and other signs of damage that may ultimately cause failure. Minor cracks can be repaired by welding, but if more extensive damage is apparent, replacement is recommended.

2   Note that crankcases are always supplied as a matched pair and should never be replaced any other way. This is important – the crankcases are line bored in pairs, so the bearing housings are aligned correctly. Replacement of only one half will result in a mismatch, which may cause the crankshaft to run out of line and absorb a surprising amount of power.

3   Make sure the mating surfaces are undamaged, otherwise oil leaks will be inevitable after reassembly. If there's any doubt about their ability to seal, use a liquid gasket sealer during reassembly.

4   Check the bearing housings to make sure they're undamaged. If they have worn as the result of a bearing rotating, it's possible to repair using a special bearing sealant such as red Loctite during reassembly. This can be used successfully only if the amount of wear is small.

5   Now is the opportunity to retap any of the threads, if they require attention. Most damage is caused by over-tightening the drain plugs. If necessary, threaded holes can be repaired by installing Helicoil thread inserts, which will permit the original drain plug to be reused. Most dealers can perform this type of repair.

24 Transmission components – inspection

1   Give the transmission components a close visual examination for signs of wear or damage, such as chipped or broken teeth, worn dogs or splines and bent shift forks. If the transmission has shown a tendency to jump out of gear look for worn dogs on the back of the gears and wear in the shift mechanism; in the former case wear will be evident in the form of rounded corners or even a wedge-shaped profile in extreme cases. On models through 1990, the corners of the cam plate tracks will wear first, being characterized by a brightly polished surface.

2   The selector arms usually wear across the fork that engages with a gear, causing a certain amount of sloppiness in the gear change movement. A bent selector will immediately be obvious, especially if overheating has blued the surface.

3   All transmission components that are worn or damaged must be replaced. There's no satisfactory method of repairing them.

4   Depending on the year of manufacture, it is possible to determine the condition of the gear shafts by direct measurement (see Specifications).

25 Transmission gear cluster – disassembly and reassembly

1970 through 1990 models (4-speed)

Refer to illustrations 13.5a, 25.1, 25.2, 25.3, 25.4a, 25.4b, 25.4c, 25.5a, 25.5b, 25.5c and 25.5d

1   Assuming the selector mechanism and cam plate have been removed (as described in Section 13), withdraw the mainshaft **(see illustration)**. Take off the tabbed thrust washer from the right-hand end and remove first gear. On the right-hand end of the mainshaft is the needle roller bearing that seats in the sleeve gear, a thrust washer and second gear, which will pull off. Before third gear can be released, remove and discard the snap-ring. The washer below should be lifted off, then the gear itself.

2   The sleeve gear will remain in position through the main bearing in the access cover and can be driven out if the access cover is supported and the gear driven through towards the inside of the cover. The bearing itself is retained by two snap-rings, which will have to be removed first **(see illustration)**. Warm the cover to simplify removal of the bearing.

3   The countershaft is disassembled by removing the thrust washers (note their number and arrangement), first gear, another washer and third gear **(see illustration)**. Reverting to the other (left-hand) end of the shaft, press the drive gear off the splined end of the countershaft. Remove the spacer, then second gear and the thrust washer.

25.1 Withdraw the mainshaft, leaving only the sleeve gear in position

25.2 Note the needle bearing inside the sleeve gear

25.3 Note the arrangement of the countershaft thrust washers

Chapter 2 Engine, clutch and transmission

25.4a Remove the snap-ring so the cam plate can be withdrawn

25.4b Access is now available to the pawl carrier and spring assembly

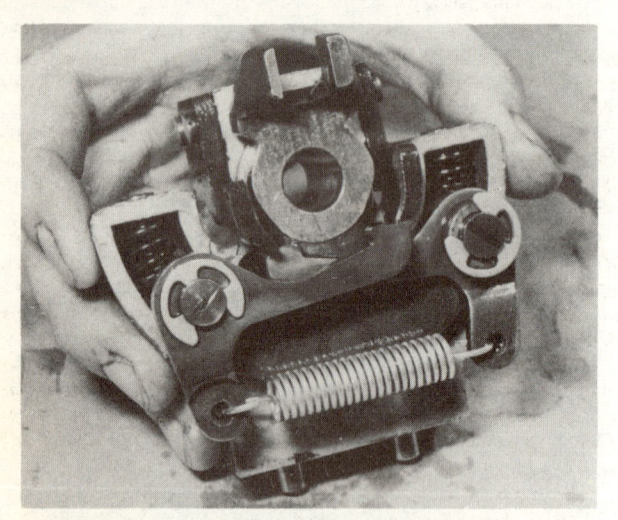

25.4c Check the pawl carrier support body for cracks

25.5a Reassemble by installing the shaft gears, thrust washers and snap-rings in the reverse order of disassembly

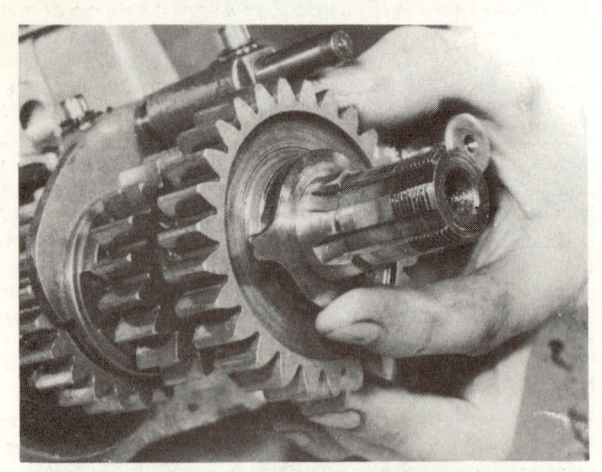

25.5b The mainshaft thrust washer has a projecting tab

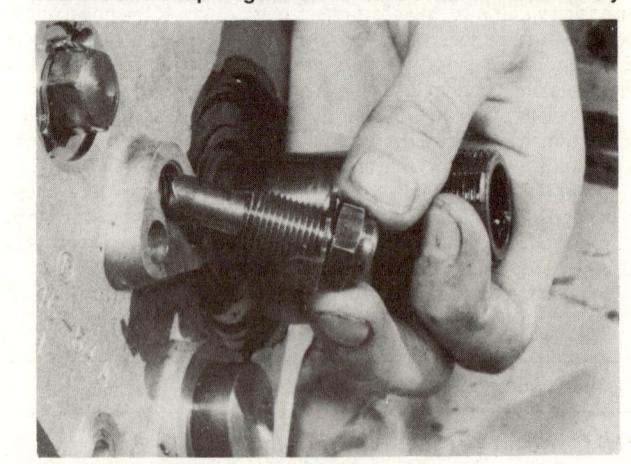

25.5c Replace the cam plate plunger, then check the action of the gear selector

4 If required, the cam plate assembly can be disassembled by removing and discarding the snap-ring on the outer end of the cam plate shaft **(see illustration)**. This will give access to the pawl carrier and associated parts, all of which can be checked for wear **(see illustration)**. Don't forget the pawl carrier support, which should be checked for surface cracks **(see illustration)**. Check the two pawl carrier springs for damage caused by acids in the oil. **Note:** *On 1979 through 1985 models, do not install cadmium-plated, 14-coil pawl carrier springs. Use cadmium-plated, 16-coil springs or black phosphatized springs (14 or 16-coil) only.*

5 Reassemble the shafts by reversing the disassembly order. Be sure to install new snap-rings and make sure all the shims and thrust washers are replaced in their original locations **(see illustrations)**. The shafts and gears must turn freely **(see illustration)**.

6 Before installing the transmission, have a Harley-Davidson dealer or

25.5d All shafts should revolve freely before installation

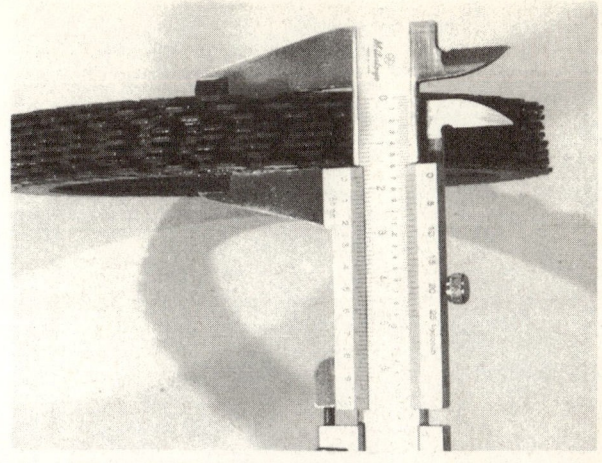

26.11 Friction plate thickness measurement – 1991-on models

a repair shop check the gear spacing and shaft end play. If it's not as specified, new pawl carrier support shims, shift forks or thrust washers may have to be installed. This is very important so let a qualified Harley-Davidson technician handle the job.

1991-on models (5-speed)
Refer to illustration 13.12

7 Disassembly of the transmission shafts is as follows. Place each component in order as it is removed as a guide to reassembly.
8 Remove the circlip from the end of the countershaft and remove its 5th gear. Next slide both the countershaft and mainshaft 2nd gears off their shafts, noting the split bearing inside the countershaft gear.
9 Slide the thrust washer off the countershaft and remove the circlip retaining the 3rd gear; slide the 3rd gear off the shaft.
10 The mainshaft 3rd gear is retained on each side by a circlip and thrust washer. First displace the circlip between 3rd and 1st gears from its groove and move it and the thrust washer towards 1st gear. Then push the 3rd gear back towards 1st gear to provide clearance to remove the circlip from its front face. Slide the thrust washer, 3rd gear, its split bearing, and second thrust washer off the mainshaft.
11 At this point, the shafts should be pressed out of the access cover. Dealing first with the countershaft, check that the Torx screw and retainer plate have been removed from its end. Place a support between the 4th and 1st gears and press on the end of the countershaft (do not press on the inner race of the bearing) to push it out of the access cover.
12 Once clear of the access cover, slide the angled spacer off its end, together with the 4th gear. The 1st gear is retained on both sides by a circlip and thrust washer; note its split bearing when removing.
13 Turning to the mainshaft, slide its 1st gear off the shaft. Place a support under its 4th gear and press the shaft from the access cover. Once free, remove the spacer, 4th gear and its split bearing. The remaining thrust washer and circlip can be removed if desired.
14 Reverse the disassembly sequence to assemble the gear shafts.

26 Clutch components – inspection

1970 through early 1984 models

1 Check the condition of the clutch sprocket, to ensure none of the teeth are chipped, broken or badly worn.
2 Clean the plates with solvent and make sure they are not warped or distorted. Remove all traces of clutch insert debris, otherwise a gradual build-up will occur and affect clutch action.
3 Visual inspection will reveal if the tongues of the clutch plates have been burred and whether corresponding indentations have formed in the slots in the clutch drum. Burrs should be removed with a file, which can also be used to dress the slots square, provided the depth of the indentations is not too great.

4 Check the thickness of the friction plates. When they have worn thin, they must be replaced. Always replace them as a complete set, irrespective of whether some may not have reached the serviceable limit. Worn friction plates promote clutch slip.
5 Check the free length of the clutch springs and compare it to the Specifications Section of this Chapter. Do not stretch the springs if they have compressed. They must be replaced when the service limit has been reached.
6 Check the clutch pushrod for straightness, if necessary, by rolling it on a sheet of glass. Heavy action is often caused by a bent rod, which may hang up in its housing. Check the action of the clutch actuating mechanism that's located on the inside of the primary chaincase cover. It should give no trouble if greased regularly.
7 The bearings in the center of the pre-1971 clutch are needle roller type and a press fit. To gain access when replacement is necessary, remove the oil seal and rivets from the center of the outer drum and press out the bearing assembly, which includes two separate needle roller races, a large diameter washer of variable thickness and the starter clutch.
8 When reassembling, press the first bearing into place, pushing on the face on which the bearing number is inscribed. Press it in to a depth of 0.010 to 0.015-inch, measured from the clutch drum to the inner face of the bearing. Then press the other bearing into position from the starter clutch side until it's flush with the first bearing. Position the roll pin and the large diameter washer, then the variable size washer. Lay the kickstart ratchet on the back plate of the clutch and press down on it while a feeler gauge is inserted between it and the variable size washer. Adjustments should be made with different thicknesses of the variable size washer until a measured clearance of from 0.001 to 0.004-inch is obtained. Allow approximately 0.001-inch for the pull of the rivets. The rivets should not protrude more than 0.010-inch above the face of the kickstart ratchet, after installing, and should be sealed off with a solvent-proof sealant. Do not forget to install the oil seal with the lip facing in.
9 Post-1970 clutches have a much simpler center bearing arrangement, composed of a roller bearing retained by a circlip. This is easily removed and replaced.

Late 1984-on models

Note: *Access to the clutch components requires disassembly of the clutch using the Harley-Davidson service tool. Due to the dangers involved in safely compressing the diaphragm spring, it is advised that the task be entrusted to a Harley-Davidson dealer. With the clutch disassembled, its components can be examined as follows.*

Refer to illustrations 26.11 and 26.12

10 Check the sprocket teeth on the clutch sprocket, together with those of the engine sprocket. Severe wear is best remedied by replacement of both sprockets and chain. Also check the teeth of the starter ring gear.
11 Check the plates as described in Steps 2 to 4 of this Section, referring to the Specifications for minimum limits. When checking the friction plates for wear on 1991-on models, wipe them free of oil then stack together

26.12 Check that spring plate is correctly positioned in plate pack

without the steel plates and spring plate and measure the overall thickness of the pack **(see illustration)**.

12 Note the position of the spring plate in the pack of plates; on models through 1990 it resides between the 3rd and 4th friction plates from the outer drum, whereas on 1991-on models it resides between the 4th and 5th friction plates from the outer drum **(see illustration)**.

13 Inspect the diaphragm spring for signs of cracking or bend tabs. Replace if either condition is found.

14 Rotate the clutch outer drum on the clutch center. If the bearing is excessively noisy it must be replaced. First remove the circlip from the rear of the outer drum and separate the clutch center from the outer drum. Remove the circlip from the front of the outer drum and press out the bearing. Use new circlips on installation.

27 Primary drive components – inspection

1 Examine the teeth on the crankshaft and clutch drum sprockets. If any are chipped, hooked, or broken the sprocket must be replaced. It's a good idea to replace the clutch drum, the crankshaft sprocket and the primary chain together as a matched set. A badly worn sprocket will cause the chain to wear more rapidly and cause the engine to lose power.

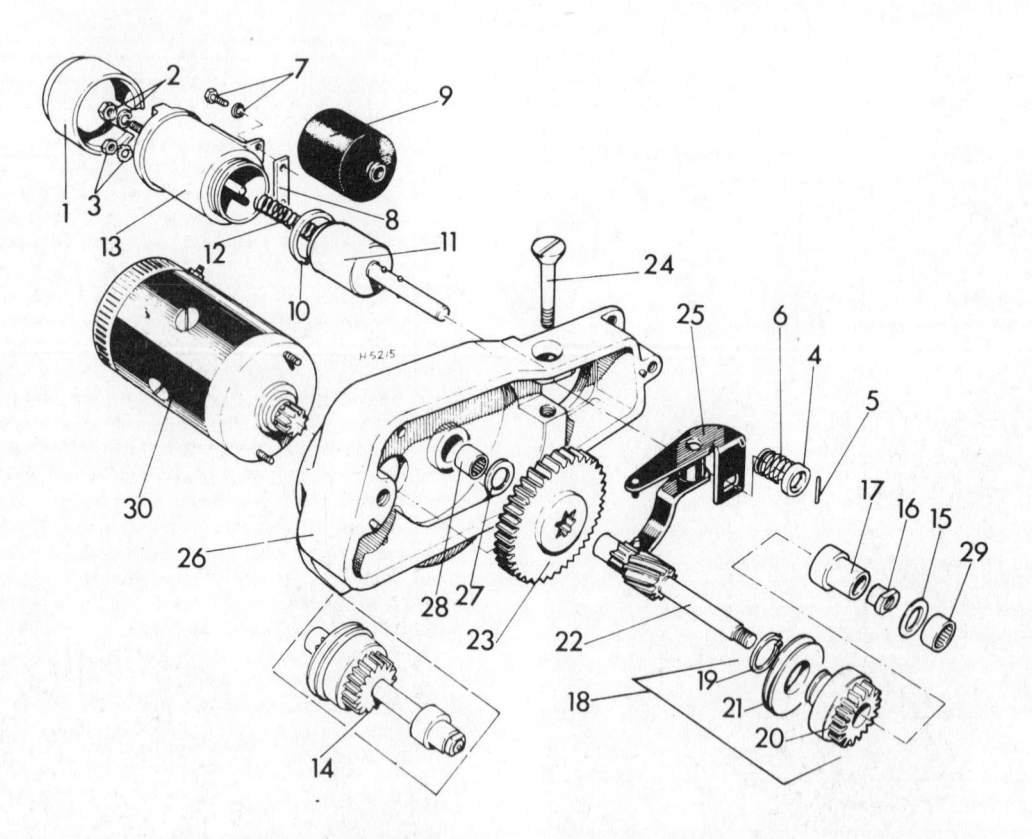

29.3 Starter, solenoid and housing – 1970 through 1980 models

1	Solenoid cover	11	Plunger	21	Reaction collar
2	Terminal nut and lock washer	12	Plunger spring	22	Shaft
3	Terminal nut and lock washer	13	Solenoid	23	Drive gear
4	Retainer cap	14	Gear and shaft assembly	24	Reaction lever screw
5	Pin	15	Thrust washer	25	Reaction lever
6	Spring	16	Gear shaft nut	26	Starter shaft housing
7	Bolt and lock washer	17	Bearing race	27	Washer
8	Spacer bar	18	Gear and reaction collar assembly	28	Needle roller bearing
9	Boot	19	Snap-ring	29	Needle roller bearing
10	Gasket	20	Gear	30	Starter motor

Chapter 2 Engine, clutch and transmission

2   On earlier models equipped with a compensating sprocket, check the condition of the splines on the sliding cam and the sprocket extension. If they are worn, the components should be replaced as a matched set. Although it's unlikely the sliding cam has worn to any great extent, it should be checked where it makes contact with the crankshaft sprocket.

3   The component most likely to wear on models with a compensating sprocket is the sprocket spring, which will compress after extended use. Increased cam action is a sure sign that the spring is in need of replacement.

28 Kickstart mechanism – inspection

1   The kickstart mechanism installed on some models is extremely rugged. The problems most likely to develop are a broken kickstart return spring, causing engagement of the mechanism while the engine is running, or slipping under load.

2   The kickstart return spring is on the outside of the cover and is easily removed and replaced after the kickstart lever has been removed **(see illustration 37.3)**. The spring is pre-tensioned, so it will return the kickstarter to an upright position when released.

3   Failure of the kickstart mechanism to disengage after the engine has started can usually be traced to a bent kickstarter shaft, worn shaft bearings or excessive shaft end play, which should normally be within 0.001 to 0.007-inch. Shims of varying thicknesses are available to take up excess play.

4   Slip is caused by worn or broken teeth on the kickstart crank gear or the kickstart ratchet gear with which it engages. It may also be caused by worn teeth on either the kickstart ratchet found on the back of the clutch outer drum or on the face of the kickstart ratchet gear. In all cases the worn parts are not repairable and must be replaced. In the case of the kickstart ratchet, this will involve removing the rivets to release the worn component and re-riveting the replacement into position. Refer to Section 26.

29 Starter drive assembly – inspection

Refer to illustration 29.3

1   The Bendix-type drive shaft and gear assembly located between the starter motor and the ring gear on the clutch engage the starter motor drive gear with the clutch ring gear when the starter motor button is depressed. It also ensures the drive is disengaged as soon as the engine starts or when the starter motor button is released.

2   Examine all the components of the starter drive assembly. Look for worn or broken teeth on the gears and damage to the worm drive gear in the drive gear. The gear must move freely on the worm, without tilting or binding at any point. All worn parts must be replaced – they cannot be repaired.

3   To disassemble the starter drive assembly on pre-1981 models, remove the bronze thrust washer and place the nut on the end of the shaft in a vise, between two soft metal clamps. The nut has a left-hand thread and will unscrew. There is a bearing race behind it. The gear and reaction collar assembly will then pull off. To disassemble this sub-assembly, remove the snap-ring from the inner end of the shaft.

4   When reassembling, give the worm drive gear a liberal coat of moly-base grease and make sure the sliding gear moves freely when the assembly is complete.

5   To remove the starter drive assembly on 1981 and later models, remove the two mounting bolts from the drive end, along with the washers, lock washers and O-rings.

6   The starter drive, idler gear and bearing can be removed from the starter housing. Check the condition of the O-ring in the groove of the starter housing.

7   Clean the drive components and apply high-temperature grease to them.

8   More information relating to the starter motor and solenoid is found in Chapter 7.

30 Oil seals – inspection and replacement

Refer to illustrations 30.3a, 30.3b and 30.4

1   Even after very careful examination it's difficult to determine whether an oil seal can be reused, especially if it has been disturbed during disassembly. Because an oil seal failure will necessitate another teardown at a later date, it's recommended that all the oil seals be replaced with new ones during an overhaul, as a precautionary measure.

2   Oil seals are very easily damaged during reassembly. Always be very careful when installing shafts and grease the lips of the seals.

3   The most important oil seals are those located around the drive side of the crankshaft assembly and around the transmission mainshaft behind the final drive sprocket (pre-Evolution engine). The former is easily pried out of position **(see illustrations)**. The latter is retained by an oil seal retainer, held to the outer face of the right-hand crankcase with four screws and lock washers.

30.3a   The crankshaft seal can be easily pried out of position

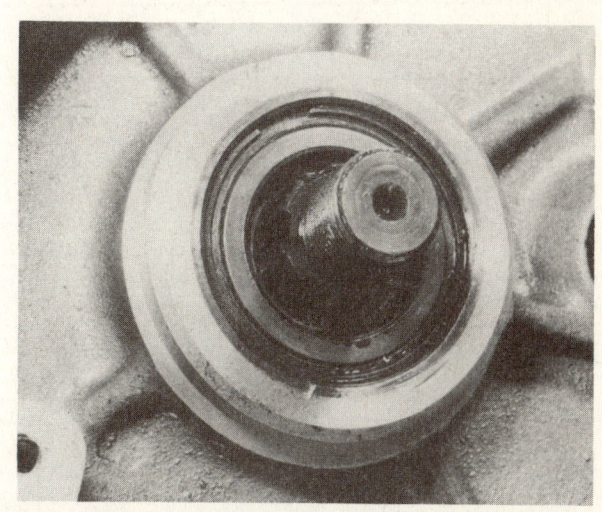

30.3b   Install new seals carefully to avoid damage that could cause oil leaks

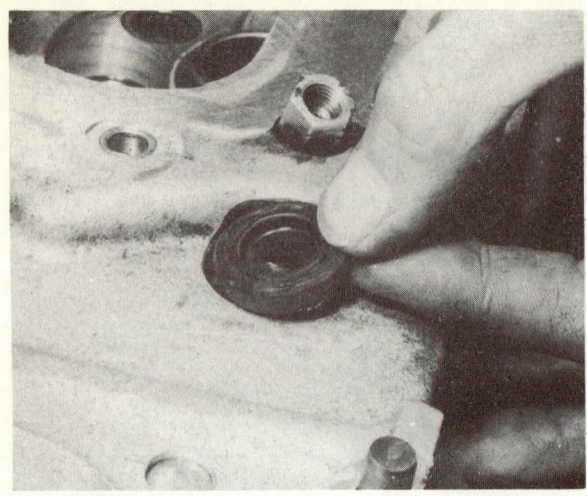

30.4 Replace the shift lever shaft seal also

32.2 Install the rear engine mount (on earlier models) before the crankcases are rejoined

4   Oil seals will also be found around the clutch pushrod, the kickstart shaft housing (if equipped) and the shift lever shaft **(see illustration)**.

31 Engine reassembly – general information

1   Before reassembly is started, the various engine components should be cleaned thoroughly and placed close to the work area.
2   Make sure all traces of old gasket material has been removed and the mating surfaces are clean and undamaged. One of the best ways to remove old gasket sealer is to apply aerosol gasket remover, available at most auto parts stores. This acts as a solvent and will ensure the sealer is removed without scraping and the accompanying risk of damage.
3   Gather up all necessary tools and have an oil can filled with clean engine oil available. Make sure all the new gaskets, oil seals and replacement parts are on hand; there's nothing more frustrating than having to stop in the middle of a reassembly sequence because a vital gasket or part has been overlooked.
4   Make sure the reassembly area is clean and well lit, with adequate working space. Refer to the torque specifications (when given). Many smaller bolts are easily sheared off if overtightened. Always use the correct size screwdriver bit for Phillips-head screws. Remember, if the heads are badly damaged during reassembly, the screws may be almost impossible to loosen later.

32 Crankcases – reassembly

Refer to illustrations 32.2 and 32.3

1   If the flywheel assembly has been removed from the left-hand crankcase, reinsert it through the left-hand main bearing assembly, which requires special equipment. This should be done by a Harley-Davidson service department with the required service tools to pull the crankshaft through the bearing assembly without risk of damage. Install the sprocket shaft extension on the left-hand end of the crankshaft on early models so equipped.
2   Before joining the right-hand crankcase, make sure the mating surfaces are perfectly clean and apply a thin coat of gasket sealant. On all models through 1990, also make sure the rear engine mount is included with the parts for reassembly **(see illustration)**. It cannot be added after the crankcases are rejoined and bolted together.
3   Reassemble the right-hand needle roller bearing on the crankshaft (on models without caged bearings) **(see illustration)**, rather than in the crankcase bearing housing. Grease can be used to retain the uncaged rollers in place during reassembly.

32.3 Assemble the right roller bearing assembly on the crankshaft – early uncaged type

4   Bolt the crankcases together, but before final tightening, make sure the crankshaft assembly revolves freely. Tighten the fasteners, then recheck it. Where fitted, replace the washer over the right-hand end of the crankshaft and the circlip in front of it.

33 Crankcase breather – timing (1976 and earlier models only)

Refer to illustrations 33.1, 33.2a, 33.2b, 33.2c and 33.2d

1   Before the gear can be installed on the crankshaft, you'll have to time the crankcase breather, which forms part of the oil pump assembly. Proceed by removing the Allen-head plug in the left crankcase **(see illustration)**. It's located above the primary chaincase, between the two cylinders. When removed it will permit the edge of the crankshaft flywheel to be seen.
2   Rotate the crankshaft until the TDC mark is exactly in the center of the hole **(see illustrations)**. Slide the oil pump spiral gear over the splines on the right-hand end of the crankshaft with the marked side facing out. The oil pump gear that meshes with the spiral gear should be arranged so when the spiral gear is seated on the crankshaft, the timing hole in the oil pump/breather sleeve drive is facing forward **(see illustration)**. Make

Chapter 2 Engine, clutch and transmission

33.1 Remove the Allen-head plug from the left side of the crankcase

33.2a Rotate the crankshaft until the TDC mark is in position

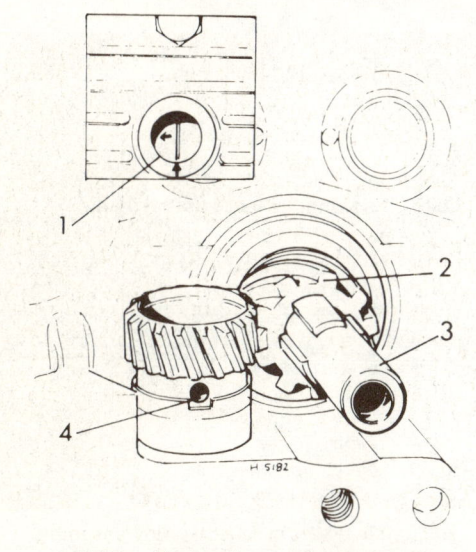

33.2b Timing the crankcase breather

1. TDC mark on flywheel
2. Spiral gear
3. Crankshaft
4. Timing hole in breather sleeve

33.2c If the breather is timed correctly, the hole in the gear boss will be in this position

33.2d Seat the crankshaft gear in front of the oil pump gear

sure all the parts mentioned are in correct alignment, then install the crankshaft gear in front of the spiral gear, making sure it's completely seated.

3 Check the position of the gear on the end of the crankshaft. The gear outer face must be 5/16-inch from the timing cover mating surface.

4 Install the plug in the left crankcase.

34 Camshafts – installation

Refer to illustrations 34.3, 34.4a, 34.4b and 34.5

1 Install any shims originally fitted to the crankshaft end, followed by the oil pump drive gear and crankshaft gear; on 1988-on models the gears must locate with the shaft's key. If the oil pump body was withdrawn, this can now be installed. **Note:** *Refer to the procedure in Section 33 for timing the crankcase breather on models through 1976.*

2 Secure the crankshaft gear with its retaining nut on later models. Note that on models through 1987 a new lock washer must be fitted under the nut. On 1988 through 1990 models apply a drop of Loctite 242 (blue) to the nut threads, and on 1991-on models apply a drop of Loctite 262 (red) to the nut threads. Tighten the nut to the specified torque and on models through 1987 secure with one of the lockwasher tabs.

Chapter 2 Engine, clutch and transmission

34.3 Make sure the bearing retainers and shims are in position

34.4a The camshaft gears are marked for correct valve timing (be sure the marks align properly) (arrows)

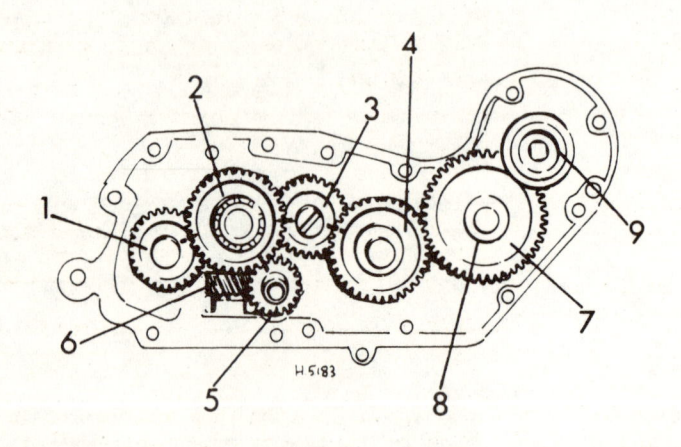

34.4b Aligning the valve timing marks

1. Rear exhaust camshaft
2. Rear intake camshaft
3. Front intake camshaft
4. Front exhaust camshaft
5. Crankshaft gear
6. Breather sleeve gear
7. Idler gear (no timing mark)
8. Fiber washer
9. Generator drive gear (through early 1984)

34.5 The oil line to the final drive chain (not all models) must be securely connected

3   Before the camshafts can be installed, you'll have to replace the bearing retainers and shims. These components should be marked, to ensure they are replaced in their correct locations **(see illustration)**. Remember to apply a liberal coat of oil to the needle roller bearings and bushings before installing the camshafts.

4   The gears on the camshafts are clearly marked and should be installed with all of the timing marks aligned **(see illustrations)**.

5   Make sure all of the gears are exactly in register to ensure the valve timing is correct. Note that the idler gear is not marked – its position isn't important. Install the rear chain oiler oil line and gland nut (if equipped); it's easier to install these components before the timing cover is installed. Make sure the oil line on the back of the timing cover is securely attached and in good condition **(see illustration)**.

6   If the check performed before removing the timing cover confirmed camshaft end play to be within the limits, refit any shims present in their original positions. Seek the advice of a Harley-Davidson dealer if adjustment need be made.

7   Install a new oil seal in the timing cover, make sure both gasket surfaces are clean and dry, then using a new cover gasket refit the timing cover. Refit all screws and bolts and tighten evenly in a criss-cross pattern.

8   Refit the ignition components as described in the next Section. If the engine is in the frame, refit all removed components and replenish the engine oil.

Chapter 2 Engine, clutch and transmission

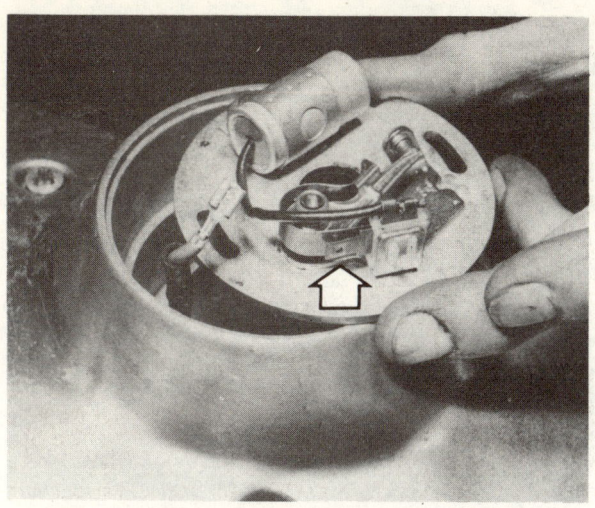

35.3a Install the contact breaker point assembly, . . .

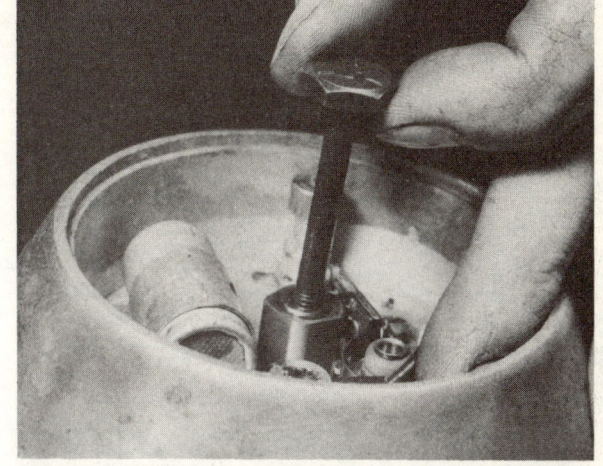

35.3b . . . then the contact breaker cam and retainer bolt

35 Ignition components – installation

Refer to illustrations 35.3a and 35.3b

1 Installation of the ignition components and the advance unit is the reverse of the removal procedure described in Section 9 of this Chapter.
2 On 1970 models, install the distributor (see Chapter 4).
3 The mechanical advance unit on 1971 through 1978 models will locate on the protruding end of the camshaft. Install the contact breaker baseplate, aligning the scribe marks made during disassembly, so the timing is approximately correct **(see illustrations)**.
4 Be sure to seat the advance assembly for the 1979 ignition system squarely on the end of the camshaft. Install the trigger rotor with the flat side next to the cam stop roll pin on the advance assembly base. It must also engage both counterweights of the advance unit.
5 When installing the rotor bolt on 1980 and later models, the threads should be coated with locking compound. If the sensor plate was completely removed on these models, it may be necessary to install new sockets, wire pins and a body receptacle.
6 Set the point gap or air gap and adjust the timing as described in Chapter 1.
7 When installing the outer cover on 1980 and later models, special rivets must be used. These rivets are specially designed so the end doesn't fall off in the timing compartment. The use of regular rivets could damage the ignition system. The recommended rivets are part number 8699.

36 Tappets – installation

1970 through 1990 models

1 On pre-Evolution engines, turn the adjustment screw assembly into the tappet body.
2 Install a new O-ring on the tappet guide and coat it with grease. Carefully push the tappet into the guide, from below. Coat the tappet assembly with fresh engine oil; be sure to apply plenty of oil to the roller needles.
3 Hold the tappet as high as possible in the tappet guide and position the assembly over the gearcase. Turn the tappet guide so the mounting hole is aligned with the hole in the case, then turn the tappet so the tappet roller is correctly aligned with the cam. **Note:** *If the tappet roller isn't aligned properly (crosswise to the camshaft) serious engine damage could result.*
4 Carefully press the tappet guide assembly into the gearcase. The guides are made of aluminum, so care must be taken not to damage them.
5 Install the tappet mounting screws and tighten them securely. On some earlier models, the rear tappet mounting screw is also used to secure the starter motor retaining strap. On these models, the bolt cannot be installed until the starter motor is installed.
6 Check the tappet for freedom of movement in the guide.

1991-on models

7 The tappets and guides are installed as part of the cylinder head installation procedure. Refer to Section 42, paragraph 21 for details.

37 Transmission gear cluster primary drive and clutch components – installation

Transmission gear cluster – 1970 through 1990 models (4-speed)

Refer to illustration 37.2

1 Turn the engine over so the left side faces up. Make sure all the transmission bearings are in position and the gear shift shaft is located in its correct position. If the gear cluster has been disassembled, reassemble it by reversing the instructions in Section 25.
2 Oil the bearings, then slide the access cover and gear cluster into position making sure the end of the gear shift shaft engages correctly with the pawl carrier **(see illustration)**. Install and tighten the four access cover bolts to the specified torque (note that the shorter bolt goes in the upper right hole). Check that the shafts are able to revolve freely.

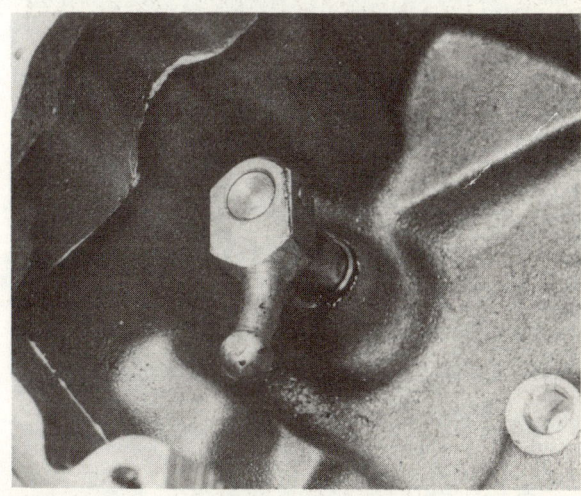

37.2 The end of the gear shift shaft must locate in the pawl carrier

Chapter 2 Engine, clutch and transmission

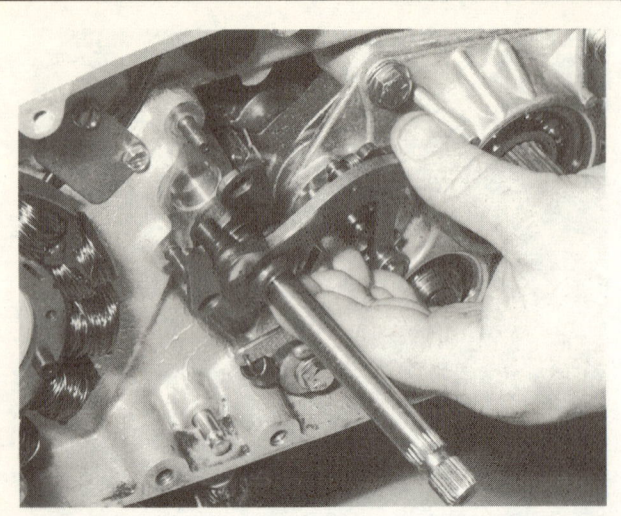

37.8 Insert shifter shaft in its crankcase bore and secure lightly with its nuts

37.10 Install cam correctly on the shifter drum end and secure with new retaining clip

3   On models with an alternator (late 1984 through 1990), refit the stator to the transmission access plate, referring to Chapter 7, Section 10 for details.

Transmission gear cluster – 1991-on models (5-speed)

Refer to illustrations 37.8, 37.10, 37.11a and 37.11b

4   If the gear cluster, shift drum/forks or their bearings have been disassembled, reassemble them as described in Section 25.
5   If the mainshaft 5th gear was removed from the crankcase it must be pressed back into its bearing until its shoulder contacts the bearing inner race. **Caution:** *Use only the correct Harley-Davidson service tool to install the 5th gear, otherwise casing damage may result.* Working from the sprocket side of the 5th gear, insert a new blind seal in the end of the gear (it must be recessed to a depth of 0.03 to 0.06 inches (0.76 to 1.52 mm). Fit a new sealing ring over the threaded part of the gear, followed by the spacer (chamfered end towards sealing ring). Lubricate the lips of the new oil seal and press it squarely into the casing until it is flush with the surface.
6   Insert the assembled access cover and gear cluster into the casing, engaging the mainshaft end in its 5th gear, and the countershaft and shifter drum ends in their respective bearings. The access cover should fit flush with the crankcase; if not, remove the shafts for investigation.
7   Inspect the five access cover bolt threads – if they show signs of having stretched in use, replace them with new ones. Apply a few drops of Loctite 242 (blue) to the bolts and tighten them evenly to the specified torque. Check that the shafts are free to rotate when in neutral.
8   Engage the shifter shaft in its bore and loosely install its washers and nuts, having engaged the claw arm over the shifter drum pins. Do not tighten the nuts at this stage **(see illustration)**.
9   Loop the detent lever spring over its anchor pin.
10   Place the cam over the shifter drum pins, aligning its blind hole with the corresponding pin **(see illustration)**. Secure the cam with a new retaining clip.
11   The shifter shaft claw must be properly aligned with the drum pins. To ensure this, shift the transmission into 3rd gear and position a No. 32 (0.116 inch) drill bit through the hole at the top of the cam and between the top part of the claw and the drum pin **(see illustrations)**. At this point press down lightly on the top of the shifter shaft crank to remove all play between the claw and drill bit. At the same time, tighten the bottom shifter shaft nut fully to the specified torque, then tighten the top nut to the same torque. Remove the drill bit.
12   If removed, install the Torx-head screw and bearing retainer in the countershaft end.

37.11a Checking shifter shaft claw alignment

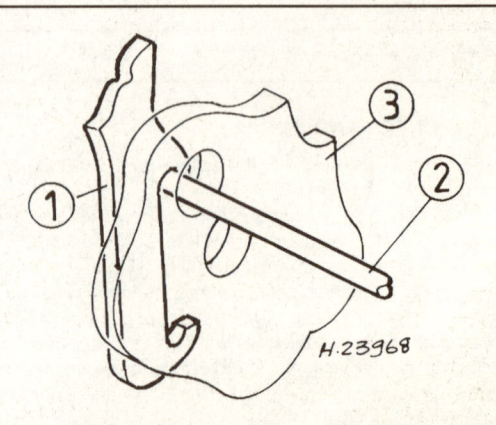

37.11b Shifter shaft claw alignment

1   Shifter shaft claw
2   Drill bit
3   Cam

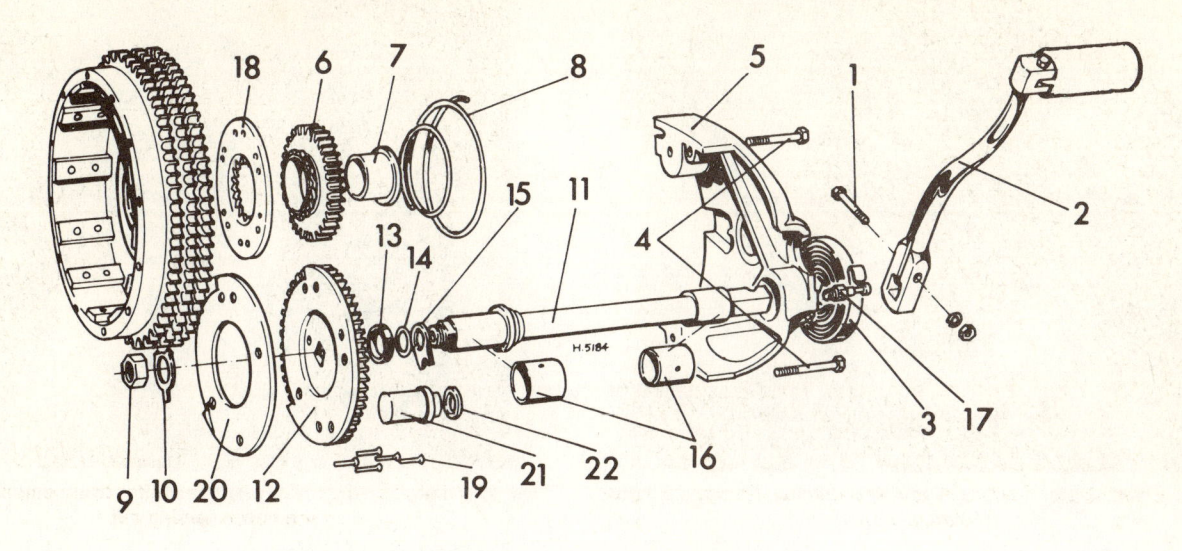

Chapter 2 Engine, clutch and transmission

37.13 Kickstarter components – exploded view

1	Pinch bolt, washer and nut	7	Clutch sprocket spacer	13	Oil seal	19	Crank gear cam plate rivet
2	Kickstarter lever assembly	8	Ratchet spring	14	Shim (0.007-inch)	20	Crank gear cam plate
3	Return spring	9	Shaft nut	15	Thrust plate	21	Crank gear stop pin
4	Sprocket cover bolt	10	Tab washer	16	Kickstarter shaft bushing	22	Crank gear stop pin washer
5	Sprocket cover	11	Kickstarter shaft	17	Spring anchor		
6	Ratchet gear	12	Crank gear	18	Ratchet plate		

Kickstart mechanism (where fitted)

Refer to illustration 37.13

13 If the machine is equipped with a kickstarter, the kickstart shaft is the next item to be replaced **(see illustration)**. Grease the shaft before it's installed and make sure the thrust plate, shims and oil seals aren't left out. Install the kickstart crank gear and tighten the center retaining nut, bending over the tab washer to secure it. Install the spring, clutch sprocket spacer and kickstart ratchet gear (the ratchet teeth should face out) over the transmission (input shaft) sleeve gear. The kickstart crank gear must be located with the segment containing the teeth in the 6 to 8 o'clock position. This will permit the kickstart ratchet gear to rest against the metal plate that covers the greater portion of the sprocket teeth.

Primary drive and clutch components – 1970 through early 1984 models

Refer to illustrations 37.14a, 37.14b, 37.14c, 37.14d, 37.15, 37.16a, 37.16b, 37.16c and 37.16d

14 Install the compensating sprocket components (if equipped) on the end of the crankshaft **(see illustration)**. Assemble the clutch outer drum and sprocket, the engine sprocket and the primary chain, and attach them to their respective shafts as an assembly **(see illustration)**. Replace the compensating sprocket spring, then the threaded collar. Lock the engine and tighten the collar or nut in a clockwise direction, using the method described in Section 8 of this Chapter **(see illustrations)**.

37.14a Install the compensating sprocket collar on the crankshaft

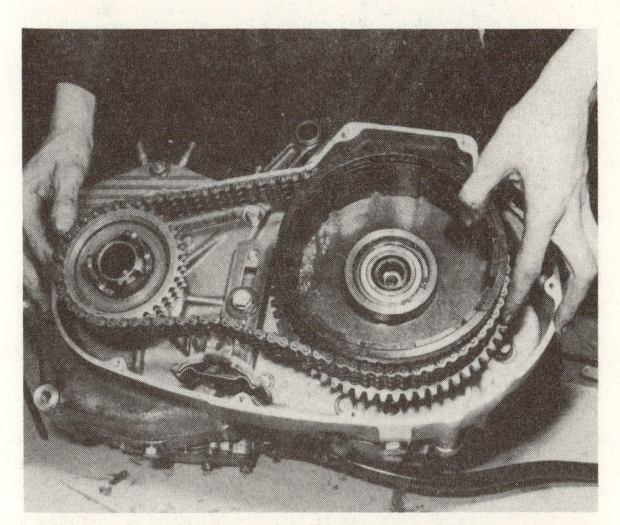

37.14b Replace the primary drive sprocket, the primary chain and the clutch drum as an assembly

37.14c Install the compensating sprocket cam, followed by the spring and outer collar

37.14d ... and the threaded collar, which must be tightened securely

37.15 Install the clutch center and tighten the retaining nut to the specified torque

37.16a Make sure the spacers are in position on the clutch studs ...

37.16b ... and the circlip is installed to retain the clutch plates

37.16c The springs and pressure plate are installed last, ...

Chapter 2 Engine, clutch and transmission

37.16d ... using a gear puller to compress the springs and start the pressure plate nuts

37.19 Apply thread-locking compound to the clutch nut on 1991-on models; install washer with OUT marking facing outwards

15 Install the clutch center, tab washer and large retaining nut. Install a new clutch pushrod seal (pre-1971 models only). Using the same metal sprag used previously, tighten the nut **(see illustration)** and bend over the tab washer. The nut must be tight. Replace the final drive sprocket temporarily and place the transmission in gear, locking the sprocket with a chain wrench wrapped around it. Reassemble the compensating sprocket and tighten the threaded collar to the specified torque while the engine is still locked.

16 The clutch can now be reassembled by reversing the procedure in Section 8 of this Chapter. Note that the pre-1971 clutch must remain oil tight, so a new gasket must be used under the domed cover. Remember to replace the shouldered clutch release rod before installing the pressure plate. In the case of the post-1970 clutch, either the appropriate Harley-Davidson service tool or a three-jaw puller is needed to compress the two clutch center springs sufficiently for the release disc retaining nuts to be started on their threads **(see illustrations)**.

17 Replace and adjust the primary chain tensioner, after checking to make sure it hasn't worn enough to require a new one (refer to Section 21 of Chapter 1). The primary chain should have the specified slack with the engine cold. When the adjustment is correct, rotate the crankshaft and check the slack at several points. Tighten the adjuster locknut when the correct adjustment has been obtained. Remove any equipment used to prevent engine rotation.

Primary drive and clutch components – late 1984-on models

Refer to illustration 37.19

Note: *This procedure describes clutch installation as a complete unit. If the clutch has been dismantled (see Note in Section 8), it must be reassembled using the Harley-Davidson service tool.*

18 On 1991-on models install the spacer on the end of the crankshaft (if it came off the shaft on removal). On all models, if the alternator rotor was detached from the clutch sprocket (models through 1990) or engine sprocket (1991-on) it should be installed as described in Chapter 7.

19 With the chain installed on the engine sprocket and clutch sprocket, install the assembly on the crankshaft and mainshaft, being careful not to damage the stator coils as the rotor is moved into position. On models through 1990 install the washer on the mainshaft and secure with a new circlip. On 1991-on models, install the conical washer on the mainshaft with the side marked 'OUT' facing outwards. Clean the nut and shaft threads, apply a drop of Loctite 262 (red) to the shaft threads and tighten to the specified torque, remembering that the nut has a left-hand thread **(see illustration)**.

20 Install the adjuster assembly in the clutch pressure plate and secure with the circlip. On 1991-on models note that the release plate ears must align with the cutouts in the pressure plate.

21 Clean the crankshaft and engine sprocket nut threads, then apply a drop of Loctite 242 (blue) thread-locking compound (Loctite 262 (red) on 1991-on models) to the shaft threads. With the engine locked, tighten the nut to the specified torque. Where a lock plate is fitted, install the plate over the nut and secure with the lock bolt.

22 Remove any equipment used to prevent engine rotation. Adjust the chain tension and clutch after the chaincase has been refitted.

---

38 Starter motor – installation

1970 through 1980 models

1 On 1970 through 1980 models, the housing containing the starter drive assembly can now be attached to the back of the chaincase. The starter drive assembly and solenoid reaction lever should be assembled first. Attach the starter solenoid and bolt the whole assembly in place. Install the spring, cap and pin.

2 Install the starter motor, securing it to the back of the chaincase with the two long through bolts. Install the mounting bracket, then the end cover.

3 Reconnect the starter motor wire to the terminal and route the cable like it was originally. Place the retaining clamp around the starter motor body before attaching the wire. It's held by the single bolt that also retains the rear tappet assembly in place.

1981-on models

4 Slip the starter motor with a new gasket into position in the primary chaincase. Insert the two mounting bolts into the front of the starter motor and through the chaincase. The short starter mounting bolt is installed at the upper left corner of the chaincase.

5 Connect the battery cable and the solenoid wire to the solenoid.

All models

6 If the engine is in the frame, refit the primary chaincase (see Section 45) and top up the transmission oil. Refit the exhaust and reconnect the battery negative cable.

7 If the starter motor is being installed as part of an engine overhaul, continue as described in the following sections.

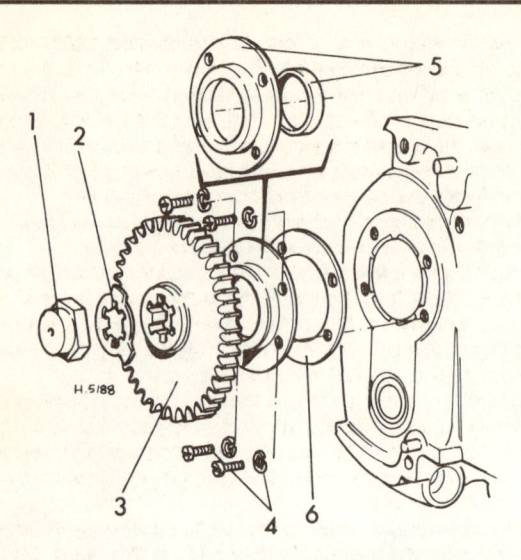

106   Chapter 2   Engine, clutch and transmission

39.1a   Final drive sprocket components – exploded view (1970 through early 1984 shown)

1. Mainshaft nut
2. Tab washer
3. Final drive sprocket
4. Oil seal retainer screws and washers
5. Oil seal and retainer
6. Gasket

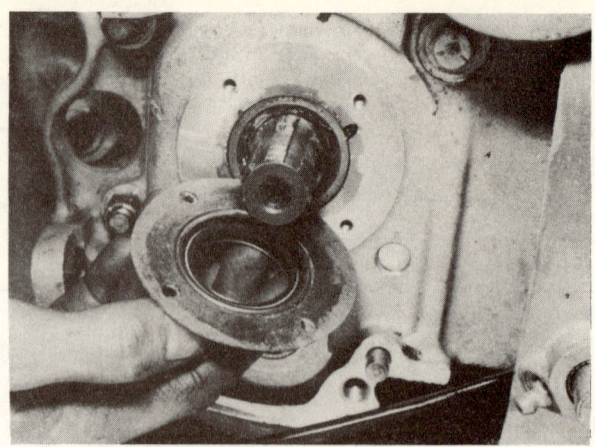

39.1b   Remove the retainer to replace the oil seal on early models

39   Final drive sprocket – installation

1970 through early 1984 models

Refer to illustrations 39.1a, 39.1b and 39.2

1   Before installing the final drive sprocket, it's a good idea to replace the oil seal behind it. Remove the four retaining screws and lift the seal retainer off (if applicable) **(see illustrations)**. Pry the oil seal out of the case or retainer carefully so the bore isn't damaged. Note the direction the seal is facing and install the new seal in the same direction.

2   Install the retainer (if applicable), then slip the final drive sprocket over the splined mainshaft. The sprocket presses into the center of the oil seal on early models **(see illustration)**.

3   Install the tab washer behind the nut that secures the sprocket. You'll have to lock the engine to tighten the nut. When the nut is tight, bend the tab washer over the nut to lock it in place **(see illustration)**. Later models have a lock screw instead of a tab washer.

Late 1984 through 1990 models

4   Install the sprocket on the mainshaft and secure with its nut. Lock the engine and tighten the nut to 35 to 65 ft-lbs (47 to 88 Nm), check if one of the three holes for the lock screw aligns with any of the nut flats. In none align, continue tightening (do not exceed maximum of 90 ft-lbs, 122 Nm) until one of the holes aligns. At this point, apply a drop of thread locking compound to the lock screw and tighten it to 50 to 60 in-lbs (6 to 7 Nm).

1991-on models

Refer to illustration 39.7

5   Install the spacer on the mainshaft with its chamfered side inwards, followed by the sprocket. Apply a few drops of Loctite 262 (red) thread locking compound to the nut threads and install the nut on the mainshaft, remembering that it has a left-hand thread. Tighten to 110 to 120 ft-lbs (149 to 163 Nm).

6   On all chain drive models and 1991 belt drive models, align one of the sprocket lock screw holes as described in Step 4, not exceeding the maximum torque of 150 ft-lbs (203 Nm). Apply a drop of Loctite 242 (blue) to the lock screw threads and tighten it to 50 to 60 in-lbs (6 to 7 Nm).

7   On 1992 through 1994 models with belt drive, fit the lock plate over the nut in such a way that it can be secured diagonally by the two lock screws (tighten the nut further if necessary as described above). Apply a drop of Loctite 242 (blue) to the lock screw threads and tighten them to 80 to 110 in-lbs (9 to 12 Nm) **(see illustration)**.

8   On 1995 and later models, install the nut as described in Step 5, but tighten it only to 50 ft-lbs (68 Nm). Draw a line across the nut and sprocket with a felt pen. Tighten another 30 to 40-degrees, aligning two of the lock plate holes with the holes in the sprocket. Try repositioning the lock plate if the holes don't line up; the lock plate holes are offset. If the holes can't be

39.2   The final drive sprocket fits into the center of the oil seal (1970 through early 1984 models)

39.7   Sprocket lock plate must be installed so that bolt holes align -- 1992-on

Chapter 2 Engine, clutch and transmission

lined up within the 30 to 40-degree range, tighten further, but not more than a total of 45-degrees (1/8-turn) from the felt pen mark. Don't loosen the nut to line up the holes. Once the holes are aligned, install the lock plate Allen bolts and tighten to 80 to 110 in-lbs (9 to 12 Nm). The Allen bolts come from the factory with a spot of Loctite on the threads and can be reinstalled three to five times. If you aren't sure how many times they've been installed, use new ones.

40 Piston rings – installation

Refer to illustrations 40.9 and 40.12

1 Before installing the new piston rings, the ring end gaps must be checked.
2 Lay out the pistons and the new ring sets so the rings will be matched with the same piston and cylinder barrel during the end gap measurement procedure and engine assembly.
3 Insert the top compression ring into the first cylinder and square it up with the cylinder walls by pushing it in with the top of the piston, The ring should be at least one inch below the top edge of the cylinder. To measure the end gap, slip a feeler gauge between the ends of the ring and compare the measurement to the Specifications.
4 If the gap is larger or smaller than specified, double-check to make sure you have the correct rings before proceeding. If the gap is too small, it must be enlarged or the ring ends may come in contact with each other during engine operation, which can cause serious damage.
5 The end gap can be increased by filing the ring ends very carefully with a fine file. Mount the ring in a vise equipped with soft jaws, holding it as close to the gap as possible, When performing this operation, file only from the outside in.
6 Excess end gap is not critical unless it's greater than the service limit in the Specifications. Again, double-check to make sure you have the correct rings for your engine.
7 Repeat the procedure for each ring that will be installed in the first cylinder and for each ring in the remaining cylinder. Remember to keep the rings, pistons and cylinder matched up.
8 Once the ring end gaps have been checked/corrected the rings can be installed on the pistons. Be sure to install the rings with their gaps staggered about 120-degrees.
9 The oil control ring (lowest on the piston) is installed first. 1970 through 1982 models have a full width slotted ring with a spring expander. 1983 and later models have rings composed of three separate components (two side rails and a spring/expander) **(see illustration)**. Place the spring/expander and the slotted oil ring in the lower groove of the piston in the same position they were in before disassembly. On 1983 and later models, slip the spacer/expander into the groove, then install the upper side rail. Do not use a piston ring installation tool on the oil ring side rails as they may be damaged. Instead, place one end of the side rail into the groove between the spacer/expander and the ring land. Hold it firmly in place and slide a finger around the piston while pushing the rail into the groove. Next, install the lower side rail in the same manner.
10 After the oil ring components have been installed, check to make sure the components can be turned smoothly in the ring groove.
11 The middle (second compression) ring is installed next. The inside chamfered edge must face the top of the piston when it's installed. If there's a dot mark on the ring, it must face upward. On 1983 and later models, the top ring is different from the middle ring. It's moly filled, beveled and barrel-faced while the second ring is tapered.
12 To avoid breaking the ring, use a piston ring installation tool and make sure the chamfered edge is facing up. Fit the ring into the middle groove on the piston. Do not expand the ring any more than is necessary to slide it into place. (An alternative method of installing the rings is shown in the accompanying illustration.)
13 Install the top ring in the same manner as the middle ring. Be sure the inside beveled edge (if equipped) faces the top of the piston. On later models, the top ring is symmetrical and can be installed with either side upward.
14 Repeat the procedure for the other piston and rings. Be very careful not to confuse the top and middle rings on 1983 and later models.
15 After all of the rings are installed, make sure they move freely in their grooves, but don't alter the positions of the ring gaps.

41 Pistons and cylinder barrels – installation

Refer to illustrations 41.4, 41.8a, 41.8b and 41.8c

1 Prior to installing the pistons on the connecting rods, install the rings on the pistons as described in the previous Section.
2 Rotate the crankshaft until the rod journal is at TDC, then position the pistons over the rods, facing in the direction marked during disassembly. If new pistons are being installed, be sure they're installed in the cylinder the rings were measured in, and that the arrow on the piston crown faces the front of the engine.
3 Place a new base gasket over the studs for each cylinder. Place clean rags around the openings in the crankcase to keep dirt and debris out. Install one new circlip in each piston to retain the piston pin. Make sure it's seated in the groove.

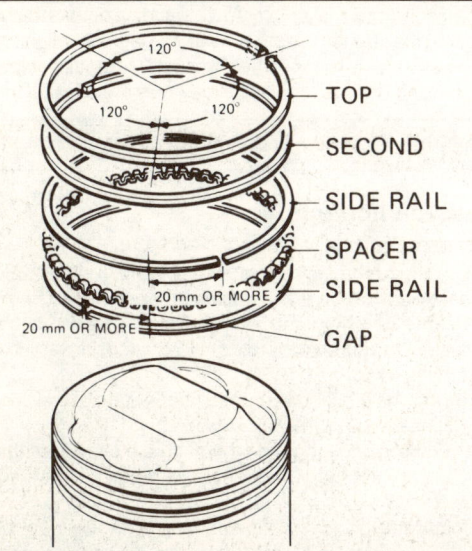

40.9 When installed, the piston ring gaps should be positioned as shown (120-degrees apart)

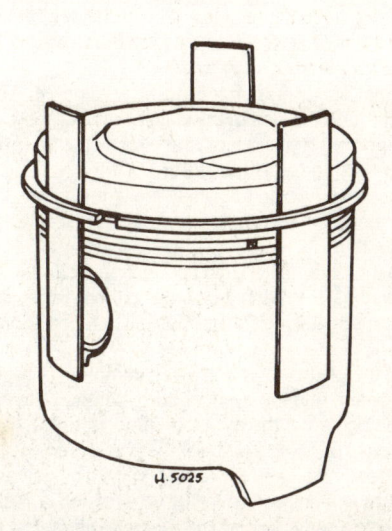

40.12 If a piston ring installation tool isn't available, the rings can be "walked" into position over thin metal strips (the strips are then pulled out)

Chapter 2 Engine, clutch and transmission

41.4 Align the piston and connecting rod and insert the piston pin

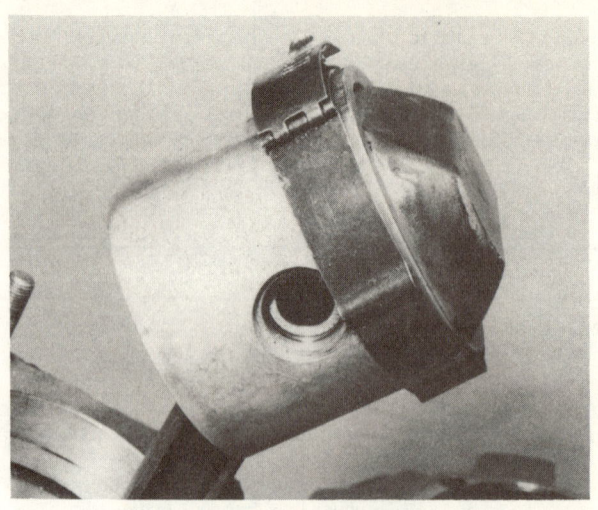

41.8a A piston ring compressor . . .

41.8b . . . makes installation of the barrels much easier

41.8c Note how the cylinder is liberally coated with clean oil

4   Coat the small end of the connecting rod with moly-base grease. Slide the piston pin through the side of the piston opposite the circlip. Align the piston and connecting rod and push the piston pin through the rod to the other side of the piston **(see illustration)**.
5   Install the other circlip on the other side of the piston. Be sure the circlips are completely seated in the grooves. The ends of the circlips should be facing up.
6   Assemble the other piston and connecting rod in the same way.
7   Coat the pistons, rings and grooves and the cylinder walls with clean engine oil. Make sure the ring end gaps are still spaced about 120-degrees apart.
8   Compress the piston rings with a ring compressor **(see illustration)** and insert the piston into the cylinder bore, carefully lowering the barrel until the piston rings enter the bore **(see illustrations)**. Remove the rag from the crankcase opening and lower the barrel until it's seated. On pre-Evolution engines, install and tighten the nuts and flat washers.
9   Install the remaining barrel over the other piston in the same manner.
10   If possible, two persons should be employed during this operation. One can insert the piston and rings while the other supports the barrel and lowers it into position.

42 Cylinder heads – installation

Pre-Evolution engine
Refer to illustrations 42.3a, 42.3b and 42.6
1   If the heads have been disassembled for repairs to the valves or valve guides, they must be reassembled as described in Section 20.
2   Place a new head gasket in position on top of each cylinder, without any sealant. Be sure the oil return holes in the gasket line up with the drain holes in the head.
3   Rotate the crankshaft until the tappets are at their lowest position. Install the pushrods in their original locations. Be sure to install new seals (cork washers or O-rings) on the pushrod tubes **(see illustrations)**.
4   Carefully set the cylinder head in position over the oil lines and the pushrod assemblies.
5   Install the cylinder head mounting bolts finger-tight.
6   Insert the previously marked pushrods in their original locations **(see illustration)**.
7   Install the remaining cylinder head following the same procedure.

Chapter 2 Engine, clutch and transmission

42.3a Reassemble the pushrod tubes using new seals

42.3b A new seal should be installed in each tappet housing

42.6 Be sure the pushrods engage in the tappets

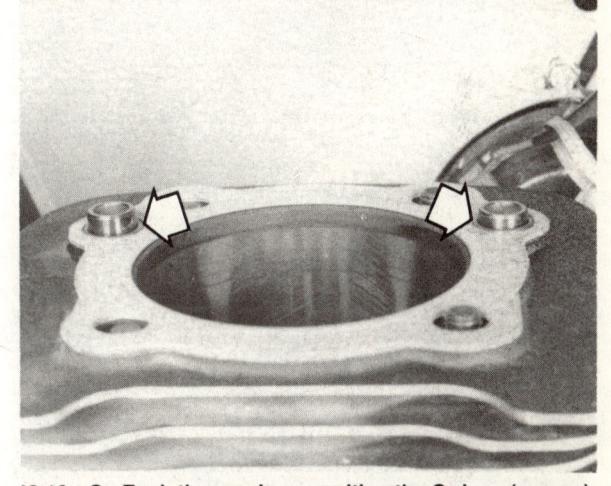

42.10 On Evolution engines, position the O-rings (arrows) over the dowel sleeves first, then install the head gasket and make sure it's aligned properly

8   Attach the intake manifold to the two cylinder heads with the manifold seals and clamps.
9   Tighten the cylinder head mounting bolts, in three steps, to the specified torque.

Evolution engine

Refer to illustrations 42.10, 42.12, 42.16 and 42.17

Note: *Before beginning this procedure, clean the cylinder head bolt threads, then lubricate them with a small amount of clean engine oil and thread the bolts onto the studs to make sure they don't bind.*

10   Install new O-rings over the dowel sleeves, then position the new head gasket on the cylinder barrel **(see illustration)**. No sealant is required on the head gaskets.
11   Make sure the bolt holes in the head are clean, then carefully lower it onto the cylinder – the dowel sleeves must enter the holes and align the head.
12   Lubricate the threads and the underside of each bolt head with a small amount of engine oil **(see illustration)**.
13   Slip the washers over the bolts (where applicable), then install the bolts finger-tight. **Caution:** *The procedure for tightening the cylinder head bolts is extremely critical to prevent gasket leaks, stud failure and cylinder or head distortion.*
14   Refer to the tightening sequence **(see illustration 6.14)** and tighten the bolts to 7 to 9 ft-lbs.
15   Tighten each bolt to 12 to 14 ft-lbs following the sequence.

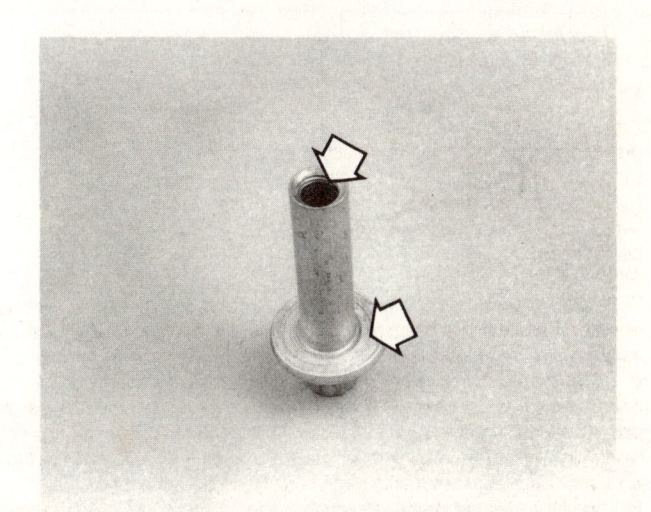

42.12 Clean and lubricate the threads and the underside of each head bolt (arrows) prior to installation

Chapter 2 Engine, clutch and transmission

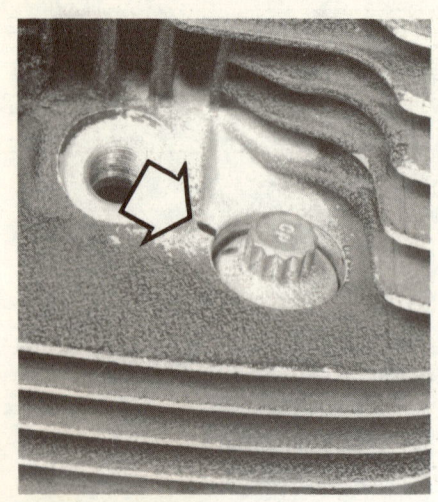

42.16 After the head bolts are all at the specified torque, make a mark on each bolt head flange and extend it onto the head (arrow)

42.17 Turn each cylinder head bolt an additional 90-degrees (1/4-turn) in an uninterrupted motion – this is very important and must be done exactly as described to prevent head gasket leaks

43.3 Make sure the O-ring seal for each pushrod tube is in place

16 Mark each bolt and the cylinder head with a felt-tip pen **(see illustration)**.
17 Turn each bolt, in the recommended sequence, an additional 90-degrees (1/4-turn) **(see illustration)**.

Models through 1990
18 Install new rocker cover gaskets with the sealant beads facing UP.
19 Install new seals and assemble the pushrod tubes, then slip the pushrods into the tubes and install them in their original locations. Make sure they're seated in the tappets. **Caution:** *The pushrods are color coded to ensure correct installation. Don't turn the pushrods end-for-end; they must be mated with the rocker arm or tappet just like they were originally. If the valves and seats were reconditioned, or if valve train components were replaced with new ones, the pushrod length requirements very likely will change. Since a special gauge is required to determine the correct length pushrod(s) to use, have a Harley-Davidson dealer service department select the parts for you.*
20 Make sure the tappets for the cylinder being reassembled are on the cam lobe base circles, not the lobes (turn the crankshaft if necessary to reposition them). **Note:** *If you're working on an early 1986 model, install the lower rocker arm cover now, but don't tighten the bolts completely (just snug them down). Don't install the two large bolts that retain the rocker arm shafts. Insert a 14-inch long piece of 0.050-to-0.060 inch wire through the hollow pushrod and into the tappet. Depress the check valve in the tappet by pushing down on the wire, then push down on the pushrod to bleed down the tappet. Hold the pushrod down, pull out the wire and install the rocker arm and shaft. Repeat the procedure for the other valve and the remaining cylinder. Tighten the lower rocker arm cover bolts to the specified torque.*

1991-on models
21 Before installing the tappets rotate the engine so that the cam followers are on their base circle (as viewed from the tappet bore). Liberally oil each tappet and insert so that its flats face the front and rear of the engine. Insert the retaining pins, followed by a new O-ring and install the triangular plate. Install the washer and screw and tighten to the specified torque.
22 Install a new seal over the top of the pushrod tube and slide it down the tube together with the retaining plate. Fit a new O-ring on top of the tube and insert the tube into its location in the cylinder head underside. Maneuver the bottom of the tube into position and secure with the retaining plate (fit over the pin in the casing); tighten its screw to the specified torque. Install the pushrods in their original bores and the original way up.

All Evolution models
23 Lubricate the rocker arm faces with moly-base grease, assemble the rocker arms and shafts in the lower rocker arm cover and slip all the bolts through the holes. The cutouts in the shafts must be positioned so the bolts will pass through the cover and the cutouts to retain the shafts.
24 Position the lower rocker arm cover on the head and thread all the bolts into place finger-tight. Make sure the pushrods are engaged in the rocker arm sockets.
25 Tighten the bolts in 1/4-turn increments, following a criss-cross pattern, to the specified torque. This will allow the tappets to bleed down slowly.
26 Make sure the pushrods spin freely. **Caution:** *To avoid pushrod and rocker arm damage, do not operate the starter or turn the engine over by hand until both pushrods can be turned with your fingers.*
27 Position new gaskets in the lower rocker arm cover, then install the middle rocker arm cover and a new gasket. On 1991-on models check the condition of the rubber umbrella valve in the middle rocker cover and renew it if it shows signs of deterioration. The valve is part of the crankcase breather system, and vents vapors passed up through the pushrod tubes into the intake system, via hollow bolts which mount the air cleaner to each cylinder.
28 Install the upper rocker arm cover and screws. Tighten the screws after making sure the middle rocker arm cover is positioned evenly on all sides.

43 Rocker boxes (pre-Evolution engine) – installation

Refer to illustrations 43.3, 43.4, 43.5 and 43.8

1 Coat the rocker shafts with clean engine oil and install new O-rings on the right end, behind the acorn nuts. Install the rocker components in their original locations.
2 Rotate the crankshaft until both valves in the head being assembled are closed (tappets on camshaft base circles). This will prevent stress on the rocker box casting as it's bolted down.
3 Place new O-ring seals in each of the recesses that the pushrod tubes are seated in **(see illustration)**.
4 Install new rubber seals on the oil lines going to the heads **(see illustration)**. Be sure the seals seat correctly, otherwise there is a risk of the oil supply being cut off to the head.
5 Place a new rocker box gasket on top of the head without sealant. Carefully install the rocker box on the cylinder head **(see illustration)**. Be sure the pushrods engage correctly with the ends of the rocker arms.

Chapter 2 Engine, clutch and transmission

43.4 External oil feed lines have a rubber seal under the union nut

43.5 Install the rocker box when the tappets are on the camshaft base circles, not the lobes (valve closed positions)

Tighten the rocker box retaining bolts in a criss-cross pattern, in three steps, to the specified torque. The valves must be closed as this is done.
6 Reconnect the external oil lines and tighten the fitting nuts.
7 Install the other rocker box in the same way.
8 Adjust the tappets as described in Chapter 1 **(see illustration)**. Be sure it's done correctly. The valves will be burned if they are adjusted too tightly; if the valves are adjusted too loose, the engine will be noisy and suffer a loss of power because the valves will not open far enough.

44 Engine installation

Refer to illustrations 44.2, 44.3, 44.4a and 44.4b
1 As you noticed during engine removal, there is very little clearance available for maneuvering the engine into place. It's recommended that three persons be available during this operation – two to lift the engine and locate it correctly, while the third person steadies the frame. Lift the engine in from the left-hand side of the frame on models through 1985, and from the right on 1986-on Evolution models.
2 When the engine is in position, attach the front engine mount plates and bolts first; do not force the bolts into the holes and risk damaging or stripping the threads. Do not tighten the bolts at this stage **(see illustration)**.

43.8 Removal of the keeper permits the pushrod tubes to be collapsed for valve adjustment

3 With the front engine bolts still loose, bolt the rear engine mount in place **(see illustration)**. Tighten the rear mount bolts, then tighten the front mount bolts.

44.2 Bolt up the front . . .

44.3 . . . and the rear engine mount (early model shown)

Chapter 2 Engine, clutch and transmission

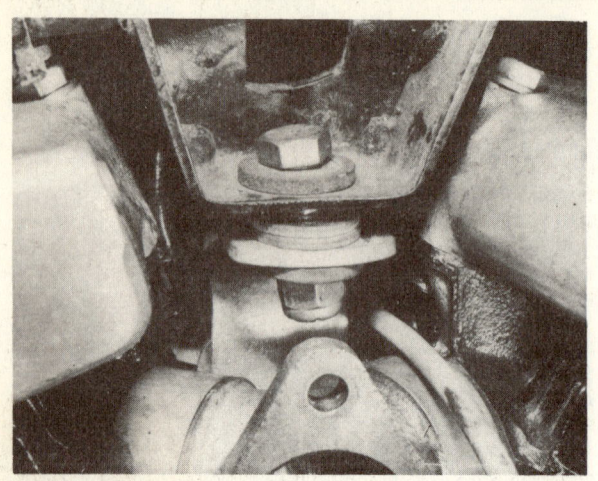

44.4a Install the cylinder head bracket with the washers in their correct locations

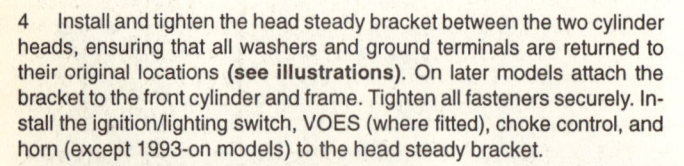

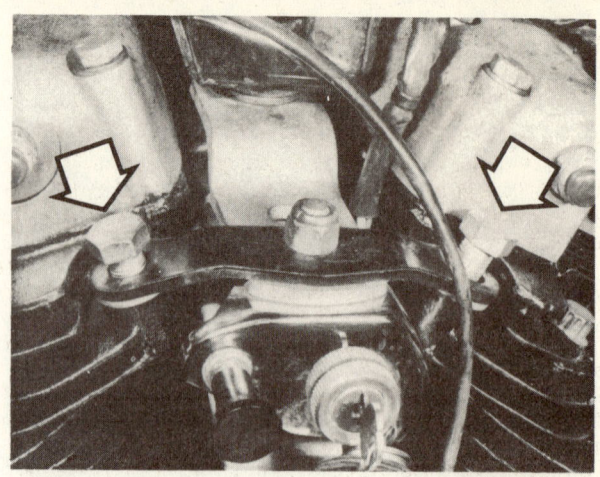

44.4b Tighten the bolts securing the bracket to the head (arrows)

4 Install and tighten the head steady bracket between the two cylinder heads, ensuring that all washers and ground terminals are returned to their original locations **(see illustrations)**. On later models attach the bracket to the front cylinder and frame. Tighten all fasteners securely. Install the ignition/lighting switch, VOES (where fitted), choke control, and horn (except 1993-on models) to the head steady bracket.

45 Generator (early models) and primary chain cover – installation

Generator (early models)
Refer to illustrations 45.1 and 45.2
1 Install the generator, tilting it as it enters the back of the timing case so the oil thrower will clear the large diameter gear **(see illustration)**.

45.1 Attach the generator to the timing case

45.2 Connect the wires to the correct terminals on the generator

46.1a Attach the choke cable to the carburetor

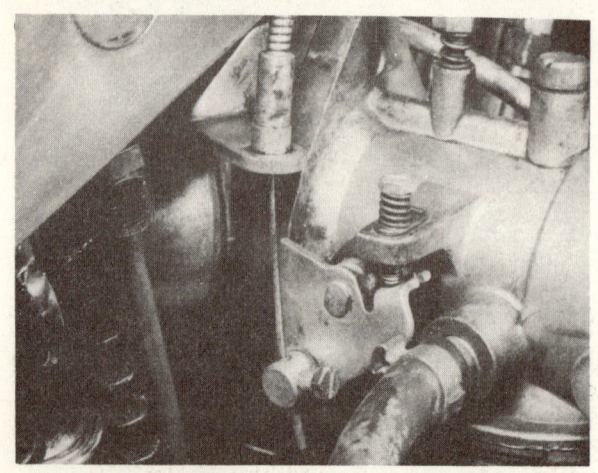

46.1b Connect the throttle cable(s)

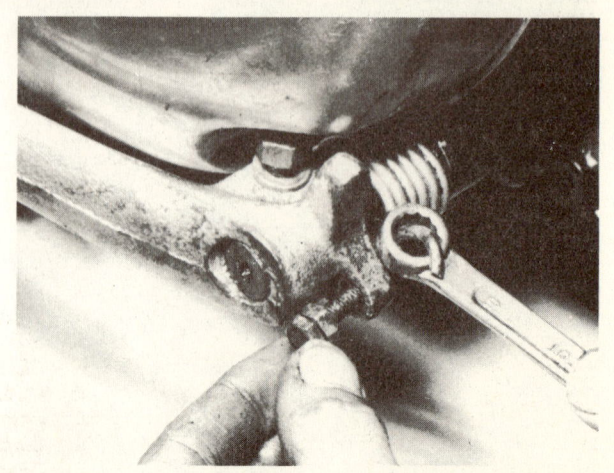

46.4 Tension the return spring while installing the brake pedal

Chapter 2 Engine, clutch and transmission

Insert the two bolts that retain the generator through the breather cover and tighten them.

2 Secure the wires to the two generator terminals **(see illustration)**.

Primary chaincase cover

3 Before installing the primary chaincase cover ensure that the clutch release mechanism is well lubricated. If the cable was disconnected on removal, reconnect it leaving slack in the in-line adjuster **(see illustrations 4.20c and 4.20d)**.

4 Check that the two cover dowels are in position. On pre-1981 models check that the thrust washer on the end of the starter shaft is installed **(see illustration 4.20b)**.

5 The new cover gasket should be installed dry, using only a smear of grease on the crankcase to hold the gasket in place while the cover is installed. Renew the gear shift shaft oil seal in the cover and make sure the seal is not damaged as the cover is installed. Return all screws to their original positions and tighten evenly in a criss-cross pattern.

6 On late 1984-on models thread the clutch adjuster nut onto the adjuster screw and carry out clutch adjustment as described in Chapter 1, Section 7. On all models, tension the primary drive chain as described in Chapter 1, Section 21, and refill the primary chaincase with oil as described in Chapter 1, Section 23.

46 Completing engine reassembly

Refer to illustrations 46.1a, 46.1b, 46.4, 26.8, 46.15a, 46.15b, 46.21a, 46.21b and 46.21c

1 Fit a new gasket or seal (as applicable) and install the carburetor and intake manifold, using the air cleaner backplate as a guide to ensure correct alignment. Reconnect its throttle and choke/enrichener cables, referring to Chapter 3 for information **(see illustrations)**.

2 Refit the engine oil tank (if removed) and using new hose clamps reconnect the oil feed, return and vent hoses to the engine.

3 If not already installed, refer to Section 45 and refit the primary chaincase. On 1971-on models adjust the clutch as described in Chapter 1, Section 7. On all models adjust the primary chain tension as described in Chapter 1, Section 21.

4 Install the left footpeg and, on 1975-on models, the gear shift lever. On models through 1974 install the brake pedal, noting that its spring must be tensioned **(see illustration)**.

5 On models through mid-1984 install the generator (see Section 45).

6 On chain drive models install the chain over the engine and rear wheel sprockets and connect its joining link. **Note:** *The closed end of the link's spring clip must face in the direction of normal chain travel.* Refer to Chapter 1, Section 5 and set the chain freeplay.

7 On belt drive models, ensure that the rear wheel is pushed forward in the swingarm to provide sufficient belt slack and then slip the belt over the engine sprocket. Refer to Chapter 1, Section 33 and set the belt tension.

8 On 1970 models install the three-part clutch pushrod through the mainshaft and attach the cable to the release mechanism in the cover **(see illustration)**. On all models, reconnect the neutral switch wire. Where a crosshaft for the gearshift is fitted, this must now be installed.

9 Install the sprocket cover, together with any dowels fitted, and tighten its retaining screws. Where the right-hand footpeg and brake pedal (or gear shift lever – models through 1974) are separate from the cover, install them. On 1970 models carry out clutch adjustment as described in Chapter 1, Section 7.

10 Where a kickstarter is fitted, install it on the cover, noting that its return spring must be tensioned so that it will return the kickstarter arm to the upright position.

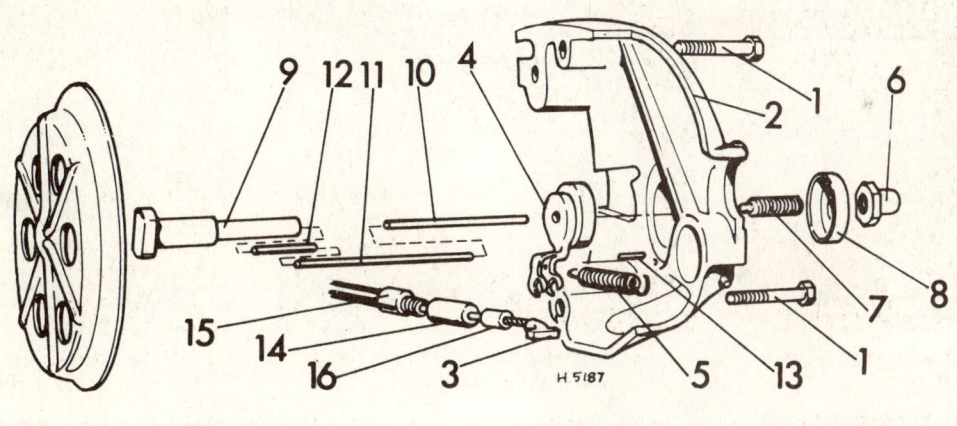

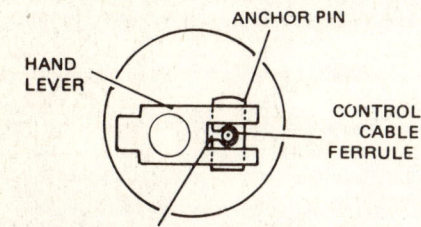

46.8 Clutch operating mechanism (1970 models) – exploded view

1	Sprocket cover bolt	9	Clutch pushrod – left
2	Sprocket cover	10	Clutch pushrod – right
3	Clutch cable end	11	Clutch pushrod – right center
4	Clutch release ramp and lever	12	Clutch pushrod – left center
5	Clutch release spring	13	Release lever stop
6	Clutch adjusting screw locknut	14	Clutch cable felt seal retainer
7	Clutch adjusting screw	15	Clutch cable ferrule
8	Clutch release ramp cover	16	Clutch cable felt seal

Chapter 2 Engine, clutch and transmission

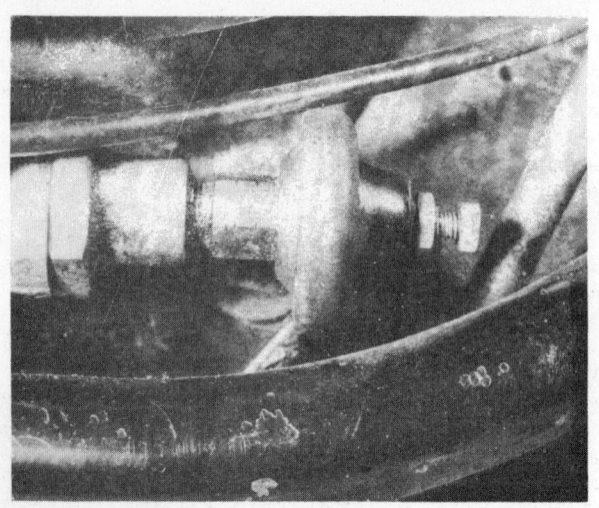

46.15a Install the oil pressure switch

46.15b Connect the wires to the regulator

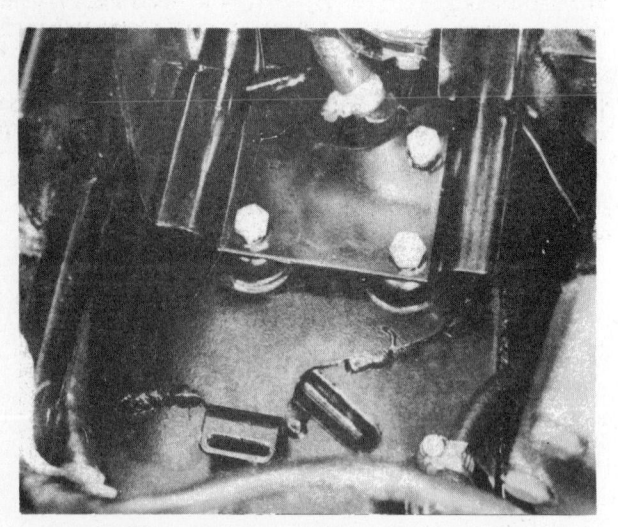

46.21a Attach the battery box to the frame (pre-1979 models), ...

46.21b ... then install the battery and retaining clamp

46.21c Tighten the clamp to hold the battery in place (note the locknuts)

11 On models with a rear drum brake, reconnect the brake linkage and adjust as described in Chapter 6, Section 12.

12 Where applicable, install the speedometer and tachometer drive gears. On early models refit the breather pipe to the domed housing on the timing cover.

13 Using new gaskets and seals at all disturbed joints, refit the exhaust system.

14 On rear disc brake models, install the master cylinder and set the pedal position as described in Chapter 5, Section 15.

15 Reconnect all electrical components and connections disturbed during engine removal (see illustrations):
   a) Oil pressure switch
   b) Distributor/contact breaker/ignition trigger
   c) Ignition coil
   d) Ignition module
   e) Spark plugs
   f) Regulator
   g) Starter solenoid
   h) Vacuum operated electric switch (if fitted)
   i) Horn
   j) Ignition/lighting switch

Chapter 2 Engine, clutch and transmission

16 Install the fuel tank, noting that on some models its front mounting also secures the ignition coil. Reconnect the fuel feed hose to the fuel valve using a new hose clamp.
17 Using a new gasket, refit the air cleaner backplate, then install the filter and cover. Refit any emission control system hoses disturbed on removal.
18 Add the specified quantity and type of transmission oil via the aperture in the primary chaincase and check the level (see Chapter 1, Section 23).
19 Install a new engine oil filter and refill the oil tank with fresh oil of the specified type (see Chapter 1, Section 17).
20 Make a final check that all fasteners are securely tightened and that all components have been installed or reconnected.
21 Install the battery tray/carrier and finally refit the battery and connect its terminals – negative terminal last **(see illustrations)**.

47 Starting and running the rebuilt engine

1 On models through mid-1984 which are equipped with a generator, the generator must be polarized after disconnection from the electrical system. Use a jumper wire to bridge the BAT and GEN terminals on the regulator for a moment. Failure to follow this procedure may cause damage to either the generator or the regulator.
2 Turn the ignition kill switch to the Off position. Check the oil level and remove the spark plugs.
3 Turn the key switch on and crank the engine over with the electric starter or kick the engine over (on kickstart models) to circulate oil through the engine. Install the spark plugs, hook up the wires and turn the ignition kill switch to the On position.
4 Make sure there is fuel in the tank, turn the fuel valve to the On position and operate the choke. Start the engine and allow it run at a fast idle until it reaches operating temperature. Check the oil return to the oil tank.
5 The engine may have a tendency to smoke initially due to the amount of oil used during reassembly of the various components. This excess oil should burn away gradually as the engine components seat themselves.
6 Check carefully for oil leaks and make sure the transmission and controls, especially the brakes, function properly before road testing the machine.
7 Refer to the following Section for the proper break-in procedure.
8 After completion of the road test and after the engine has cooled down, on pre-Evolution engines retorque the cylinder head bolts and recheck the valve adjustment.

48 Recommended break-in procedure

1 Any rebuilt engine needs time to break-in, even if parts have been installed in their original locations. For this reason, treat the machine gently for the first few miles to make sure oil has circulated throughout the engine and any new parts installed have started to seat.
2 Even greater care is necessary if the engine has been rebored or a new crankshaft has been installed. In the case of a rebore, the engine will have to be broken in as if the machine were new. This means greater use of the transmission and a restraining hand on the throttle until at least 500 miles (800 km) have been covered. There's no point in keeping to any set speed limit – the main idea is to keep from lugging the engine and to gradually increase performance until the 500 mile (800 km) mark is reached. These recommendations can be lessened to an extent when only a new crankshaft is installed. Experience is the best guide, since it's easy to tell when an engine is running freely.
3 If a lubrication failure is suspected, stop the engine immediately and try to find the cause. If an engine is run without oil, even for a short period of time, irreparable damage will occur.

Chapter 3 Fuel system

Contents

Fuel system – general information 1
Fuel tank – removal and installation 2
Fuel control valve – removal and installation 3
Carburetor – removal and installation 4
Carburetor overhaul – general information 5
Carburetor – disassembly, inspection and reassembly 6
Carburetor – adjustments 7
Air cleaner – removal and installation 8
Evaporative emission control system – general information 9
Throttle cable – adjustment 10

Specifications

Carburetor type

1970 and 1971	Tillotson
1972 through 1976	Bendix
1977 through 1987	Keihin
1988-on	Keihin CV

Idle speed

	RPM
Tillotson	900 to 1100
Bendix	700 to 900
Keihin	
1977 and 1978	700 to 900
1979 only	900
1980 through 1987	900 to 950
1988 through 1990	1000 to 1050
1991-on	950 to 1050

Fast idle speed

1979 through 1985	1500 rpm
1986 and 1987	1500 to 1550 rpm

Main jet sizes

Tillotson	0.055, 0.057, 0.059, 0.061 and 0.063
Bendix	90, 95, 100, 105, 110, 115, 120 and 125
Keihin	
1976 through 1978	1.60, 1.65, 1.70, 1.75, 1.80 and 1.85 mm
1979	No. 165
1980 through 1985	No. 160
1986	No. 155
1987	
883	No. 155
1100	No. 150
1988	
California only	
883	No. 165
1200	No. 180

Chapter 3 Fuel system

All others
   883 .... No. 170
   1200 .... No. 200
1989 – California only
   883 .... No. 155
   1200 .... No. 160
1990 and 1991 – California only
   883 .... No. 160
   1200 .... No. 160
1989 through 1991 (all others)
   883 .... No. 175
   1200 .... No. 175
1992 through 1994
   883 .... No. 160
   1200 .... No. 170
1995-on
   883
      California .... No. 170
      49 states .... No. 160
      International .... No. 190
      Switzerland
         1995 .... No. 160
         1996-on .... No. 190
   1200 (except 1998 and 1999 XL1200S)
      California .... No. 185
      49 states
         1995 on .... No. 170
      International .... No. 190
      Switzerland
         1995 .... No. 160
         1996 through 1998 .... No. 190
         1999 .... No. 200
   1998 and 1999 XL1200S
      California .... No. 195
      49 states .... No. 195
      International .... No. 195
      Swiss .... No. 195

Slow jet sizes *(1979 and later models only)*

1979 only .... No. 65
1980 through 1982 .... No. 68
1983 through 987 .... No. 52
1988 only .... No. 35
1989 through 1991
   California only .... No. 42
   All others .... No. 45
1992 through 1994 .... No. 40
1995
   US and Switzerland .... No. 40
   International .... No. 42
1996-on .... No. 42

Torque specifications

	Ft-lbs (unless otherwise indicated)	Nm
Tillotson carburetor		
Inlet needle valve seat	40 to 45 in-lbs	4.5 to 5
Diaphragm cover plug	23 to 28 in-lbs	2.6 to 3
Keihin carburetor (1977 through 1985)		
Carburetor mounting nuts	19	26
Keihin carburetor (1986 and 1987)		
Carburetor-to-intake manifold bolts	15 to 17	20 to 23
Intake manifold flange-to-cylinder head nuts/bolts	72 to 120 in-lbs	8 to 14
Air cleaner backplate		
To engine bolts	10 to 12	14 to 16
To carburetor bolts	36 to 60 in-lbs	4 to 7
Air cleaner cover screws	36 to 60 in-lbs	4 to 7
Keihin carburetor (1988-on)		
Intake manifold flange-to-cylinder head nuts/bolts	72 to 120 in-lbs	8 to 14
Carburetor clamp screw	10 to 15 in-lbs	1.1 to 1.7

Chapter 3 Fuel system

1 Fuel system – general information

The fuel system consists of the gas tank, the control valve, the fuel line and the carburetor.

Gas is fed by gravity to the carburetor through the control valve, which contains a built-in filter. The control valve has three positions; On, Off and Reserve. The reserve position provides a small amount of fuel after the main supply has run out, so the engine will still run for a short time.

Three different makes of carburetors were installed on the motorcycles covered by this manual. Tillotson carburetors were used on 1970 and 1971 models, Bendix carburetors were used on 1972 through 1976 models and Keihin carburetors were used on 1977 and later models. Beginning in 1988, a constant velocity (CV) carburetor was standard equipment. All carburetors have a butterfly throttle (CV carburetors also have a slide, but it's vacuum operated and responds to throttle butterfly movement) and incorporate an accelerator pump. The Bendix and Keihin carburetors have an integral float chamber. Each carburetor has a manual choke to facilitate easy starting in low outside temperatures.

A large capacity air cleaner is attached to the carburetor intake on all models. Refer to Chapter 1 for filter maintenance instructions.

All 1986 and later California models are equipped with an Evaporative emission control system to reduce air pollution that stems from evaporation of gasoline in the fuel tank when the motorcycle is parked.

Several fuel system routine maintenance procedures are included in Chapter 1. They include *Fuel system – check*, *Idle speed – adjustment*, *Fuel filter – cleaning and replacement* and *Throttle – check and lubrication*.

2 Fuel tank – removal and installation

Refer to illustrations 2.1 and 2.5

Warning: *Gasoline is extremely flammable and highly explosive under certain conditions – safety precautions must be followed when working on any part of the fuel system! Don't smoke or allow open flames or unshielded light bulbs in or near the work area. Don't do this procedure in a garage with a natural gas appliance (such as a water heater or clothes dryer). Also, before starting work, disconnect the negative battery cable from the battery.*

Removal

1 Turn the fuel control valve to the Off position and disconnect the fuel line from the carburetor. Some hose clamps installed at the factory must be cut or pried off and can't be reused **(see illustration)**.

2 Insert the end of the fuel line into a clean gasoline container and turn the fuel valve to the Reserve position to drain the tank. Use a funnel to direct the gasoline into the container.

3 On some tanks, the crossover hose, that connects the two lower portions of the tank, must be disconnected from one of the fittings. Release the clamp and slide it down the hose. Some clamps have to be cut off with wire cutters and can't be reused – install a worm-drive clamp when the hose is reinstalled.

4 On 1983 and later XLS models, remove the three screws, disconnect the tachometer, remove the gas tank cap and detach the center panel.

5 On most models, the fuel tank is attached to the upper frame rail at the front and rear by a nut and bolt. The bolts pass through flanges on the tank and lugs on the frame **(see illustration)**. On some models, the rear of the tank is secured by a strap and spring-loaded clip.

6 On some models, the front tank mount is also used to secure the ignition coil and horn with brackets and spacers. When the front mounting bolt is removed, the coil assembly should be held in place to prevent it from dropping down onto the engine, possibly causing damage.

7 Remove the nuts/bolts securing the fuel tank to the frame. Support the ignition coil, if necessary. Lift the fuel tank away from the frame and store it in a safe place, away from sparks and flames. **Note:** *On later models with an evaporative emission system, the vent hose must be detached from the fitting on the tank.*

Installation

8 When installing the tank, reverse the removal procedure. Make sure the coil and horn mounting brackets and spacers are correctly reinstalled at the front fuel tank mount. Don't pinch the wiring harness between the tank and frame. Position the horn so it doesn't touch the coil bracket or frame.

3 Fuel control valve – removal and installation

Refer to illustration 3.2

Warning: *Gasoline is extremely flammable and highly explosive under certain conditions – safety precautions must be followed when working on any part of the fuel system! Don't smoke or allow open flames or unshielded light bulbs in or near the work area. Don't do this procedure in a garage with a natural gas appliance (such as a water heater or clothes dryer). Also, before starting work, disconnect the negative battery cable from the battery.*

1 If the control valve is leaking or the filter must be cleaned, the valve must be detached from the bottom of the tank. Drain the fuel into a clean gasoline container first (see Section 2), then detach the fuel line from the valve. Most hose clamps installed at the factory must be pried or cut off and can't be reused.

2.1 Remove the hose clamp and detach the hose from the carburetor fitting – if it's stuck, twist it back-and-forth with pliers, then pull on it

2.5 On most models, the tank is held in place by nuts/bolts at both ends

Chapter 3 Fuel system

3.2 Unscrew the fuel control valve with a wrench on the large hex nut (arrow) – make sure the tank is empty first!

4.4 Loosen the set screw (arrow) securing the choke cable to the carburetor lever

2 Unscrew the large nut and detach the valve from the bottom of the fuel tank **(see illustration)**.
3 Refer to Chapter 1 for instructions on cleaning and replacing the fuel filter.
4 If the valve leaks badly or doesn't work correctly, it must be replaced with a new one. The control valve on later models can be disassembled by removing the two screws from each side of the lever.
5 The lever, spring and nylon valve can be removed after the screws are removed.
6 Before installing the valve in the fuel tank, apply teflon tape or sealant that's resistant to gasoline to the tank threads. On 1975 and later models, the valve has a left-hand thread and the fitting on the bottom of the fuel tank has a right-hand thread. As the large nut is tightened, the fuel valve and the tank are drawn together.
7 When the fuel valve is securely fastened to the tank, connect the fuel line to it. Install a new hose clamp if necessary.

4 Carburetor – removal and installation

Refer to illustration 4.4

Warning: *Gasoline is extremely flammable and highly explosive under certain conditions – safety precautions must be followed when working on any part of the fuel system! Don't smoke or allow open flames or unshielded light bulbs in or near the work area. Don't do this procedure in a garage with a natural gas appliance (such as a water heater or clothes dryer). Also, before starting work, disconnect the negative battery cable from the battery.*

Note: *Although it isn't absolutely necessary, you probably will find it easier to remove the carburetor if the fuel tank is removed first (see Section 2).*

1 Refer to Section 8 and remove the air cleaner assembly.
2 Turn the fuel control valve to the Off position and disconnect the fuel line from the carburetor. Most hose clamps installed at the factory, must be pried or cut off and can't be reused **(see illustration 2.1)**.
3 Disconnect the throttle cable(s) from the carburetor. Tillotson and Bendix carburetors use set screws to secure the throttle cable. To disconnect the cable(s) from a Keihin carburetor, turn the throttle valve open by hand and pull the cable ferrule out of the hole in the throttle lever. On 1981 and later models, there are two cables attached to the throttle lever.
4 Disconnect the choke cable from the carburetor – it's attached with a set screw on all carburetors **(see illustration)**. **Note:** *The CV carburetor used on 1988 and later models doesn't have a choke. It has an enrichener valve that's cable-operated just like the choke. To detach it, unscrew the*

fitting and pull the valve out of the left (rear) side of the carburetor. Be careful not to damage the end of the valve while it's exposed.
5 Disconnect the vacuum and EVAP (emission) system hose(s) from the carburetor (some later California models only).
6 On all but 1988 and later models (CV carburetor), remove the nuts/bolts that secure the carburetor to the intake manifold. **Note:** *On 1983 and later models, remove the nut and washer and detach the VOES bracket from the carburetor mounting stud. Remove the stud and lower bolt.* Carefully separate the carburetor from the intake manifold. On 1977 and 1978 Keihin carburetors, an O-ring seal is installed between the carburetor and intake manifold. All other models have a gasket.
7 On 1988 and later models, simply loosen the hose clamp and pull the carburetor out of the intake manifold.
8 Installation is the reverse of the removal procedure, but be sure to install a new gasket or O-ring between the intake manifold and carburetor.
9 Replace the gasket between the air cleaner baseplate and the carburetor with a new one.

5 Carburetor overhaul – general information

1 Poor engine performance, hesitation and little or no engine response to idle fuel/air mixture adjustments are all signs that major carburetor maintenance is required.
2 Keep in mind that many so-called carburetor problems are really not carburetor problems at all, but mechanical problems in the engine or ignition system faults. Establish for certain that the carburetor needs maintenance before assuming an overhaul is necessary.
3 For example, fuel starvation is often mistaken for a carburetor problem. Make sure the fuel filter, the fuel line and the gas tank cap vent hole are not plugged before blaming the carburetor for this relatively common malfunction.
4 Most carburetor problems are caused by dirt particles, varnish and other deposits which build up in and block the fuel and air passages. Also, in time, gaskets and O-rings shrink and cause fuel and air leaks which lead to poor performance.
5 When the carburetor is overhauled, it's generally disassembled completely and the metal components are soaked in carburetor cleaner (which dissolves gasoline deposits, varnish, dirt and sludge). **Caution:** *Don't soak any rubber parts (especially the vacuum piston diaphragm on the CV carburetor) in carburetor cleaning solvents. They will be damaged if you do.* The parts are then rinsed thoroughly with solvent and dried with compressed air. The fuel and air passages are also blown out with compressed

Chapter 3 Fuel system

air to force out any dirt that may have been loosened but not removed by the carburetor cleaner. Once the cleaning process is complete, the carburetor is reassembled using new gaskets, O-rings, diaphragms and, generally, a new inlet needle and seat.

6 Before taking the carburetor apart, make sure you have a rebuild kit (which will include all necessary O-rings and other parts), some carburetor cleaner, solvent, a supply of rags, some means of blowing out the carburetor passages and a clean place to work.

7 Some of the carburetor settings, such as the sizes of the jets and the internal passageways are predetermined by the manufacturer after extensive tests. Under normal circumstances, they won't have to be changed or modified. If a change appears necessary, it can often be attributed to a developing engine problem.

6 Carburetor – disassembly, inspection and reassembly

Warning: *Gasoline is extremely flammable and highly explosive under certain conditions – safety precautions must be followed when working on any part of the fuel system! Don't smoke or allow open flames or unshielded light bulbs in or near the work area. Don't do this procedure in a garage with a natural gas appliance (such as a water heater or clothes dryer).*

1 Before disassembling the carburetor, clean the outside with solvent and lay it on a clean sheet of paper or a shop towel.

2 After it's been completely disassembled, submerge the metal components in carburetor cleaner and allow them to soak for approximately 30 minutes. Do not place any plastic or rubber parts in it – they'll be damaged or dissolved. Also, don't allow excessive amounts of carburetor cleaner to get on your skin.

3 After the carburetor has soaked long enough for the cleaner to loosen and dissolve the varnish and other deposits, rinse it thoroughly with solvent and blow it dry with compressed air. Also, blow out all the fuel and air passages in the carburetor body. **Note:** *Never clean the jets or passages with a piece of wire or drill bit – they could be enlarged, causing the fuel and air metering rates to be upset.*

Tillotson carburetor

Refer to illustrations 6.4 and 6.19

4 Carefully turn the idle mixture adjustment screw in until it bottoms, while counting the number of turns, then remove it, along with the spring **(see illustration)**. Record the number of turns – you'll need to refer to it later. Remove the intermediate mixture adjusting screw in the same manner. By counting the number of turns until they bottom, the adjustment screws can be returned to their original positions and adjustments will be kept to a minimum.

5 Note the position of the throttle valve before removing it to ensure it's reinstalled in the same position. Remove the two screws securing the throttle valve to the shaft and detach the valve.

6 Remove the screw securing the throttle shaft and accelerator pump. Pull the throttle shaft out of the carburetor body along with the throttle shaft spring and washers, then remove the dust seals from both sides of the carburetor body.

7 Invert the carburetor and remove the screws securing the diaphragm cover. Carefully lift off the cover, then remove the diaphragm and gasket. Separate the diaphragm from the gasket by peeling them apart.

8 Take out the screw that secures the accelerator pump plunger, then withdraw the plunger.

9 Remove the plug screw from the diaphragm cover.

10 Remove the inlet control lever screw. This will permit the control lever pin, the control lever and the inlet needle to be removed. These parts are very small and easily lost if you aren't careful. Remove the control lever tension spring from below the assembly.

11 Remove the inlet needle valve seat and gasket with a thin-wall 3/8-inch socket. Note the position of the seat insert with the smooth side toward the inside of the cage. The gasket can be lifted out with a scribe.

12 Unscrew the main jet plug, then remove the main jet and gasket.

13 Drill a 1/8-inch hole in the center of the main nozzle welch plug. Be careful not to drill beyond the welch plug, since damage to the main nozzle can result. Insert a small punch through the hole and carefully pry the plug out of the casting.

14 Remove the idle port welch plug as described in the previous Step.

15 Using a small punch, remove the welch plug over the economizer check ball and let the check ball roll out.

16 Remove the screws securing the choke valve and lift out the bottom part of the valve.

17 Slide the choke shaft assembly out of the carburetor (when this is done, the upper part of the choke valve will be released and can be removed). Remove the choke spring and the choke shaft friction ball and spring. Pry the choke shaft dust seal out of the carburetor body.

18 Clean and inspect the parts as described in Steps 2 and 3 of this Section. Inspect the carburetor body for cracks and make sure the throttle shaft turns freely without excessive play. If it's sloppy, a new carburetor will be needed (although sometimes the throttle shaft bores can be reamed out and bushings installed – check with a dealer service department).

19 Make sure the inlet control lever rotates freely on the pin and the forked end of the lever engages with the slot in the inlet needle valve. Check the end of the control lever for wear and burrs. Check the spring to be sure it isn't stretched or distorted. Check the inlet needle valve and seat for nicks and a pronounced groove or ridge on the tapered end of the valve **(see illustration)**. If there is one, a new needle and seat should be used when the carburetor is reassembled.

20 Examine the rest of the parts for wear and damage.

21 Reassemble the carburetor by reversing the disassembly sequence. Use a new diaphragm as well as new gaskets and seals and don't overtighten any of the small fasteners or they may break off.

22 Seat the new welch plugs by striking them with a punch slightly smaller than the plug itself. When the plug is seated, it should be flat, not concave. This will ensure a tight fit around the edge of the casting opening.

23 The inlet control lever tension spring should be installed in the carburetor body. Be sure it attaches to the protrusion on the inlet control lever. Bend the diaphragm end of the control lever so when the lever is installed, it's flush with the floor of the metering chamber.

24 Be sure to tighten the inlet needle valve seat and the diaphragm cover plug to the torque figures listed in this Chapter's Specifications.

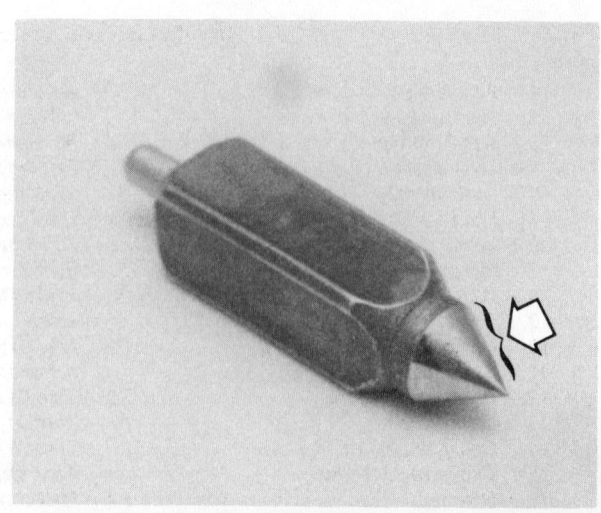

6.19 Check the inlet needle valve for a groove or ridge in the tapered area (arrow)

Chapter 3 Fuel system

6.4 Tillotson carburetor components – exploded view

1	Accelerator pump	18A	Accelerator pump check ball retainer	32	Inlet needle valve and seat
2	Accelerator pump lever			33	Gasket
3	Accelerator pump screw	18B	Accelerator pump check ball	34	Inlet control lever tension spring
4	Lock washer	19	Diaphragm cover plug	35	Intermediate mixture adjusting screw
5	Channel plug	20	Diaphragm cover screws		
6	Main nozzle welch plug	21	Diaphragm cover gasket	36	Intermediate mixture adjusting screw gasket
6A	Welch plug	22	Economizer check ball		
7	Idle port welch plug	23	Fuel filter screen	37	Washer
8	Welch plug	24	Idle mixture adjustment screw	38	Washer
9	Choke shaft friction ball			39	Main jet
10	Choke shaft friction spring	25	Idle mixture adjustment screw spring	40	Main jet plug
11	Choke valve (top)			41	Main nozzle check valve
12	Choke valve spring	26	Throttle stop screw	42	Throttle shaft
13	Choke shaft	27	Throttle stop screw cup	43	Throttle lever cable screw
14	Choke shaft dust seal	28	Throttle stop screw spring	44	Dust seal
15	Choke valve (bottom)	28A	Lock washer	45	Washer
16	Screws	29	Inlet control lever	46	Throttle shaft spring
17	Diaphragm	30	Inlet control lever pin	47	Throttle valve
18	Diaphragm cover	31	Inlet control lever screw	48	Screws

Chapter 3 Fuel system

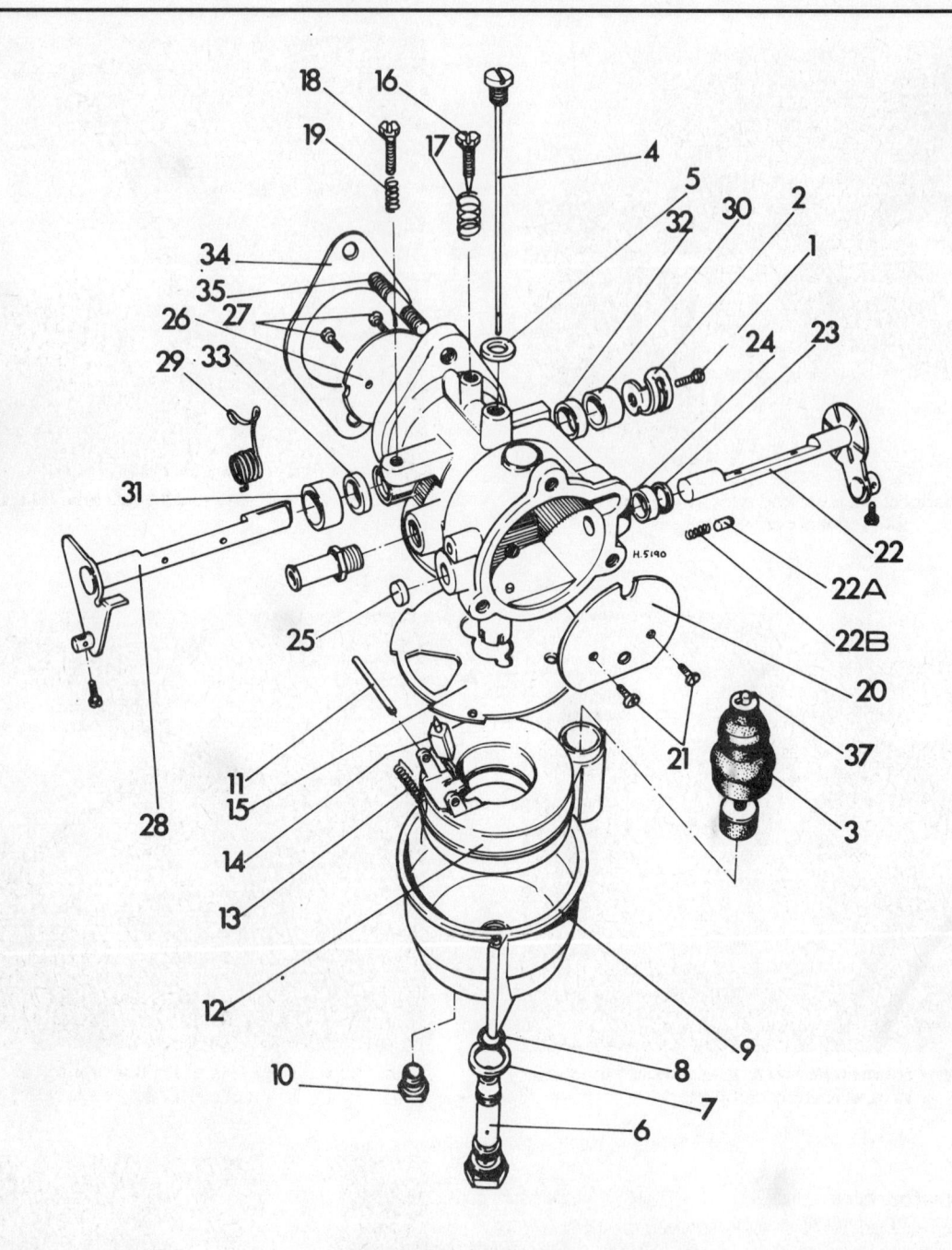

6.25a Bendix carburetor components – exploded view

1 Screw
2 Accelerator pump lever
3 Accelerator pump
4 Idle tube
5 Idle tube gasket
6 Main jet and tube assembly
7 Fiber washer
8 O-ring
9 Float bowl
10 Float bowl drain plug
11 Float pivot pin
12 Float assembly
13 Float spring
14 Inlet needle valve
15 Gasket
16 Idle mixture adjusting screw
17 Spring
18 Throttle stop screw
19 Spring
20 Choke valve
21 Screw
22 Choke shaft and lever
22A Plunger
22B Spring
23 Choke shaft seal retainer
24 Choke shaft seal
25 Choke shaft cup plug
26 Throttle valve
27 Screw
28 Throttle shaft and lever
29 Throttle shaft spring
30 Throttle shaft seal retainer
31 Throttle shaft seal retainer
32 Throttle shaft seal
33 Throttle shaft seal
34 Manifold gasket
35 Stud
36 Accelerator pump shaft pin

Chapter 3 Fuel system 123

6.25b Disconnect the lever and remove the accelerator pump (Bendix carburetor)

6.27 Unscrew and withdraw the idle tube (Bendix carburetor)

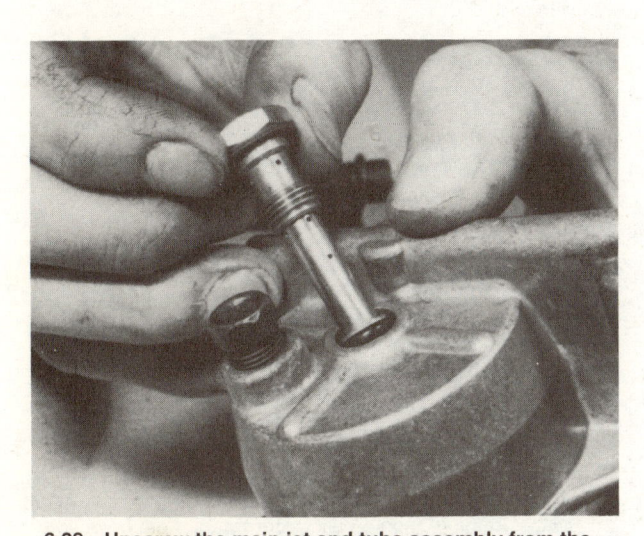

6.28 Unscrew the main jet and tube assembly from the float bowl (Bendix carburetor)

6.29a Note the position of the float spring before removing the pivot pin (Bendix carburetor)

Bendix carburetor

Refer to illustrations 6.25a, 6.25b, 6.27, 6.28, 6.29a, 6.29b, 6.40 and 6.42

25 Remove the screw securing the accelerator pump lever to the throttle shaft **(see illustration)**. Disconnect the accelerator pump boot from the float bowl. Remove the accelerator pump and lever **(see illustration)**.
26 Compress the spring on the accelerator pump shaft, rotate the pump lever 1/4-turn and disengage the pin at the top of the shaft from the lever.
27 On 1972 through 1974 models, unscrew the idle tube from the top of the carburetor body and detach the gasket at the same time **(see illustration)**.
28 Unscrew the main jet and tube assembly from the bottom of the float bowl **(see illustration)**. This will release the float bowl. Remove the O-ring and fiber washer from the main jet assembly.
29 Note how the float spring is positioned **(see illustration)**, then push the float pivot pin out of the throttle body. You may have to use a small punch to push the pin out. Remove the float assembly along with the inlet needle valve and the float spring **(see illustrations)**. Remove the float bowl gasket.
30 Carefully screw the idle mixture adjusting screw in until it bottoms,

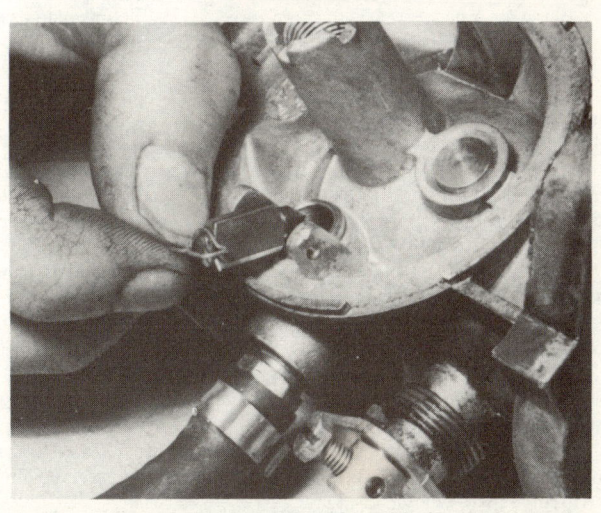

6.29b Lift out the inlet needle valve with the retaining clip attached (Bendix carburetor)

6.40 Be sure the inlet needle valve is correctly seated during reassembly – the retaining clip fits over the tab on the float (arrow) (Bendix carburetor)

6.42 The long end of the float spring (arrow) must be positioned as shown (Bendix carburetor)

while counting the number of turns, then remove it along with the spring. Remove the throttle stop screw and spring in the same manner. Counting and recording the number of turns required to bottom the screws will enable you to return them to their original positions and minimize the amount of adjustment required after reassembly.
31 Close the choke and remove the screws securing the choke valve to the shaft. Pull the choke shaft and lever out, releasing the spring and the plunger.
32 Pry the choke shaft seal and retainer out of the carburetor body. Remove the choke shaft cup plug only if it's damaged and must be replaced.
33 Close the throttle valve and remove the two screws securing it to the shaft. Remove the valve, then slide the shaft out of the carburetor body. Release the spring from the throttle shaft.
34 Pry the throttle shaft retainers and seals out of both sides of the carburetor body.
35 Clean and inspect the parts as described in Steps 2 and 3 of this Section. Make sure the throttle shaft turns freely without excessive play. If it's sloppy, a new carburetor will be needed (although sometimes the throttle shaft bores can be reamed out and bushings installed – check with a dealer service department). Check the inlet needle valve and seat for nicks and a pronounced groove or ridge on the tapered end of the valve **(see illustration 6.19)**. If there is one, a new needle and seat should be used when the carburetor is reassembled. Check the float pivot pin and its bores for wear – If the pin is a sloppy fit in the bores, excessive amounts of fuel will enter the float bowl and flooding will occur. Shake the float to see if there's gasoline in it. If there is, install a new one.
36 Reassembly is the reverse of disassembly. Use new gaskets and seals. Whenever an O-ring or seal is installed, lubricate it with grease or oil. Don't overtighten any of the small fasteners or they may break off.
37 Position the throttle shaft so the flat section is facing out and install the throttle valve (leave the screws slightly loose). Open and close the throttle a few times to center the valve on the shaft. Hold the valve firmly in place while tightening the screws.
38 Insert the choke shaft seal and seal retainer into the choke shaft hole and stake the retainer in place with a small punch.
39 Connect the choke valve to the choke shaft in the same manner as the throttle valve.
40 When the inlet needle valve assembly is installed, be sure the clip is attached to the float tab **(see illustration)**. Check and adjust the float level as described in Section 7.
41 Turn the idle mixture adjusting screw and the throttle stop screw in until they bottom and back each one out the number of turns required to restore them to their original positions.
42 Invert the carburetor and rotate the long end of the float spring up, against the float. Position the float bowl carefully over the throttle body, releasing the float spring so the long end of the spring is pressed against the

6.44a Remove the screws securing the float bowl (Keihin carburetor)

side of the float bowl **(see illustration)**.
43 Install the main jet and tube assembly through the bottom of the float bowl and into the throttle body. Tighten it securely.

Keihin carburetor (except CV type)

Refer to illustrations 6.44a, 6.44b, 6.45, 6.47, 6.48, 6.49, 6.50a, 6.50b, 6.51 and 6.52

44 Remove the screws securing the float bowl to the bottom of the throttle body and detach the float bowl **(see illustrations)**.
45 Loosen the float retaining screw and slide the pivot pin out **(see illustration)**, then carefully separate the float assembly from the carburetor.
46 Remove the inlet needle valve and retaining clip from the float assembly.
47 Separate the rubber boot from the float bowl, then disengage the accelerator pump rod from the rocker arm **(see illustration)**.
48 Remove the plug from the bottom of the throttle body to gain access to the low speed jet **(see illustration on page 122)**. On some models, the low speed jet is accessible only after removing the main jet and main nozzle as described below.
49 Unscrew the main jet **(see illustration on page 122)**. On 1976 through 1978 models, tip the throttle body to remove the main nozzle, then unscrew the low speed jet.

Chapter 3 Fuel system

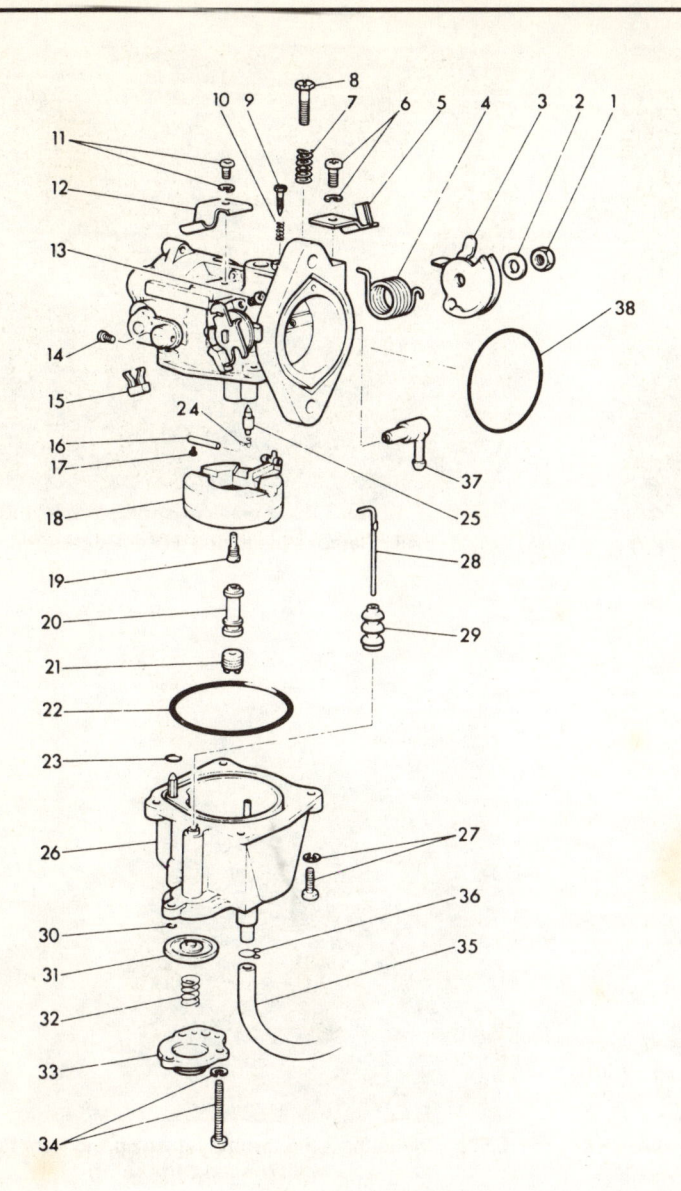

6.44b Typical non-CV type Keihin carburetor components – exploded view

1. Nut
2. Plain washer
3. Throttle lever cable pulley
4. Return spring
5. Throttle cable bracket
6. Screw/washer
7. Spring
8. Throttle stop screw
9. Low speed mixture adjusting screw
10. Spring
11. Screw/washer
12. Choke cable bracket
13. Throttle pump stroke adjusting screw
14. Screw
15. Spacer clip
16. Float pivot pin
17. Float retaining screw
18. Float assembly
19. Low speed jet
20. Main nozzle (not used on all models)
21. Main jet
22. O-ring
23. O-ring
24. Retaining clip
25. Inlet needle valve
26. Float bowl
27. Screw/washer
28. Accelerator pump rod
29. Rubber boot
30. O-ring
31. Diaphragm
32. Spring
33. Accelerator pump cover
34. Screw/washer
35. Overflow tube
36. Clip
37. Union
38. O-ring

6.45 Loosen the retaining screw to release the float pin (Keihin carburetor)

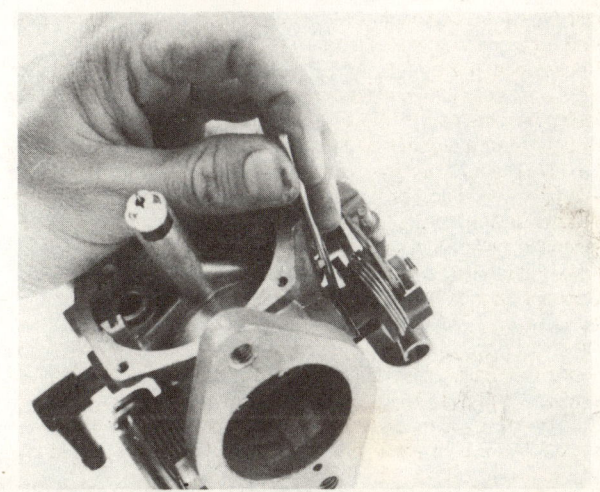

6.47 Disconnect the accelerator pump rod from the rocker arm (Keihin carburetor)

Chapter 3 Fuel system

6.48 Remove the low speed jet plug (Keihin carburetor)

6.49 Remove the main jet (Keihin carburetor)

6.50a Remove the throttle cable pulley mounting nut (Keihin carburetor)

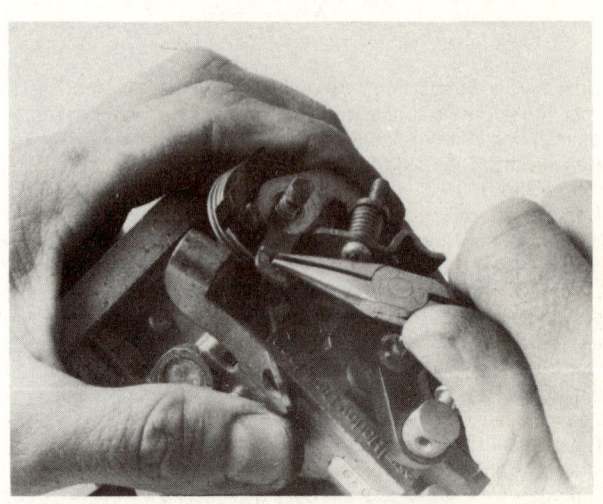

6.50b Disconnect the throttle return spring from the throttle shaft (Keihin carburetor)

6.51 Locations of the throttle cable (1) and choke cable (2) brackets (Keihin carburetor)

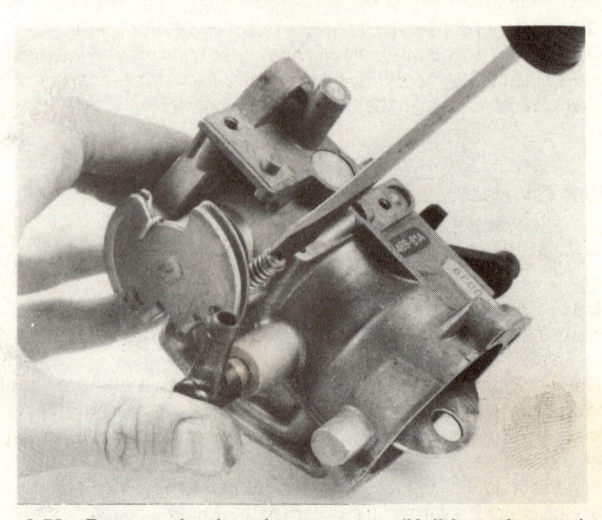

6.52 Remove the throttle stop screw (Keihin carburetor)

Chapter 3 Fuel system

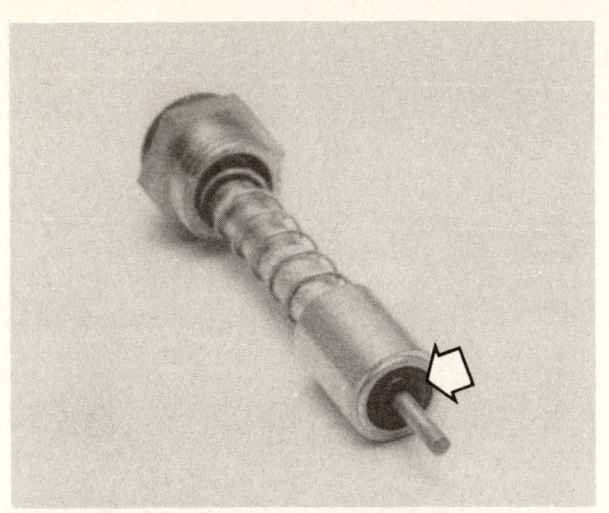

6.64 The enrichener valve seal (arrow) must be in good condition or fuel may leak past it and cause an overly rich mixture (Keihin carburetor)

50 Unscrew the nut securing the throttle cable pulley or the fast idle cam assembly in place **(see illustration)**. Remove the throttle cable pulley and the return spring from the throttle shaft **(see illustration)**.
51 Remove the screws and detach the choke and throttle cable brackets **(see illustration)**.
52 Remove the low speed mixture adjusting screw (1975 through 1979 models only) and the throttle stop screw as described in Step 30 **(see illustration)**. **Note:** *Late 1977 through 1979 models have a limiter cap on the low speed mixture screw that must be pried off to remove the screw. Models built after 1979 don't have a low speed mixture screw – the idle mixture is preset and can't be changed.*
53 Don't disassemble the throttle valve – it's matched to the carburetor and isn't replaceable. If there's a problem with the valve, the carburetor must be replaced with a new one.
54 Clean and inspect the parts as described in Step 35.
55 Check the accelerator pump boot for cracks, the rod for distortion and the diaphragm for holes, cracks and other defects. If any wear or damage is evident, replace the pump components with new ones.
56 Reassembly is the reverse of disassembly. Use new gaskets and seals. Whenever an O-ring or seal is installed, lubricate it with grease or oil. Don't overtighten any of the small fasteners or they may break off. Check the float level (Section 7) before installing the float bowl.
57 Turn the low speed mixture adjusting screw (if used) and the throttle stop screw in until they bottom and back each one out the number of turns required to restore them to their original positions.

Keihin CV carburetor

Refer to illustration 6.64

58 Remove the screws and detach the vacuum chamber cover and spring.
59 Lift out the vacuum piston and rubber diaphragm assembly with the jet needle attached. Be careful not to bend or nick the jet needle – store the vacuum piston assembly in a safe place.
60 Refer to Step 44 – the rest of the disassembly procedure is the same as the one for the early Keihin carburetor. After removing the float assembly, unscrew the main jet and invert the carburetor to remove the needle jet holder and needle jet.
61 In addition to the inspection procedures outlined for the early Keihin carburetor, check the following items: Hold the vacuum piston diaphragm up to a strong light. Look for pinholes and small tears.
62 Make sure the vacuum passage in the bottom of the piston is clear. The piston itself must be smooth and perfectly clean.
63 Check the jet needle for wear, nicks, scratches and distortion. It should be straight and the surface of the taper should be smooth and even. Clean the needle jet and jet holder.
64 Check the end of the enrichener valve **(see illustration)**. Make sure the rubber pad is in good condition. If it's cracked, distorted or deeply grooved, replace the valve with a new one.
65 Reassembly is the reverse of disassembly (see Steps 56 and 57).

7 Carburetor – adjustments

Tillotson carburetor

1 Carburetor adjustments should be made with the engine at normal operating temperature. Also, be sure the air filter element is clean and the air cleaner assembly is installed securely. Adjustments cannot be done accurately unless the air cleaner assembly is in place.
2 Make sure the twist grip closes the throttle lever on the carburetor completely. Turn the idle mixture screw in until it seats lightly, then back it out 7/8-turn. Adjust the intermediate mixture screw in a similar fashion.
3 Start the engine and adjust the throttle stop screw until the engine is running at approximately 2000 rpm. Turn the intermediate mixture screw in both directions until the highest engine speed is obtained without any misfiring or surging. Turn the screw an additional 1/8-turn counterclockwise.
4 Turn the throttle stop screw to adjust the idle to the recommended speed.

Bendix carburetor

Refer to illustrations 7.5 and 7.6

5 The float level must be adjusted before anything else is done to the carburetor. The carburetor must be removed and the float bowl detached to adjust the float level. Refer to Section 4 for carburetor removal and Section 6 for removal of the float bowl. Invert the carburetor and measure the distance between the lower edge of the float, opposite the pivot pin, and the gasket mating surface. The distance should be 3/16-inch (insert a 3/16-inch drill bit between the gasket surface and the float as a gauge) **(see illustration)**. If adjustment is necessary, bend the tab that contacts the inlet needle valve with needle-nose pliers. Reassemble and install the carburetor.

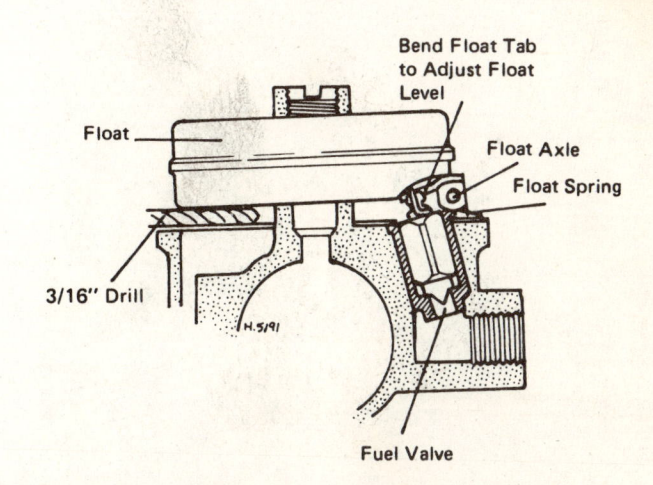

7.5 Checking the float level with a 3/16-inch drill bit as a gauge (Bendix carburetor)

7.6 Locations of the throttle stop screw (1) and the idle mixture adjusting screw (2) (Bendix carburetor)

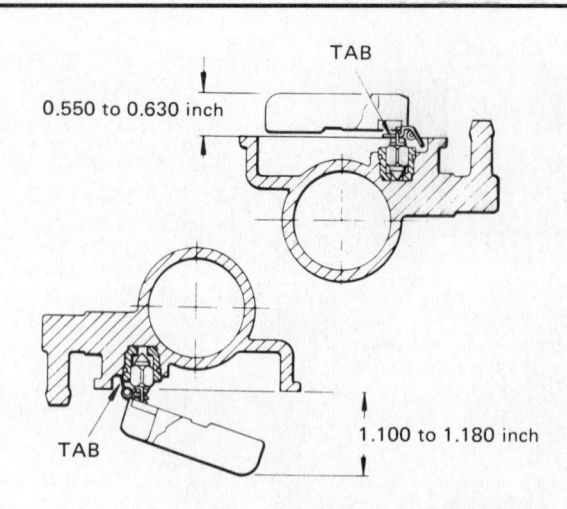

7.9 Checking the float level on 1977 and 1978 Keihin carburetor

6 Run the engine until it reaches normal operating temperature, then shut it off. Carefully turn the idle mixture adjusting screw in until it seats lightly **(see illustration)**. On 1972 and 1973 models, back it out 1-1/2 turns. On 1974 through 1976 models, back it out 2-1/4 turns. Adjust the engine speed by turning the throttle stop screw until it runs at 700 to 900 rpm with the twist grip closed.

7 Adjust the idle mixture screw until the engine will run smoothly at idle speed and accelerate crisply. If necessary, adjust the throttle stop screw until the engine idles at 700 to 900 rpm.

8 Remember, all adjustments must be made with the engine at normal operating temperature and the air cleaner securely in place.

Keihin carburetor (except CV type)
Float level (1977 and 1978 models)
Refer to illustration 7.9

9 These models must have the float level measured in both the open and closed positions. Hold the carburetor upside-down and measure the distance between the gasket surface and the upper edge of the float. The distance should be 35/64 to 5/8-inch (14 to 16 mm) **(see illustration)**.

10 Hold the carburetor right side up and measure the distance between the gasket surface and the lower edge of the float while it's suspended. The distance should be 1-3/32 to 1-3/16-inches (28 to 30 mm). Bend the tabs on the float, as necessary, to obtain the proper float levels.

Float level (1979 through 1987 models)
Refer to illustrations 7.11a and 7.11b

11 Hold the carburetor vertically with the float pivot pin at the top. Measure the distance between the outside edge of the float and the gasket surface **(see illustration)**. The distance should be 0.630 to 0.670-inch (16 to 17 mm). If necessary, bend the tab on the float until the desired float level is attained **(see illustration)**.

Low speed mixture adjustment (1977 and 1978 models)

12 Adjustments should be made with the engine at normal operating temperature and the air cleaner securely mounted.

13 Turn the low speed mixture adjusting screw **(see illustration 6.44a)** in (clockwise) carefully until it seats lightly, then back it out 7/8-turn.

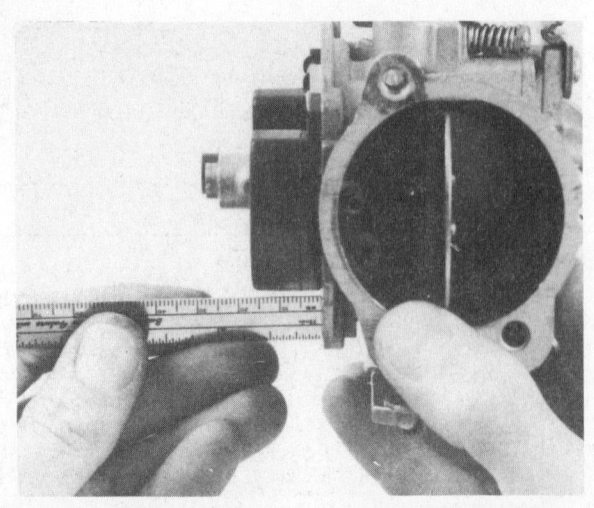

7.11a Measure the float level with the carburetor held vertically (1979 through 1987 Keihin carburetor)

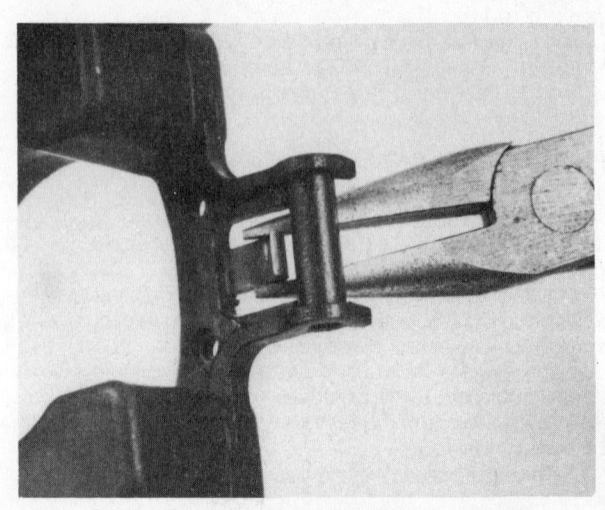

7.11b Bend the tab to adjust the float level (Keihin carburetor)

Chapter 3 Fuel system

7.21 Fast idle adjusting screw (arrow) (1981 through 1987 Keihin carburetor) (1980 slightly different)

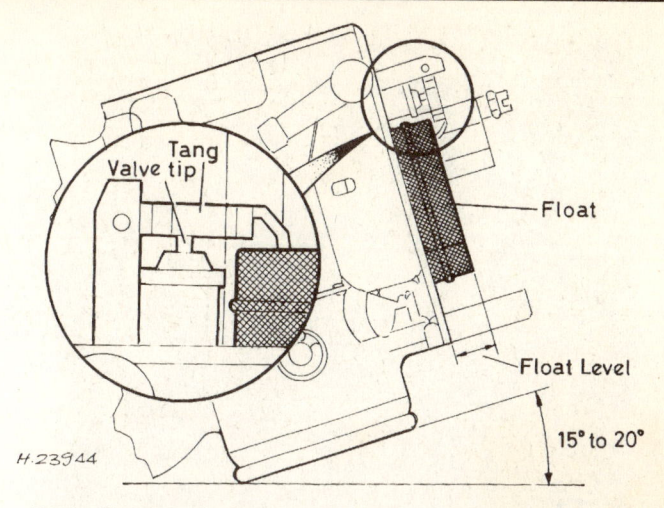

7.27 Checking the float level on 1992-on Keihin CV carburetor

14 Adjust the throttle stop screw until the engine runs at 700 to 900 rpm. Turn the low speed mixture screw until the engine runs smoothly or until it reaches its highest rpm. Readjust the throttle stop screw, if necessary, to obtain the specified idle speed.

Low speed mixture adjustment (1979 models)

15 There's a limiter cap installed over the low speed mixture adjusting screw on these models. Normally the limiter cap shouldn't be removed. The low speed mixture screw should be turned only within the limits of the cap. If necessary, the limiter cap can be removed and the mixture can be altered.
16 Turn the mixture screw in until it seats lightly, then back it out 1-1/4 turns. Reinstall the limiter cap in the center position on the adjusting screw.
17 Adjust the idle speed with the throttle stop screw to 900 rpm. Turn the limiter cap to the leanest setting that still permits the engine to run smoothly. **Note:** *Turn the limiter cap counterclockwise for a richer mixture and clockwise for a leaner mixture.*
18 Readjust the idle speed, if necessary, to the specified setting.

Low speed mixture adjustment (1980 and later models)

Refer to illustration 7.21)

19 The low speed mixture for 1980 and later models is set at the factory and sealed – it's not possible to adjust it.
20 Adjust the idle speed with the choke completely open. Turn the throttle stop screw until the engine is idling at the speed listed in this Chapter's Specifications.
21 Pull the choke out to the second position and turn the fast idle adjusting screw **(see illustration)** until the engine is idling at the specified speed.

Keihin CV carburetor

Float level (1988 through 1991 models)

22 The carburetor must be removed and the float bowl detached when checking the float level. Make sure the floats are aligned with each other – bend them carefully to realign them if necessary.
23 Invert the carburetor and measure the distance from the float bowl mounting surface to the very bottom (curved) side of the float(s) with a dial or vernier caliper. Don't push down on the float(s) as this is done. It should be 0.725 to 0.730 inch (18.4 to 18.5 mm) on 1988 through 1990 models, and 0.690 to 0.730 inch (17.5 to 18.5 mm) on 1991 models.
24 If the level is incorrect, carefully bend the tab on the float that contacts the inlet needle valve until it is.
25 Reinstall the float bowl and the carburetor.

Float level (1992-on models)

26 Remove the carburetor and detach the float bowl. Place the carburetor body face down on a flat surface, on its engine manifold side.

27 Tilt the body at an angle of 15 to 20 degrees until the float tang is seen to just contact the needle valve tip, but not compress it **(see illustration)**. Do not tilt any more or less than the specified angle or the reading will be inaccurate.
28 Measure the distance from the top of the float to the bowl mounting surface; it should be within 0.413 to 0.453 inches (10.5 to 11.5 mm).
29 If the level requires adjustment carefully bend the tab on the float that contacts the inlet needle valve, then recheck the setting.
30 Reinstall the float bowl and carburetor.

Enrichener valve cable

31 The enrichener valve control knob should open, remain open and close without binding. The knurled plastic nut behind the knob controls the amount of resistance the cable offers.
32 If adjustment is required, loosen the locknut at the back of the cable bracket, then detach the cable from the bracket.
33 Grip the flats on the cable housing with an adjustable wrench, then turn the knurled plastic nut until the enrichener operates as described in Step 26.
34 Reattach the cable to the bracket and tighten the locknut. Do not lubricate the cable or housing – it must have a certain amount of resistance to work properly.
35 When the knob is closed, make sure the valve at the carburetor is closed completely.

Slow idle

36 With the engine at normal operating temperature and the enrichener valve fully closed, turn the throttle stop screw until the idle speed is correct (see this Chapter's Specifications). If the motorcycle doesn't have a tachometer, a hand-held instrument will be needed to measure the engine rpm.

8 Air cleaner – removal and installation

Refer to illustrations 8.2, 8.3 and 8.4

Note: *Although the air cleaners used on the models and years covered by this manual differ slightly in some details, they are all basically the same. They consist of a backplate, fastened to the carburetor with screws and a gasket, mounting bracket(s), a filter element and a cover. The following procedure is typical of what must be done to completely remove the air cleaner assembly – take notes, label parts and make a simple sketch of the mounting bracket(s) if the air cleaner on the machine you have appears different from the one described in the text. Make sure the gasket between the backplate and carburetor is in place and in good condition and hook up all hoses during installation. Use thread locking compound on the backplate mounting screws/bolts to prevent the screws from backing out during engine operation.*

8.2 After the cover is removed, the filter can be detached from the air cleaner backplate

8.3 Remove the bolts holding the backplate to the carburetor (arrows) – some models also have nuts (arrows) holding the backplate to brackets

1 Remove the screws or bolts securing the outer cover to the air cleaner assembly. Pull the cover off.
2 Remove the filter **(see illustration)**.
3 Unscrew the bolts securing the air cleaner backplate to the carburetor. Some models also have two nuts attached to backplate mounting brackets that must be removed **(see illustration)**. Later models have two Allen-head bolts which attach the backplate directly to each cylinder head; from 1991 these bolts are hollow and form part of the crankcase ventilation passage.
4 Some models have a vent hose attached to the rear of the backplate, which must be disconnected **(see illustration)**.
5 Installation is the reverse of removal. Be sure to install a new gasket between the backplate and carburetor.

9 Evaporative emission control system – general information

Refer to illustration 9.3

This system, installed on all 1986 and later California models only, is virtually maintenance-free and shouldn't be tampered with unless a new canister, hoses or valves are required because of leaks or deterioration. An occasional check to make sure the hoses are routed properly, secured to the fittings and not kinked or blocked should be sufficient. Make sure the hoses don't come too close to or touch the exhaust system components. Also, check all canister and valve mounting fasteners to see if they're tight. On 1988 through 1991 models, check the reed valves in the air cleaner assembly backplate to make sure they aren't cracked or broken off.

The system is designed to prevent fuel vapor from escaping from the tank into the atmosphere when the engine is off. The vapor is directed from the tank through a hose and valve to a charcoal-filled canister (mounted on a frame front tube or under the swingarm, depending on the model), where it's absorbed by the charcoal. When the engine is started, the vapor is drawn from the canister into the carburetor and then into the engine, where it's burned in the cylinders. A large diameter hose also purges the canister with fresh air from the air cleaner during engine operation. The vapor valve prevents raw fuel from entering the vent hose when the motorcycle is at extreme angles.

1988 through 1991 models also have a Vacuum Operated Electric Switch (VOES – see Chapter 4) and a vacuum operated valve. The VOES directs vacuum to the vacuum operated valve, which seals off the carburetor float bowl vent when the engine's off (it also vents it to the atmosphere when the engine's running). If the diaphragm in the vacuum valve starts to leak, the fuel/air mixture would be upset (leaned out) at high speeds. To check the valve, apply a small vacuum (1 to 2 in-Hg only) to port A **(see**

8.4 After the backplate has been separated from the carburetor, detach the hose from the fitting on the back side

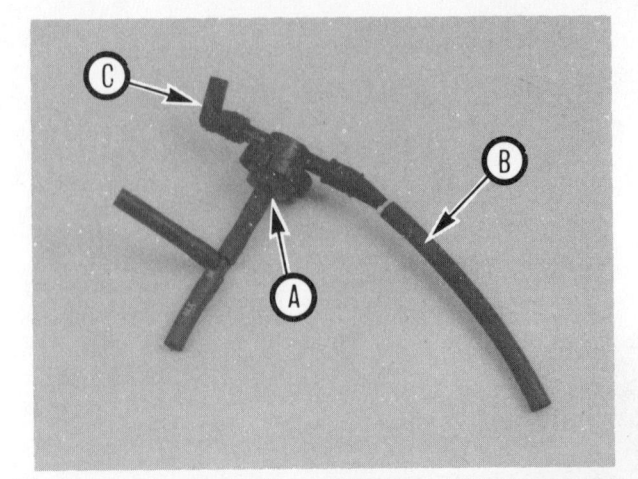

9.3 Vacuum operated valve port locations (1988 through 1991 California models) – see text

illustration). The vacuum should remain steady and the valve should be open (you should be able to blow through port B or C – air should pass through). Release the vacuum and make sure the valve is closed (no air should pass through when blowing into port B or C). If the valve doesn't function as described, install a new one.

1992-on models retain the VOES described above, but venting of the carburetor vent hose is controlled electrically by a solenoid-operated butterfly valve, clamped to the rear of the air cleaner baseplate. The butterfly valve itself resides in a housing at the bottom of the air cleaner and if functioning correctly, should remain closed when the engine is stopped, yet open when the starter circuit operates and stay open until the engine is stopped. If failure is suspected, check first the mechanical linkage from the valve pivot to the solenoid plunger. If this is in order, check the solenoid operation as described in Chapter 7.

10 Throttle cable – adjustment

Warning: *Do not over-tighten the throttle grip friction screw. Riding with the friction screw too tight is not recommended because of the danger involved when the engine won't return to idle automatically in an emergency.*
Caution: *This adjustment is necessary to prevent excess stress on and potential failure of the throttle cable.*

1979 and 1980

1   When turned by hand and released, the throttle grip must return to the closed (idle) position. If it doesn't return freely, back off the friction screw until it does. If the throttle grip turns stiffly, it should be disassembled, cleaned and inspected (see Chapter 1).
2   Adjust throttle cable free play by turning the adjusting nut. The cable should not pull on the carburetor lever when the handlebars are turned to the left and right stops.
3   With the front wheel pointed straight ahead, adjust the throttle grip travel limit. This is done by turning the throttle to the wide open position and adjusting the set screw with a two millimeter Allen wrench. The throttle grip should be all the way open as the carburetor lever reaches the fully open position.

1981-on

Note: *The throttle cable has a 5/16-inch x 18 threaded adjuster and is attached to the right side of the throttle grip. The idle cable has a 1/4-inch x 20 threaded adjuster and is attached to the left side of the throttle grip.*

4   Turn the cable adjusters and locknuts clockwise as far as possible.
5   Point the front wheel straight ahead. Turn the throttle grip so the throttle is wide open and hold it there. Now turn the throttle cable adjuster until the throttle cam stop barely touches the stop on the carburetor. Tighten the locknut against the adjuster and release the throttle.
6   Turn the front wheel all the way to the right. Turn the idle cable adjuster until the cable housing just touches the spring in the cable support sleeve. Open and close the throttle grip to make sure the throttle cable returns to the idle position. If it doesn't, continue to turn the adjuster until it does. Tighten the locknut against the idle cable adjuster. **Note:** *The throttle grip must operate freely without binding. With the friction adjusting screw backed off, the carburetor lever must return to the closed (idle) position.*

Chapter 4 Ignition system

Contents

General information 1	Electronic ignition system – check 7
Spark plugs – check and replacement 2	Ignition components – removal and installation 8
Ignition coil – check, removal and installation 3	Ignition timing – check and adjustment 9
Contact breaker points – check and replacement 4	Vacuum Operated Electric Switch (VOES) – check
Condenser – removal and installation 5	and replacement 10
Air gap – check and adjustment 6	Distributor (1970 models only) – removal and installation 11

Specifications

Note: *Additional ignition system specifications are included in Chapter 1, Tune-up and routine maintenance.*

Ignition system type

1970 through 1978	Mechanical contact breaker point
1979-on	Breakerless inductive discharge (electronic)

Air gap *(1979 only)* .... 0.004 to 0.006 in (0.102 to 0.152 mm)

Ignition coil resistance

1970 through 1978	
Primary	4.7 to 5.7 ohms
Secondary	16,000 to 20,000 ohms
1979	
Primary	4.5 to 5.7 ohms
Secondary	16,500 to 20,000 ohms
1980 through 1985	
Primary	3.3 to 3.7 ohms
Secondary	16,500 to 19,500 ohms
1986 on	
Primary	
All except 1998 and 1999 XL1200S	2.5 to 3.1 ohms
1998 and 1999 XL1200S	0.4 to 0.6 ohms
Secondary	
1986 through 1990	Not specified
1991 and 1992	11,250 to 13,750 ohms
1993 on (except 1998-on XL1200S)	10,000 to 12,500 ohms
1998-on XL1200S	11,700 to 12,700 ohms

1 General information

In order for the engine to run correctly, an electrical spark must ignite the fuel/air mixture in the combustion chambers at exactly the right moment in relation to engine speed and load. The ignition system operation is based on feeding low tension (primary) voltage from the battery to the coil where it's converted to high tension (secondary) voltage by a process known as induction. The secondary voltage is powerful enough to jump the spark plug gap in the cylinders many times a second under high compression pressures, provided the system is in good condition and all adjustments are correct.

The ignition system installed on 1970 through 1978 models as standard equipment is a conventional mechanical contact breaker point type. The system installed on 1979 and later models is a breakerless inductive discharge system (electronic ignition).

Both systems are divided into two circuits: the primary (low tension) circuit and the secondary (high tension) circuit. The primary circuit consists of the battery, the contact breaker points (early models), the control module (1979 only) or the computerized control module (1980-on), the primary coil, the ignition switch and the wires connecting the components.

The secondary circuit consists of the secondary coil, the spark plugs and the wires.

Mechanical flyweights are incorporated in the ignition systems on pre-1980 models to advance the ignition timing mechanically. On 1980 and later models, the computer (control module) advances the timing electronically. On 1983 and later models, a Vacuum Operated Electric Switch (VOES) senses vacuum in the intake manifold and sends a signal to the computer, which advances or retards the ignition timing as needed.

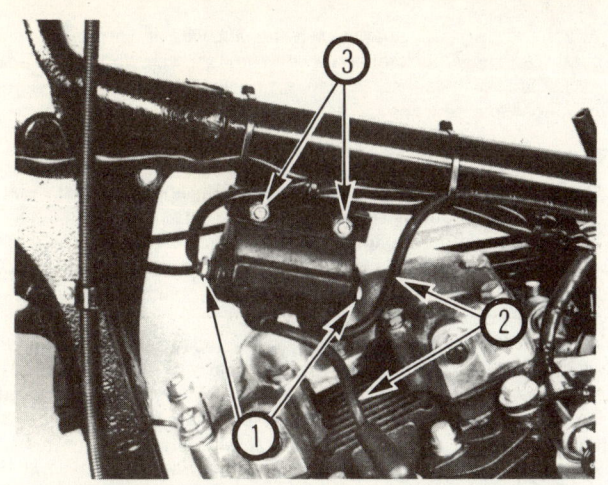

Chapter 4 Ignition system

3.3 The ignition coil is located under the fuel tank

1 Primary wire terminals
2 Spark plug wires
3 Mounting nuts

7.1 A simple spark test tool can be made from a block of wood, a large alligator clip, some nails, screws, and wire and the end of an old spark plug – leave about 1/4-inch between the electrode tips (attach the plug wire to the protruding electrode [arrow] and grip the cylinder head fins with the alligator clip)

2 Spark plugs – check and replacement

Refer to Chapter 1, *Tune-up and routine maintenance,* for spark plug check and replacement procedures.

3 Ignition coil – check, removal and installation

Refer to illustration 3.3

1 Detach the wires from the coil and connect an ohmmeter to the coil primary wire terminals. The primary resistance should be as listed in this Chapter's Specifications – if it isn't, the coil is probably defective. If an ohmmeter isn't available, take the coil to a dealer service department to have it checked or temporarily install a known good coil. **Caution:** *Be sure to connect the wires to the correct terminals to avoid damaging other ignition components. If the ignition system trouble is eliminated by the temporary installation of the new coil, install it permanently.*
2 Pre-1979 models with contact breaker point ignitions should be checked to make sure the condenser is good before replacing the coil. A bad condenser can act just like a defective coil, but the condenser is much more likely to fail. **Note:** *The easiest way to check for a defective condenser is to substitute a known good component. If the ignition system problem goes away when the new condenser is installed, the original condenser is defective.*
3 The coil is mounted under the front of the fuel tank **(see illustration)**. On 1988 and later XL1200S models, remove the fuel tank (see Chapter 3).
4 Label the primary (small) wires and coil terminals, then disconnect the wires. Detach each spark plug wire from the coil by pulling the boot back, grasping the wire as close to the coil as possible and pulling the wire out of the coil. Note the routing of the dual plug wires on 1998 and later XL1200S models.
5 Remove the bolts/nuts and detach the coil.
6 If a new coil is being installed, make sure it's the right one. **Caution:** *1980 and later models must have a coil marked "Electronic Advance". Installing the wrong type of coil could result in failure of the electronic components.*
7 Installation of the coil is the reverse of removal. Be sure the spark plug wire boots are seated on the coil towers to keep dirt and moisture away from the terminals. Install new boots if the originals are hardened, cracked or torn.

4 Contact breaker points – check and replacement

Refer to Chapter 1, *Tune-up and routine maintenance,* for contact breaker point replacement and adjustment procedures.

5 Condenser – removal and installation

1 A condenser is included in the primary circuit of 1970 through 1978 models to prevent arcing across the contact breaker points as they open. If it fails, the ignition system will malfunction.
2 If the engine is difficult to start, or if misfiring occurs, it's possible the condenser is defective. To check it, separate the contact breaker points by hand when the ignition switch is on. If a spark occurs across the points and they appear to be discolored and burned, the condenser is defective.
3 It isn't possible to check the condenser without special test equipment. Since the cost is minimal, install a new condenser to see the effect on engine performance.
4 Because the condenser and contact breaker points supply a spark to both cylinders, it's virtually impossible for a faulty condenser to cause a misfire on one cylinder only.
5 The condenser is attached to the contact breaker baseplate by a clamp and screw. If the wire is disconnected and the screw removed, the condenser can be detached from the baseplate.
6 When installing the new condenser, be sure the clip around the body (which forms the ground connection) makes good contact with the baseplate. Tighten the mounting screw securely.

6 Air gap – check and adjustment

1 On 1979 models, the air gap between the trigger rotor and sensor is adjustable.
2 Remove the ignition component covers from the right side of the engine (see Chapter 2).
3 Remove the spark plugs and rotate the crankshaft until the wide lobe on the trigger rotor is centered in the sensor.
4 Insert a feeler gauge of the specified thickness between the lobe of the rotor and the sensor. If the air gap isn't correct, loosen the screws securing the sensor and move it until the specified gap is obtained. Rotate the crankshaft 180-degrees until the other lobe of the rotor is centered in the sensor and measure the gap to be sure it's the same.

7 Electronic ignition system – check

All models

Refer to illustration 7.1

1 Check the condition of the spark plugs and the spark plug wires as described in Chapter 1. One way to check the ability of the ignition system to produce a strong enough spark is to construct a simple home-made test tool **(see illustration)** and hook it up to the plug wires, one at a time (with

Chapter 4 Ignition system

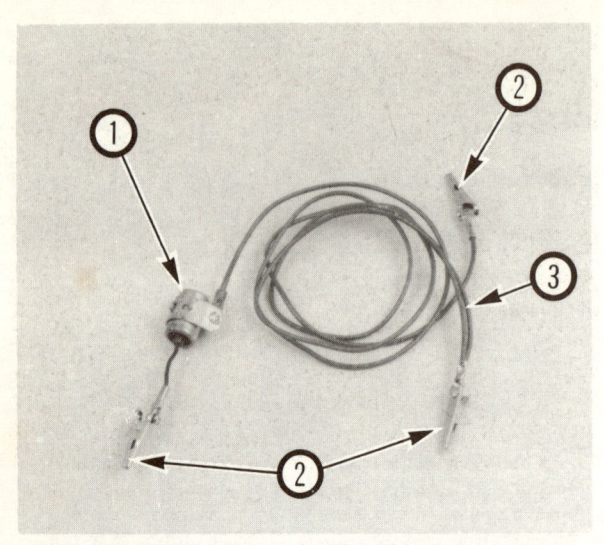

7.12 Fabricate a jumper wire like this one to check the ignition system on 1984 and later models

1. 0.33 MF capacitor
2. Alligator clip
3. 16-gauge wire (approx. 18-inches long)

the alligator clip grounded on the engine), to see if the spark will jump the gap. Another way, though less conclusive, is to attach a spark plug that's known to be good to the spark plug wires and ground the plug against the engine. Crank the engine over and check for spark. If there is a spark, the coil and ignition system are working properly. Check the choke and the carburetor and replace the spark plugs with new ones if the engine doesn't run right.

2 If no spark is obtained, attach a voltmeter to the battery (black lead to the negative terminal, red lead to positive) and turn the ignition switch and the engine stop switch to the On position. The reading on the voltmeter should be at least 11.5 volts.

3 Measure the air gap (1979 models only) between the sensor and both trigger rotor lobes (refer to Section 6). If the gap between both rotor lobes can't be adjusted within the range given, replace the trigger rotor and/or the timer mechanism.

4 Make sure the ground cable from the battery is secure and making good contact.

5 Check the ground for the module at the timer plate on the engine (1979 models) or at the frame (1980 and later models). It should be clean and tight to ensure a good connection.

6 Attach a voltmeter between the positive terminal of the coil and ground (red lead to the coil, black lead to ground). Rotate the crankshaft until the lobes of the trigger rotor (1979 models) or the slots in the rotor (1980 and later models) are equal distances from the center of the sensor. Turn the ignition switch and the engine kill switch to the On position and read the voltage on the meter. It should be within 1/2-volt of the battery voltage. If not, the trouble is somewhere between the battery and the coil. Check the connections at the ignition switch, the engine kill switch and the circuit breaker.

7 Disconnect the blue wire (pink wire 1991-on) from the negative terminal of the ignition coil and attach a voltmeter between the terminal and ground. With the ignition and kill switches on, the voltage should be the same as the battery. If the voltage is different, the coil should be replaced with a new one (after checking the resistance as described in Section 3).

1979 models only

8 Reconnect the blue wire to the negative terminal of the coil, then connect the voltmeter between ground and the negative terminal of the coil. The voltmeter should read 1.0 to 2.0 volts. Place the blade of a screwdriver against the face of the sensor and read the voltmeter. The reading should be between 11.5 and 13.0 volts. By removing the screwdriver, the voltage should drop to 1.0 to 2.0 volts. If not, the ignition module is defective and must be replaced with a new one.

9 Attach a known good spark plug to the plug wire and ground the plug against the engine. Check for spark each time the screwdriver is placed against the face of the sensor. If there is no spark, the ignition coil should be replaced with a new one.

1980 through 1985 models only

10 Unplug the connector between the module and the sensor plate. Connect a voltmeter between the red and black wires in the connector on the module side (positive voltmeter lead to the red wire, negative lead to black). Turn the ignition and the engine kill switches on and read the voltmeter. It should indicate 4.5 to 5.5 volts. If not, the ignition module is defective and should be replaced with a new unit.

11 A special jumper cable test adapter is needed to do the following test. It's available from a harley-davidson dealer (part number HD 94465-81). Attach the test adapter to the connector halves between the module and sensor. Be careful not to let the exposed wires touch a grounded component or each other or damage to the module will result. Recheck the voltage between the red and black wires. Connect the voltmeter between the green and black wires to test the sensor output (voltmeter positive lead to the green wire, negative lead to the black wire). The voltmeter should read 4.5 to 5.5 volts with the rotor slots away from the sensor, and 0 to 1.0 volt with the slot aligned with the sensor. If either of these voltage readings are not attained, the sensor plate must be replaced with a new unit.

1986 and later models only

Refer to illustration 7.12

12 Assemble a jumper wire from 16-gauge wire, a 0.33 microfarad capacitor and three alligator clips **(see illustration)**. A known good condenser from a breaker point ignition system can be used if a capacitor isn't available.

13 Attach a known good spark plug (or the home-made test tool) to one of the plug wires and ground the plug or tool on the engine.

14 Connect the jumper wire with the capacitor in it to the negative coil primary terminal (the blue wire [pink wire 1991-on] was disconnected from the terminal in Step 7). Attach the jumper wire common alligator clip to a good ground.

15 Momentarily touch the remaining jumper wire alligator clip to the negative terminal on the ignition coil – when you do, a spark should occur at the plug or test tool.

16 If not, replace the ignition coil – it's defective. If a spark occurred, proceed to Step 17.

17 Follow the instructions in Step 10 above. If the voltage readings are not as specified, check the ignition module power and ground wires for loose and dirty connections. On 1991-on models note that the black wire described is black/white, and that the expected result is 12V ± 0.5V.

18 If the wires and connections are okay, turn the ignition and kill switches on and momentarily connect the black (black/white 1991-on) and green wire connector pins in the module wire connector with a jumper wire or a screwdriver. If a spark occurs at the plug or test tool when the screwdriver or jumper wire is disconnected, the sensor is probably defective (check the sensor resistance as described below before buying a new one).

19 If no spark occurred during the test in Step 18, check the module resistance (as described below) and install a new one if the resistance isn't as specified.

Intermittent ignition problem check

20 Check the battery terminals and the module ground connection to make sure they're clean and tight.

21 Disconnect the white wire from the ignition coil primary terminal (not the white wire that goes to the module from the same terminal).

22 Connect a 16-gauge jumper wire with alligator clips to the positive battery post and the terminal on the coil the white wire was disconnected from.

23 Start the engine and see if the problem is eliminated. If it is, the problem is possibly in the starter safety switch connections (they're probably loose). **Caution:** *The engine won't stop running until the jumper wire is removed.*

24 If the intermittent problem still exists, refer to Chapter 2 and remove the ignition component covers to gain access to the ignition sensor.

Chapter 4 Ignition system

25 Start the engine and spray canned compressed air (available from electronics or computer supply stores) on the sensor – be sure to wear safety glasses or a face shield as this is done! **Note:** *Turn the can upside down for the maximum effect.* **Warning:** *Avoid skin contact and don't breathe the vapors.*

26 If the engine dies, the sensor is temperature sensitive (to cold) and should be replaced with a new one.

27 If the engine keeps running, allow it to reach normal operating temperature, then use a blow dryer to apply heat to the sensor. If the engine stops, the sensor is temperature sensitive (to excess heat) and should be replaced with a new one.

28 If the engine doesn't stop, apply heat to the ignition module. If the engine stops, the module is defective and should be replaced.

Sensor resistance check

Note: *The following resistance test applies to 1986 through 1990 models – no resistance test details are available for 1991-on models.*

29 Position the ohmmeter selector switch on the Rx1 scale. Unplug the sensor wire harness connector and attach the positive (red) lead from the ohmmeter to each of the sensor harness terminals in the connector, one at a time, with the negative (black) ohmmeter lead connected to a good ground (use the sensor plate). If each terminal reading indicates infinite resistance, the sensor is good. If any of the terminal connections produce a resistance reading, the sensor is bad.

30 Connect the positive ohmmeter lead to the green sensor wire and the negative lead to the black sensor wire. If the ohmmeter indicates infinite resistance, the sensor is good. If a resistance reading is produced, the sensor is bad.

31 Reverse the ohmmeter leads. If the ohmmeter reads 300-to-750 K-ohms, the sensor is good. If it indicates infinite resistance, the sensor is defective.

Ignition module resistance check

32 Position the ohmmeter selector switch on the Rx1 scale. Unplug the ignition sensor-to-module wire harness connector and on 1986 through 1990 models, attach the positive (red) lead from the ohmmeter to the black wire terminal in the module side of the connector – the negative (black) ohmmeter lead should be connected to the black module ground wire. On 1991-on models attach the positive meter lead to the black/white wire terminal and the negative lead to ground on the frame. If the ohmmeter indicates 0-to-1 ohm, the module is good. If it indicates more than 1 ohm, replace the module.

33 The following resistance tests apply to 1986 through 1990 models only. No similar data is available for 1991-on models; if a fault is suspected with the module and all other ignition components check out, including the wiring, then replacement of the module is the only solution.

34 Connect the positive ohmmeter lead to the white module wire at the ignition coil (disconnect it from the coil first) and check the meter reading again. If it's 800-to-1300 K-ohms, the module is good. If it indicates infinite resistance, the module is defective.

35 Reverse the ohmmeter leads (red lead to ground, black lead to the white module wire at the coil). If the ohmmeter indicates infinite resistance, the module is good. If it indicates any resistance, the module is bad.

36 Attach the positive ohmmeter lead to the blue module wire at the coil (disconnect it from the coil first) and attach the negative lead to the black module ground wire. If the meter indicates infinite resistance, the module is good. If it indicates any resistance, the module is bad.

37 Reverse the ohmmeter leads. If the ohmmeter indicates 400-to-800 K-ohms, the module is good. If infinite resistance is indicated, the module is defective.

38 If the module fails any of the tests, replace it with a new one.

8 Ignition components – removal and installation

Refer to illustration 8.1

Refer to Chapter 2 to remove and install the engine-mounted ignition components. The electronic ignition system module is mounted as follows:

a) 1980 and 1981 – on the bottom of the battery carrier/oil tank.
b) 1982 through 1997 – on the frame next to the battery (under the side cover) **(see illustration)**.

8.1 Ignition module location – 1982 through 1987 models

c) 1998 and later (except XL1200S) – on the right side of the engine, in the rear case cover.
d) 1998 and later XL1200S – under the seat.

9 Ignition timing – check and adjustment

Refer to Chapter 1 for the ignition timing check and adjustment procedures.

10 Vacuum Operated Electric Switch (VOES) – check and replacement

1 The Vacuum Operated Electric Switch is controlled by intake manifold vacuum, sensed through an opening in the carburetor body. Under low vacuum conditions, such as hard acceleration, the switch is open, which causes the ignition module to retard the timing, reducing detonation. When high vacuum is present in the intake manifold, the ignition module switches to the advanced timing mode for better fuel economy and performance.

Check

2 Set the ignition timing with a timing light (see Chapter 1).

3 With the engine idling, detach the VOES vacuum hose from the carburetor and plug the carburetor fitting. The ignition module will select the retard mode and engine rpm should drop.

4 Reconnect the vacuum hose. Engine speed should increase and the timing mark should reappear in the timing plug hole.

5 If the engine speed doesn't change when the hose is disconnected and reconnected, check the wire between the VOES and the ignition module and the VOES ground wire.

Replacement

6 Detach the air cleaner cover and backplate, then remove the rear fuel tank mounting bolt and raise the rear of the tank to get at the VOES (see Chapter 3).

7 Disconnect the wire from the VOES to the ignition module.

8 Disconnect the VOES ground wire (to the engine).

9 Detach the vacuum hose from the VOES (and the carburetor or T-fitting, if necessary).

10 Remove the nut, bolt and washer and detach the VOES from the bracket.

11 Installation is the reverse of removal. Make sure the ground wire is securely attached to the engine bracket with the VOES mounting bolt. Position the high-temperature conduit sleeve over the wire connectors and secure them to the frame with the nylon strap. The VOES and wires must not come in contact with the engine rocker arm covers – heat and vibration from the engine could damage them.

11.2 Distributor components – exploded view (1970 models only)

1. Point cover
2. Screw and lock washer
3. Primary wire terminal screw
4. Nut and lock washer
5. Moving contact breaker point
6. Point adjusting lock screw
7. Fixed contact point and base
8. Condenser mounting clip and screw
9. Condenser
10. Contact breaker base plate
11. Washer
12. Primary wire
13. Fiber washer
14. Terminal post
15. Insulator
16. Distributor body
17. O-ring
18. Contact breaker point cam
19. Clip
20. Flyweight
21. Base nut/washer
22. Flyweight spring
23. Washer
24. Shaft
25. Roll pin
26. Gear
27. Washer
28. Eccentric breaker point adjusting screw
29. Clamp
30. Clamp bolts

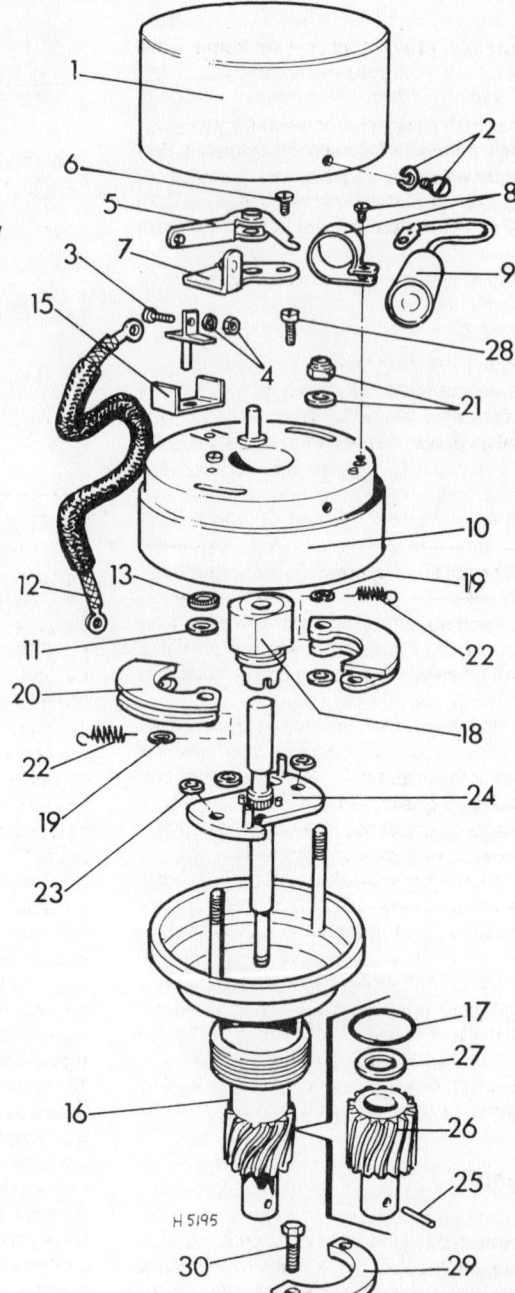

11 Distributor (1970 models only) – general information

Refer to illustration 11.2

1 The distributor should operate reliably for a long time and require inspection at infrequent intervals. Problems will occur if the drive gear is damaged or if excessive end or side play develops as the result of worn bearings, which will allow oil to contaminate the contact breaker points. End play must not exceed 0.008-inch.

2 The mechanical advance flyweights must be free to move on the pivot pins and the springs must be in good condition **(see illustration)**. Clean and lubricate the flyweight bores and pivot pins and make sure the springs are properly attached. Make sure the point cam is free on the shaft.

3 While it's possible to disassemble and repair a worn distributor, it isn't recommended. Replace the distributor with a new one instead.

4 To remove the distributor, detach the primary wire from the terminal on the outside of the distributor and remove the clamp bolts.

5 When installing the distributor in the engine, make sure the external primary wire terminal faces the rear of the machine so the contact breaker points are positioned toward the outside of the engine. This will allow easier access for adjustments. Always recheck the ignition timing before and after tightening the distributor clamp bolts.

12 Bank angle sensor (1998 and later models) – check and replacement

1 The bank angle sensor is mounted on the backside of the battery tray. It consists of a magnet in a fluid-filled channel. If the bike's lean angle exceeds 80 degrees, the magnet moves from its normal position, causing an open in the ignition module circuit. When the module detects this open, it shuts off the ignition system.

Chapter 4 Ignition system

Check

2 Carefully pull off the left side cover (it's clipped onto the frame; there are no fasteners).
3 Start the engine.
4 Locate the bank angle sensor on the backside of the battery tray. Put a magnet on top of the bank angle sensor. If the sensor is operating correctly, the magnet inside the sensor will move, cause an open circuit, shut off the ignition system and stop the engine. Remove the magnet and turn the ignition switch to OFF. This resets the bank angle sensor (unless it was defective).
5 If the bank angle sensor fails to operate as described, replace it.

Replacement

6 Remove the bank angle sensor retaining screw.
7 Disconnect the bank angle sensor electrical connector.
8 Installation is the reverse of removal. Make sure that the locating pin on the sensor body is correctly aligned with the hole before tightening the sensor retaining screw.
9 Install the left side cover.

13 Ignition system (XL1200S models) – general information

1 The ignition system on XL Sport models is different from other late-model Sportsters. It consists of a timing rotor, a cam position sensor, a manifold absolute pressure (MAP) sensor, an ignition module, a dual-coil unit, and a bank angle sensor (see Section 12).
2 Ignition timing is handled by the timing rotor, the cam position sensor, the module and the MAP sensor. The rotor, which is mounted on the end of the camshaft, rotates at one-half crankshaft speed. As it rotates, slots in the rotor's circumference break the magnetic field of a Hall-effect device in the ignition module. The voltage output of the Hall-effect device provides accurate ignition timing even at very slow crank speed.
3 The MAP sensor, which is mounted on a bracket on the frame backbone under the fuel tank, monitors pressure in the intake manifold and provides an analog signal to the ignition module in response to changes in the intake manifold pressure.
4 The ignition module, which is located under the seat, turns battery voltage to the ignition coil on and off in response to the signal it receives from the Hall-effect device. The module also computes the correct spark advance in response to changes in the manifold absolute pressure.
5 The dual-coil unit fires a pair of spark plugs in each cylinder in "single-fire" mode; there is no "waste spark," i.e. both plugs fire in one cylinder but not in the other.

Diagnostic trouble codes

Refer to illustration 13.18

6 The XLH 1200S ignition system has some self-diagnostic capabilities. When the engine is running, the module monitors the MAP sensor, battery voltage, the ignition coils, the tachometer, the cam sensor, the bank angle sensor, random access memory/read-only memory (RAM/ROM), electronic erasable programmable read-only memory (EEPROM), and the module itself. If a malfunction occurs in one of these circuits, the module stores a diagnostic trouble code and turns on the Check Engine Lamp.
7 When the ignition key is turned to ON (after it's been turned to OFF for 10 or more seconds), and the engine kill switch in the RUN position, the Check Engine Lamp lights up for about four seconds, then goes out.
8 If the Check Engine Lamp does not come on when the ignition key is turned to ON, or if it fails to go out after four seconds, there is a malfunction in one or more of the monitored circuits in the ignition system.
9 There are two ways to output any diagnostic trouble code(s) stored in the ignition module. Harley-Davidson dealers use a portable scan tool known as a "Scanalyzer" which interfaces directly with the ignition module through the Data Link Connector. This method is beyond the scope of the home mechanic. The other method, which can be done at home, requires no fancy tools. This method is simply a matter of determining the two-digit trouble code(s) by counting the number of flashes of the Check Engine Lamp.

Code	Description
12	MAP sensor
16	Battery voltage
24	Front coil
25	Rear coil
35	Tachometer
41	Cam sync failure
44	Bank angle sensor
52	RAM/ROM failure
54	EEPROM failure
55	Module failure

13.18 Trouble codes (XL1200S models)

10 Fabricate a two-inch jumper wire from 18-gauge wire with the correct terminal (Harley part no. 72191-94) on each end.
11 Locate the Data Link Connector on the electrical bracket under the left side cover. The Data Link Connector has four wires (light green/red, black, violet/red, white/black) on one side of the connector, and none on the other side. Bridge terminals 1 and 2 (light green/red and black wires) with the jumper wire.
12 Turn the ignition switch key to IGNITION and wait about eight seconds for the Check Engine Lamp to begin flashing.
13 Each stored code is preceded by a series of rapid flashes (about three per second), followed by a two-second pause (lamp is off), followed by the code. (If the lamp continues to flash at the faster rate, no codes are stored).
14 The lamp indicates the first digit of the trouble code by flashing one or more times. Each flash is about one second in duration, followed by a one-second pause (lamp off). To determine the first digit, simply count the number of times the lamp flashes (it could be one, two, three, four or five flashes). Write down this number.
15 After the lamp has flashed the first digit of the code, there is another two-second pause (lamp off). The lamp then flashes one or more times to indicate the second digit of the code (two, four, five or six flashes).
16 After the lamp has flashed the second digit of the code, there is another two-second pause. Then, if there are no other codes stored, the lamp will repeat this trouble code again. If there is more than one code stored, the lamp will flash out the second stored code, in the same manner as the first. Again, jot down the code. This will be followed by another two-second pause, then the lamp will either repeat the first two codes again or, if there is a third code, it will display that code.
17 After the lamp has completed its output of all stored codes, it repeats, starting with the first code again. Once you note that the lamp is repeating itself, there is no need for further observation. (You may wish to record them a second time just to be sure that you counted the correct number of flashes for each digit of each code the first time.)
18 Compare the displayed trouble codes to the accompanying table **(see illustration)**. You now know which circuit(s) has/have a problem.
19 The first things to check in any circuit with an apparent problem are the connectors and the wires. Make sure that all connectors are clean, dry, corrosion-free and tight. Inspect each wire in the circuit and make sure that it's not grounded, open or shorted. If all the connectors and wires are in good shape, take the bike to a Harley-Davidson dealer service department. No further diagnosis is possible at home.
20 To take the ignition module out of diagnostic mode, remove the jumper wire from the Data Link Connector and turn the ignition switch to OFF. Once any problems have been corrected, the trouble codes can be cleared in one of two ways. The convenient method requires the Scanalyzer. The other method is not convenient, but can be done at home if necessary. It requires 50 "start-and-run" cycles (each start-and-run cycle consists of starting the engine, allowing it to run for at least 30 seconds, then turning it off).

Chapter 5 Frame and suspension

Contents

General information ... 1	Swingarm – removal, inspection and installation ... 6
Frame – inspection and repair ... 2	Forks – removal and installation ... 7
Instrument cluster – removal and installation ... 3	Forks – disassembly, inspection and reassembly ... 8
Shock absorbers – removal and installation ... 4	Steering head bearings – maintenance ... 9
Swingarm bearings – check ... 5	Seat – removal and installation ... 10

Specifications

Fork oil capacity .... See Chapter 1

Torque specifications

	Ft-lbs	Nm
Upper yoke stem and fork pinch bolts		
1986 and 1987	21 to 27	28 to 37
1988 through 1990	25 to 30	34 to 41
1991-on	30 to 35	41 to 47
Lower yoke fork pinch bolts		
1986 and 1987	30 to 35	41 to 47
1988-on 883 models	30 to 35	41 to 47
1988 through 1990 1200 model	25 to 30	34 to 41
1991-on 1200 model	30 to 35	41 to 47
Swingarm Allen-head bolt (1986-on)	50	68

1 General information

The Harley-Davidson models covered by this manual have a duplex tube, full cradle frame with a single top tube running from the steering head to the seat tube. Unlike many other designs, the twin tubes are very close together so the engine tends to straddle the frame rather than sit in it.

The front suspension is conventional, consisting of oil damped, telescopic forks. The rear suspension is also conventional, composed of a swingarm and two hydraulically-damped shock absorbers with adjustable spring rates.

2 Frame – inspection and repair

1 The frame is unlikely to require attention unless accident damage has occurred. In most cases, frame replacement is the only satisfactory remedy for such damage. A few frame specialists have the jigs and other equipment necessary for straightening frames to the required standard of accuracy, but even then there's no sure way of determining exactly how much the frame was overstressed.

2 After the machine has accumulated a lot of miles, it's a good idea to examine the frame closely for cracks at the welded joints. Rust can also cause weakness at the joints. Loose engine mount bolts can cause enlargement of the holes and cracks at the mounting tabs. Minor damage can often be repaired by welding, depending on the extent and nature of the damage.

3 Remember, an out-of-alignment frame will cause handling problems. If misalignment is suspected as the result of an accident, you will have to strip the machine completely so the frame can be thoroughly checked.

3 Instrument cluster – removal and installation

1 Before either of the instruments can be removed, the drive cable must be disconnected. Unscrew the coupling nut from the underside of each instrument and detach the cables. Some later models have an electronic tachometer so there will be no cable to unscrew. Instead, disconnect the wire.

2 Detach the baseplate nuts, and withdraw the baseplate and instrument housing to gain access to the bulbs. Pull the bulbholders from the base of the instrument. Early models use bayonet fitting bulbs which are twisted counterclockwise to remove, and later models use capless bulbs which are pulled from position.

Chapter 5 Frame and suspension

4.3 Typical rear shock absorber (early models) – exploded view

1. Mounting stud nut
2. Washer
3. Stud cover
4. Cup washer
5. Rubber bushing
6. Spring keeper
7. Shroud (long)
8. Shroud (short)
9. Washer
10. Spring
11. Seal washer (low ground clearance)
12. Adjusting cup
13. Adjustment cam
14. Damper unit

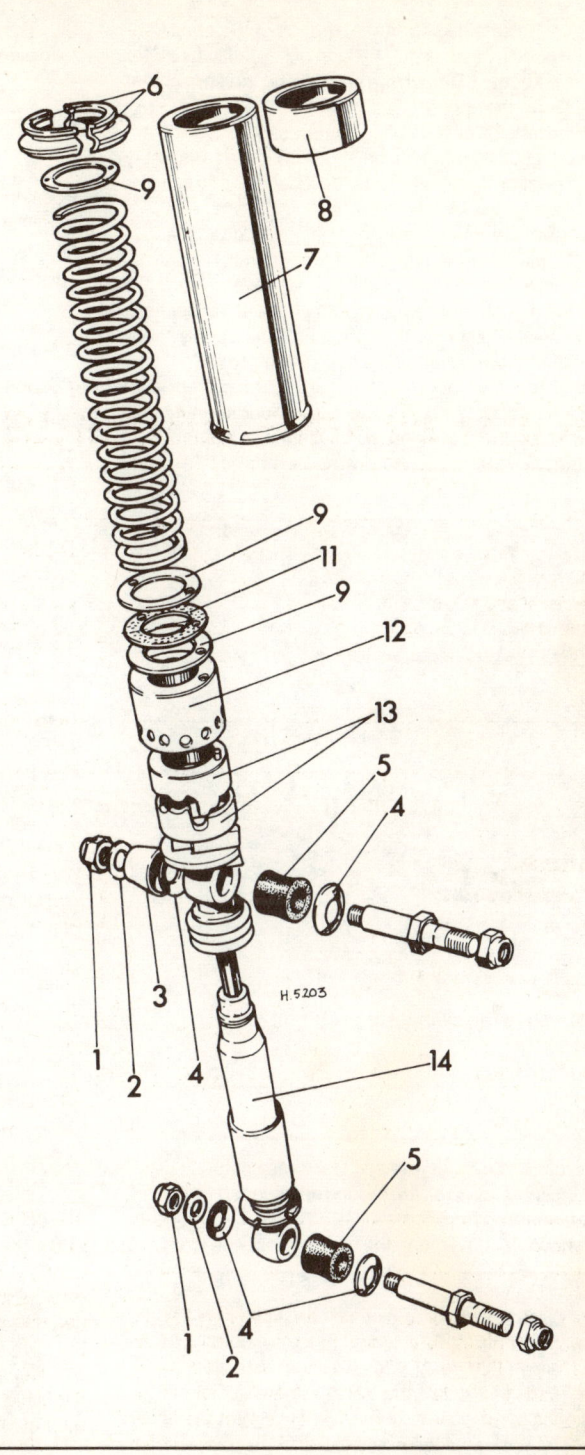

3 Apart from defects in either the drive or the drive cable, a speedometer or tachometer that malfunctions is difficult to repair – install a new or known good used one.

4 Shock absorbers – removal and installation

Refer to illustration 4.3

1 Raise the motorcycle and support it securely on blocks with the rear wheel slightly off the ground. If blocks aren't available, remove one shock absorber at a time. The other shock will hold the rear of the motorcycle in place.

2 Remove the passenger footpegs on pre-1979 models.

3 Remove the mounting nuts (later models use bolts and nuts) securing the shock absorbers to the swingarm and frame **(see illustration)**. Carefully separate the shock absorbers from the frame and swingarm. When the shock absorbers are removed, the rear wheel and swingarm will drop to the ground if not supported.

4 The rear shock absorbers can't be disassembled or rebuilt. If they're worn out or leaking fluid, they must be replaced with new ones.

5 It is possible to replace the springs or transfer the springs to new shock absorbers. Compress the spring with an approved spring

Chapter 5 Frame and suspension

compressor and remove the spring keeper(s) at the top of the shock absorber assembly (early models have a split key type of keeper, while later models have a one-piece keeper). Detach the shroud and the spring seat (if equipped) from the top of the spring.

6 Carefully release the tension on the spring compressor until the compressor can be removed. Lift the spring off the shock absorber assembly and remove all of the mounting hardware. Note the locations of the hardware to simplify reassembly.

7 Clean and inspect all of the components for wear and damage, especially the rubber parts. Replace any defective components with new parts.

8 Reassemble the shock absorber and spring in the reverse order of disassembly. Be sure the adjustment cams on both shock absorbers are positioned at the same level. Compress the spring until the spring keeper(s) can be installed, then slowly release the spring compressor.

9 Have an assistant lift the rear wheel and swingarm to the correct level to install the first shock absorber. With one shock absorber in place, the swingarm should be in the correct position to install the other shock.

10 Tighten the nuts (and bolts) securely and lower the motorcycle off the support blocks.

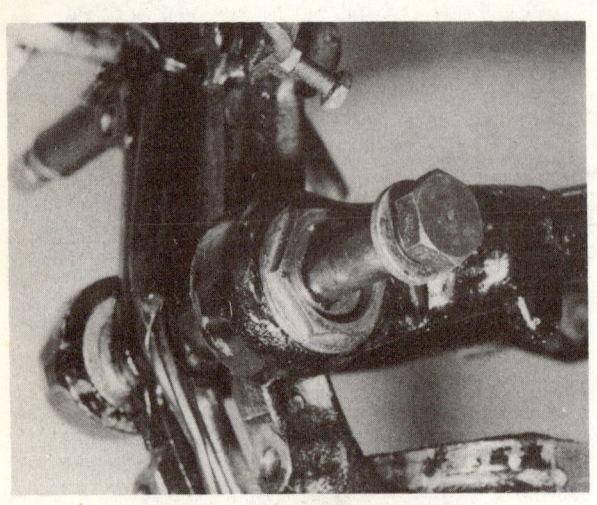

6.10a Remove the swingarm pivot bolt . . .

5 Swingarm bearings – check

1 The swingarm pivots on bearings, which rarely wear out.
2 Raise the rear of the motorcycle off the ground and support it securely on blocks. Grab the swingarm in front of the axle and move it from side-to-side. If there's any discernible movement, the bearings may be worn out.

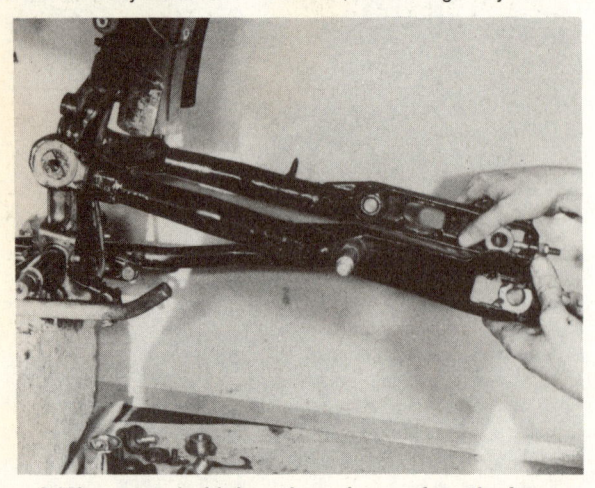

6.10b . . . and withdraw the swingarm from the frame

3 The swingarm should be removed and inspected as described in the following Section.

6 Swingarm – removal, inspection and installation

1 Block the motorcycle up so the rear wheel is slightly off the ground. Position the blocks under the frame.
2 Disconnect the final drive chain by removing the master link. Move the chain out of the way of the swingarm, but don't remove it from the front sprocket. On belt drive models, slacken the axle nut and belt adjusters to allow the wheel to be moved fully forwards in the swingarm; slip the belt off the rear sprocket.
3 Remove the rear wheel as described in Chapter 6.

Removal
Refer to illustrations 6.10a and 6.10b

1970 through 1978
4 Remove the rear brake assembly from the swingarm as described in Chapter 6.
5 Remove the rear exhaust pipe and muffler. Some models require the removal of the complete exhaust system.
6 Unscrew the passenger footpegs to release the lower shock absorber mounts. Support the swingarm and remove the shock absorbers from it.

1979-on
7 Detach the rear disc brake caliper from the swingarm. You don't have to disconnect the brake line or drain the brake fluid. Hang the brake caliper from the frame with a piece of wire, away from the swingarm. **Warning:** *Do not let the caliper hang by the brake line.*
8 Support the swingarm and remove the lower shock absorber mounting bolts.
9 Remove the chain guard from the swingarm on 1979 through 1981 models. On 1982 models, remove both the front and rear chain guards. On 1983 and later chain drive models, remove only the rear chain guard. On all belt drive models, remove the rear belt guard and the debris deflector.

All models
10 On 1970 through 1985 models, unscrew the pivot bolt. On 1986 and later models, hold the pivot bolt with a wrench and remove the Allen-head bolt from the left end of the pivot bolt. Support the swingarm and pull out the pivot bolt **(see illustration)**. Pull the swingarm to the rear, away from the frame **(see illustration)**.

Inspection
Refer to illustrations 6.11, 6.12, 6.14, 6.15a, 6.15b and 6.15c

1970 through 1981
11 Disassembly of the swingarm should be done on a clean workbench. Remove the screw securing the lockplate on the right side of the swingarm **(see illustration)**.

6.11 Remove the screw securing the lockplate to the bearing locknut (early models only)

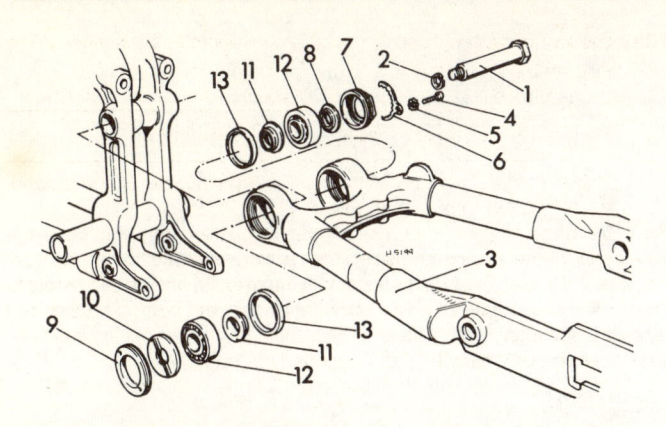

Chapter 5 Frame and suspension

6.12 Typical swingarm components – exploded view (1970 through 1981 models shown)

1. Pivot bolt
2. Bearing lock washer
3. Swingarm
4. Screw
5. Lock washer
6. Lockplate
7. Bearing locknut (right-hand side)
8. Outer spacer
9. Bearing locknut (left-hand side)
10. Pivot nut
11. Bearing spacer
12. Tapered roller bearing
13. Dust shield

6.14 Carefully pry out the bearing dust shield

12 Remove the bearing locknut and the outer spacer behind it **(see illustration)**. This will give access to the right-hand bearing.

13 Working from the left side of the swingarm, loosen the bearing lock ring with a punch until it can be unscrewed by hand.

14 Remove the pivot nut. This will give access to the left-hand bearing. Pry the bearing dust shields out from the inside of the swingarm **(see illustration)**.

15 Working inside the swingarm, drive or press out the bearing and bearing spacer by applying pressure to the spacer **(see illustrations)**. Although it isn't necessary unless the bearing is being replaced, the bearing outer race can be pressed out from the inside **(see illustration)**.

16 If the bearing is replaced, the race should be replaced with it as a matched set.

1982-on

17 The bearing components should not be interchanged. If any bearings are defective, the complete bearing assembly should be replaced with a new one.

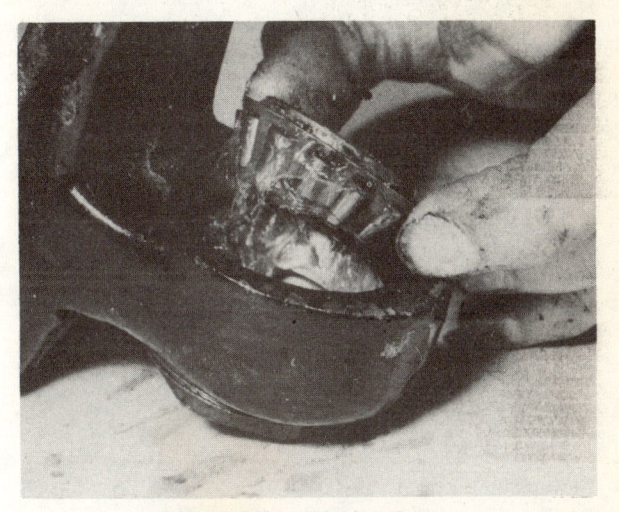

6.15a Lift out the bearing . . .

6.15b . . . and the bearing spacer

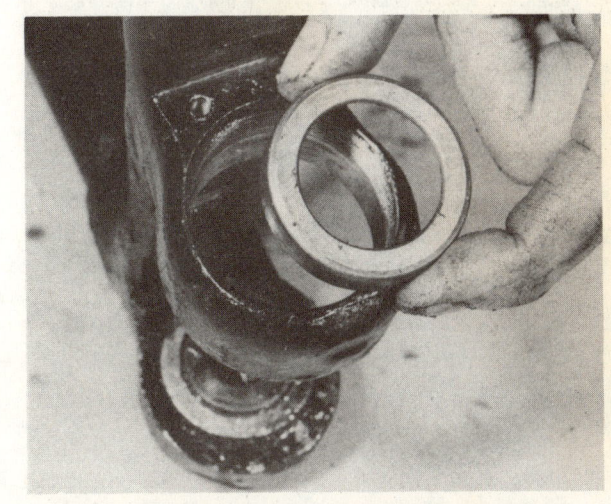

6.15c The bearing outer race must be pressed out

Chapter 5 Frame and suspension

18 You'll have to take the swingarm to a Harley-Davidson dealer or an automotive machine shop to have the bearing races or bushings pressed out and the new bearing races or bushings installed. Make sure the spacer is installed between the inner bearing races or bearing failure could occur.

All models
19 Clean and examine the tapered roller bearings for signs of wear and damage (which should be evident). If there is any doubt about the suitability of the bearings for further use, replace them with new ones. Most bearings fail as a result of pitting, which occurs on the inner and outer races.

Installation
1970 through 1981
20 Reassemble the swingarm in the reverse order of disassembly. Make sure the bearings are liberally packed with grease. When inserting the bearing spacers, the shouldered portion must face IN. The wide side of each bearing outer race must face OUT. Tighten the right-hand bearing locknut first, then loosen it one-half turn. On the left-hand side, insert the pivot bolt nut so it will align with the area in the back of the primary chaincase that it's recessed into after the swingarm has been replaced. Install and tighten the lock ring and stake it in three places.

21 After the pivot bolt has been installed, you'll have to pre-load the tapered roller bearings. This is done by attaching a spring balance to the extreme end of one of the arms, before the shock absorbers are installed. Take the reading (in pounds) necessary to make the swingarm turn on the pivot. Loosen the lock plate of the right-hand bearing locknut and tighten the locknut until the drag on the spring balance is increased by about two pounds. Install the lock plate and tighten the retaining screw.

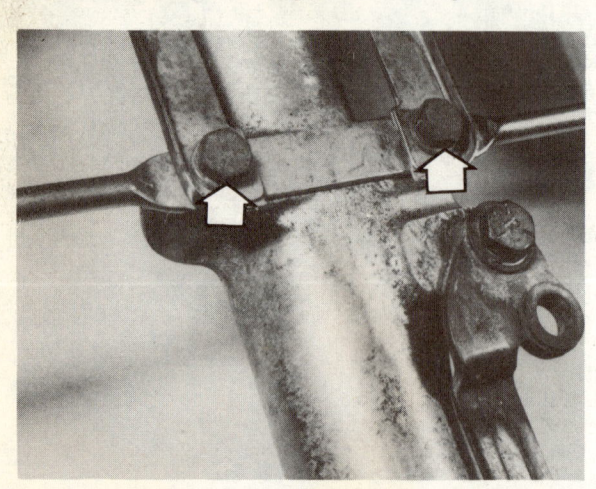

1982-on
22 Lubricate the components with grease as they're assembled.
23 The chamfered end of the spacer in the bushing (left-hand side) must face out. Insert the main pivot bolt in through the swingarm and, having applied a drop of thread-locking compound to its threads, install the Allen-head bolt in the end of the pivot bolt.
24 Tighten the Allen-head bolt to the specified torque.

All models
25 Install the shock absorbers and the rear wheel brake components.
26 Install the rear wheel and connect the drive chain or slip the belt over the sprocket (as applicable). Be sure the closed end of the master link faces the direction of chain travel. Adjust the chain or belt (as applicable) as described in Chapter 1.
27 Readjust the brake on drum brake models and make sure all of the nuts and bolts are tight.

---

7 Forks – removal and installation

Removal
Refer to illustrations 7.4, 7.6, 7.8, 7.9, 7.10a, 7.10b, 7.11 and 7.13

1 The forks can be removed very simply by separating the fork tubes from the yokes. Don't disassemble the steering head components unless the bearings require attention (Section 9).
2 Raise the front of the machine until the front wheel is off the ground. Support the machine securely on blocks. This will require a little forethought and care because the lower frame tubes are close together and the machine won't be very stable unless it has some side support.
3 Remove the front wheel as described in Chapter 6.
4 Remove the front fender, which is attached to the inside of the fork legs by two bolts on each side **(see illustration)**.
5 On models with front disc brakes, remove the clamp securing the hydraulic lines to the fork legs. Disconnect the caliper(s) from the fork leg(s) and tie the caliper(s) to the frame, out of the way.
6 Remove the finned cap that retains the handlebars. It's held in place with four Allen-head bolts **(see illustration)**.
7 The handlebars can usually be moved out of the way without disconnecting any of the control cables or electrical wires. If necessary, disconnect the control cables and the electrical wires from the handlebar controls, or remove the control cables and electrical wires with their respective controls still attached. The shape of the handlebars and the length of the cables will dictate the best approach. It's a good idea to disconnect the battery before starting this operation. Be sure to keep the front brake master cylinder (on front disc brake models) level to keep air from entering the system.
8 Unscrew the drive cables from the speedometer and tachometer. Remove the instruments and mounts by unscrewing the bolt that retains each mount to the top of each fork leg **(see illustration)**. Disconnect the electrical wires and store the instruments in a safe place.

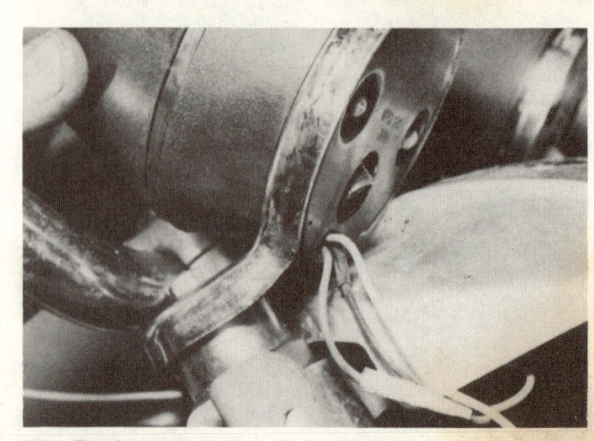

7.4 The front fender is attached to the inside of the forks with two bolts (arrows) on each side

7.6 Remove the four Allen-head screws (arrows) securing the handlebar clamp

7.8 The instruments are bolted to the top of each fork leg

Chapter 5 Frame and suspension

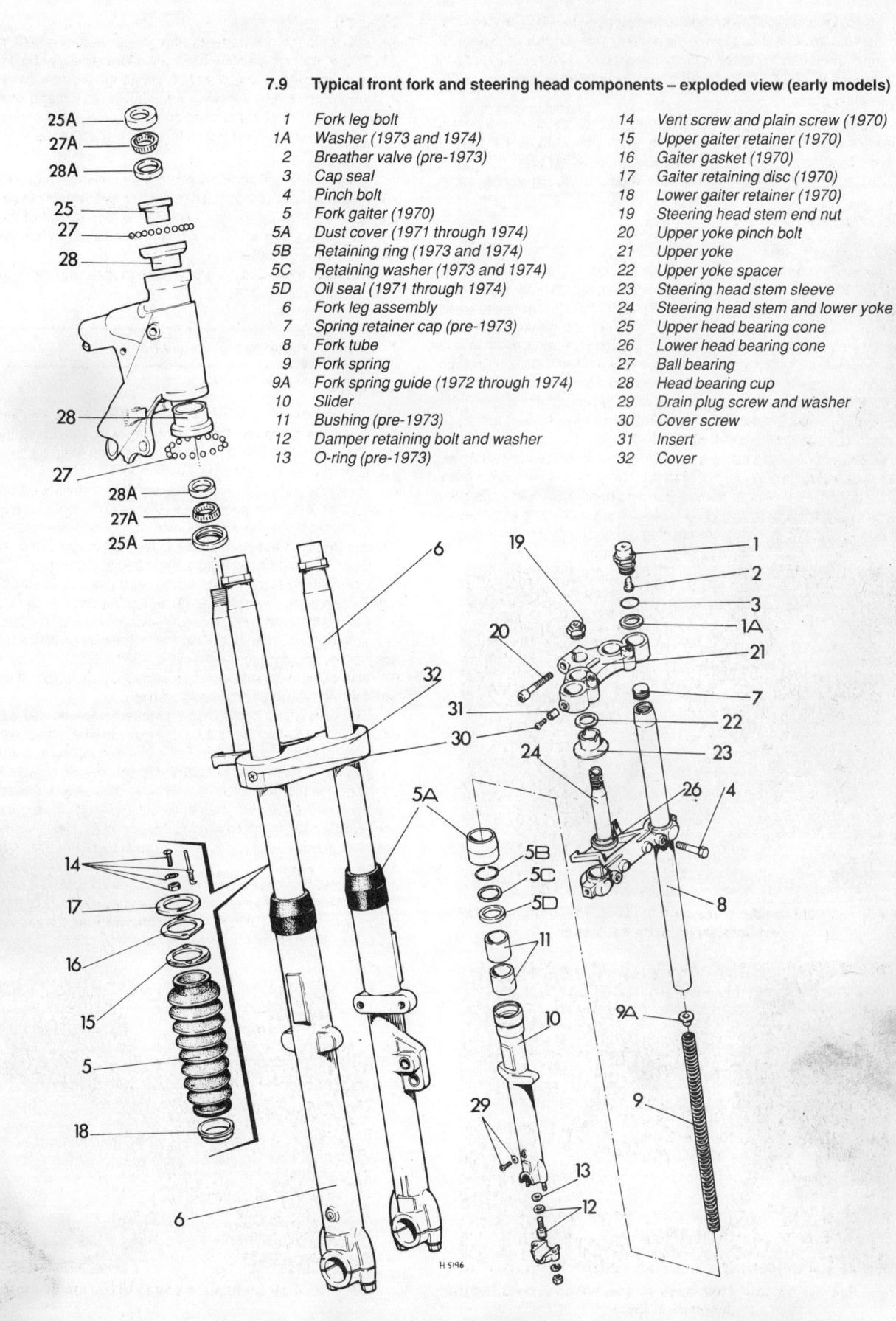

7.9 Typical front fork and steering head components – exploded view (early models)

1	Fork leg bolt	14	Vent screw and plain screw (1970)
1A	Washer (1973 and 1974)	15	Upper gaiter retainer (1970)
2	Breather valve (pre-1973)	16	Gaiter gasket (1970)
3	Cap seal	17	Gaiter retaining disc (1970)
4	Pinch bolt	18	Lower gaiter retainer (1970)
5	Fork gaiter (1970)	19	Steering head stem end nut
5A	Dust cover (1971 through 1974)	20	Upper yoke pinch bolt
5B	Retaining ring (1973 and 1974)	21	Upper yoke
5C	Retaining washer (1973 and 1974)	22	Upper yoke spacer
5D	Oil seal (1971 through 1974)	23	Steering head stem sleeve
6	Fork leg assembly	24	Steering head stem and lower yoke
7	Spring retainer cap (pre-1973)	25	Upper head bearing cone
8	Fork tube	26	Lower head bearing cone
9	Fork spring	27	Ball bearing
9A	Fork spring guide (1972 through 1974)	28	Head bearing cup
10	Slider	29	Drain plug screw and washer
11	Bushing (pre-1973)	30	Cover screw
12	Damper retaining bolt and washer	31	Insert
13	O-ring (pre-1973)	32	Cover

Chapter 5 Frame and suspension

7.10a Loosen the pinch bolts under the cover in the lower yoke (remove the screws so the cover can be raised to gain access to the pinch bolts)

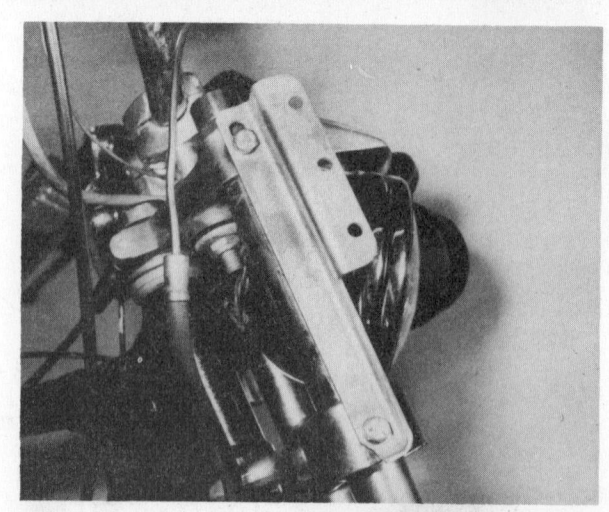

7.10b Remove the bracket (if equipped) from the fork yokes

7.11 Drive the forks through the yokes with a hammer and wood block

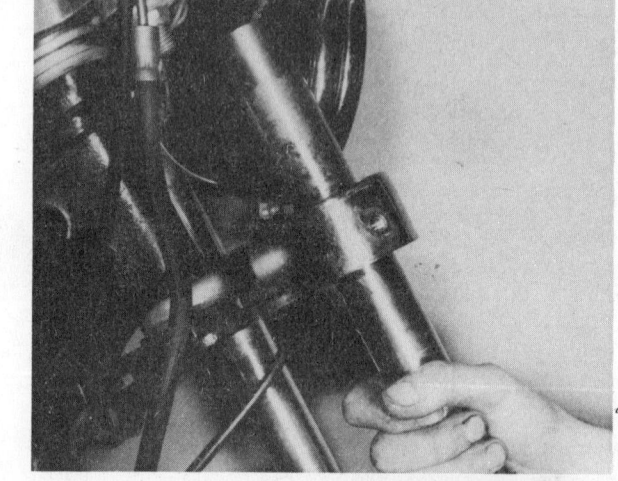

7.13 Pull the fork legs through the lower yoke (don't compress the tube and slider or oil may spurt out)

9 Unscrew the fork leg bolt from the top of each fork leg **(see illustration)**, but don't remove it at this point.
10 Loosen the pinch bolts in the lower fork yoke, under the plated cover **(see illustration)**. Remove the two bolts that secure the mounting bracket between the upper and lower yokes and remove the bracket (not used on all models) **(see illustration)**. On later models, remove the upper yoke fork pinch bolts.
11 The upper ends of the fork legs are tapered and mate with a matching taper in the upper fork yoke. To break the taper, give the loosened fork leg bolts a sharp rap with a hammer and block of wood **(see illustration)**. Do not use excessive force or the internal threads in the fork tube may be damaged. If the fork legs are a tight fit, try tapping up on the upper fork yoke at the same time.
12 After the forks are free from the upper yoke, remove the fork leg bolts from the top of each fork leg (except 1995 and later XL 1200S models).
13 Remove the fork legs from the lower yoke by pulling only on the fork tubes **(see illustration)**. **Caution:** *Do not compress the fork tube and slider or oil may spurt out of the open end of the tube. Keep the fork leg upright until the bolts (caps) are installed (also, store them in an upright position).*

Installation

14 Install the fork legs in the steering head yokes by reversing the removal procedure. Where tapered fork tubes are fitted, ensure that they mate fully with the matching taper in the fork yoke. Later models have parallel-sided fork tubes, which must be installed to a height of 0.42 to 0.50 inch (11 to 13 mm) from the top of the top bolt to the fork yoke surface. Be sure to push only on the fork tubes as the fork legs are slipped into the yokes and don't compress the fork tube and slider until the bolts (caps) are in place.
15 Tighten the lower yoke pinch bolts temporarily, then add the specified amount of fork oil to each fork leg (see Chapter 1).
16 Install the fork leg bolts, loosen the lower yoke pinch bolts, tighten the fork leg bolts securely, then tighten the lower yoke pinch bolts (and on later models, the upper yoke fork pinch bolts) to the specified torque (where given).
17 Install the instruments and handlebars, then attach the fender to the forks. Refer to Chapter 6 and install the front wheel and related components.

8 Forks – disassembly, inspection and reassembly

1 Remove the fork legs from the steering head yokes as described in Section 7.
2 Always disassemble one fork leg at a time to avoid mixing up parts.

Chapter 5 Frame and suspension

8.4 Remove the fork spring and guide (1972 through 1974 models) from the fork tube

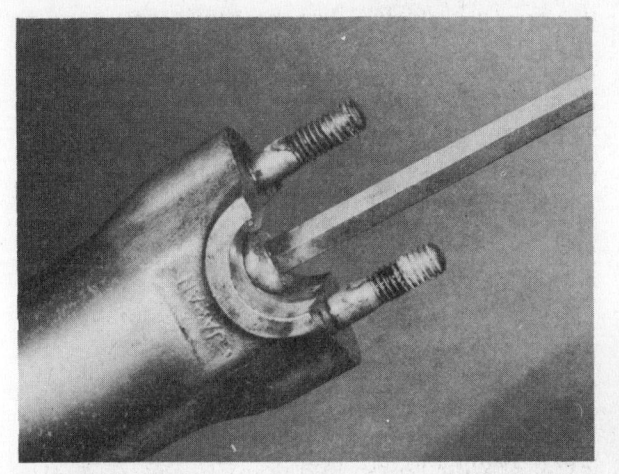

8.5 Remove the damper retaining bolt from the bottom of the fork slider

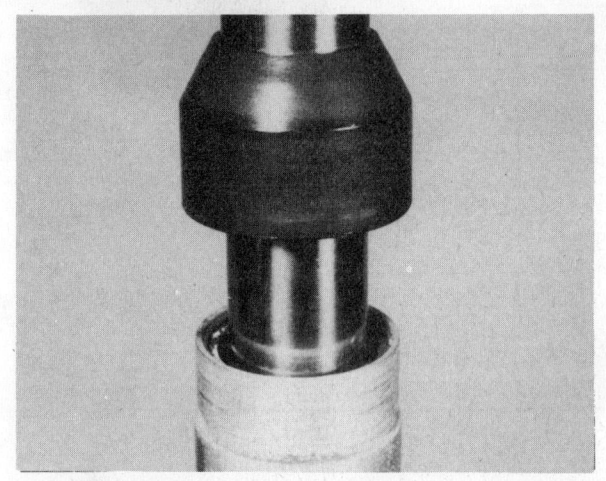

8.6 Separate the dust cover from the fork leg

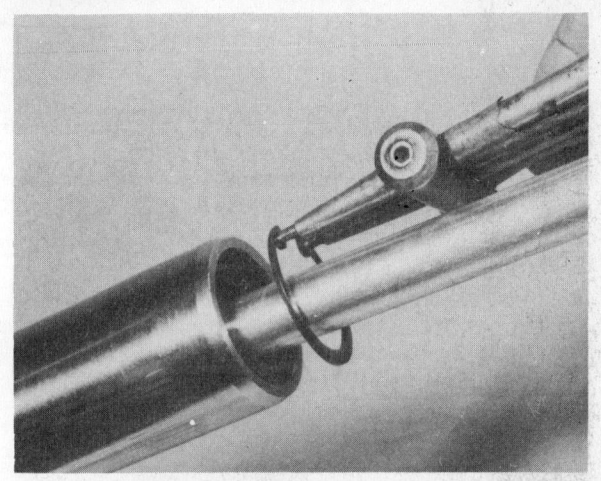

8.7 The damper unit is held in the slider with a snap-ring

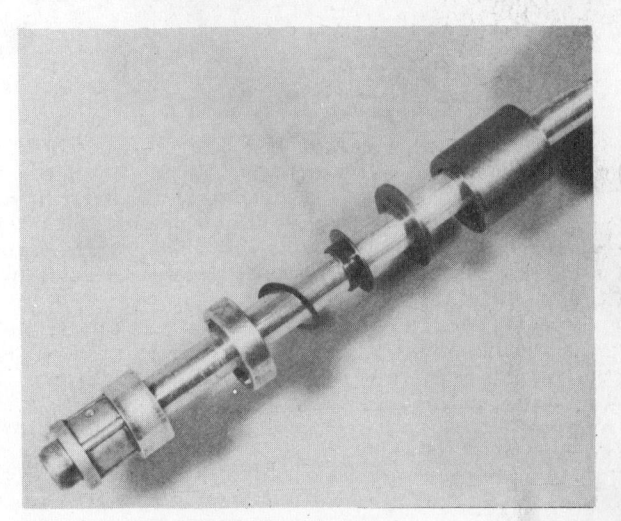

8.8a The damper unit will pull out as a complete assembly

Disassembly (except 1995-on XL 1200S models)

All models

Refer to illustrations 8.4, 8.5 and 8.6

3   Turn the fork leg upside-down and drain the oil into a container. Keep in mind that when the fork is inverted the spring will slide out – don't let it fall. Compress the fork tube and slider a few times to ensure that it drains completely.
4   When the majority of the oil has drained, slide out the spring (and spring guide on 1972 through 1974 models) and wipe off as much of the oil as possible **(see illustration)**.
5   Secure the slider in a vise equipped with soft jaws. Clamp it very lightly – just enough to prevent it from rotating while the Allen-head screw is removed from the end of the slider **(see illustration)**. You may find that once loosened, the damper tube turns inside the fork tube, preventing removal of the Allen screw. Try installing the fork spring and cap and compressing the fork to apply pressure to the damper rod head to hold it still. If this fails, obtain either the Harley-Davidson service tool, or a similar length of rod which will pass down through the fork tube and locate in the head of the damper tube; with the exposed section of the tool held tightly, unscrew the Allen screw.
6   Withdraw the dust cover or gaiter from the slider **(see illustration)**. Separate the fork tube from the slider, on 1984-on models noting the procedure described in Steps 12 and 13.

1970 through 1974

Refer to illustrations 8.7, 8.8a, 8.8b, 8.8c, 8.9a and 8.9b

7   To detach the damper unit, remove the snap-ring from the lower end of the fork tube **(see illustration)**.
8   The damper unit is then free to be drawn out. It's a built-up assembly that can be disassembled if parts are worn and require replacement **(see illustrations)**.
9   There's an oil seal in the upper portion of each slider. If the retaining

8.8b Front fork damper unit – 1970 through 1972 models

1. Snap-ring
2. Lower valve body
3. Valve washer
4. Upper body valve
4A. Spring (1971 only)
5. Piston retaining ring
6. Piston
7. Damper tube
8. Fork tube

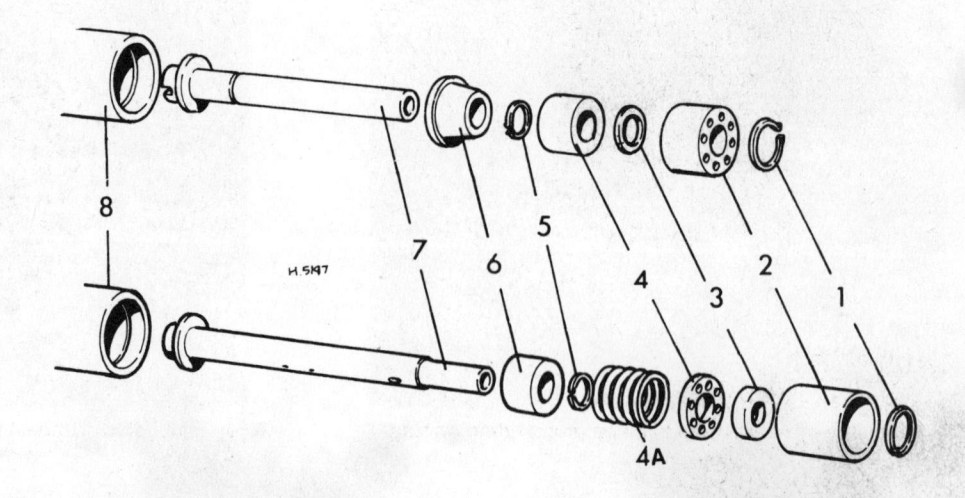

ring is removed, the oil seal can be pried out of position **(see illustrations)**. Do not disturb it unless it has to be replaced or if access is required to the two inner bushings (pre-1973 models), which will have to be replaced when play develops in the fork legs.

1975 through 1983
Refer to illustration 8.10

10 Invert the fork tube to remove the damper. Remove the fiber wear rings from the grooves at the top of the damper **(see illustration)**. Invert the slider to remove the damper tube sleeve after removing the boot from the top of the lower fork leg.

11 If the fork was leaking, replace the oil seal. Remove the retaining ring from the fork leg and carefully pry out the seal. The seal may be easier to remove if the fork leg is heated slightly.

1984-on
Refer to illustration 8.12

12 Having pulled the dust cover off the slider, carefully hook the dust seal out of its location in the lower leg. Remove the retaining ring from its groove below **(see illustration)**.

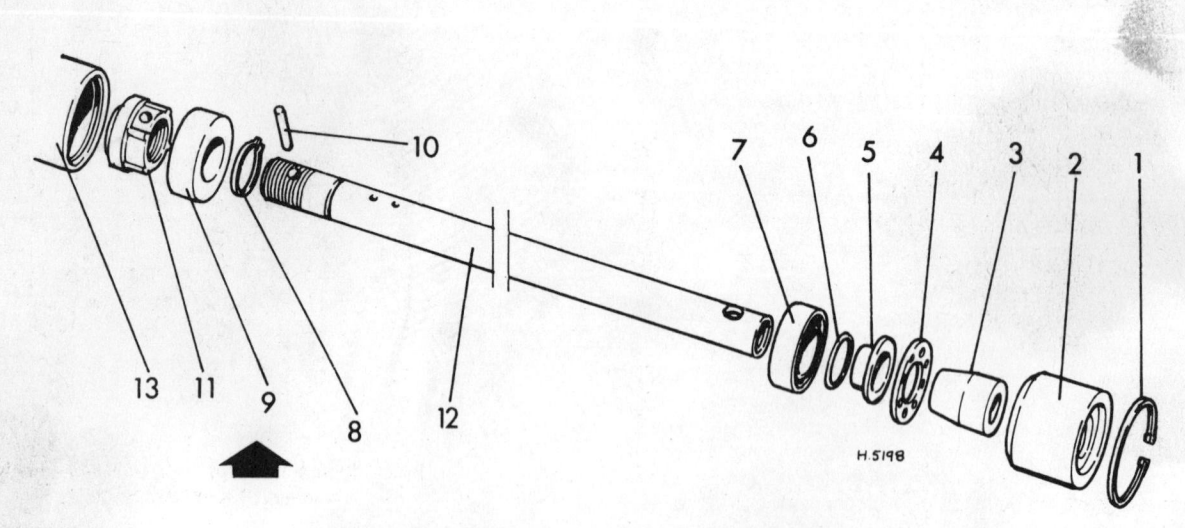

8.8c Front fork damper unit – 1973 and 1974 models

1. Snap-ring
2. Lower piston
3. Lower stop
4. Orifice washer
5. Valve
6. Spring washer
7. Valve body
8. Retaining ring
9. Upper piston
10. Roll pin
11. Upper stop
12. Damper tube
13. Fork tube

Chapter 5 Frame and suspension

147

8.9a The oil seal is held in place with a retaining ring

8.9b Don't disturb the oil seal unless it's going to be replaced with a new one

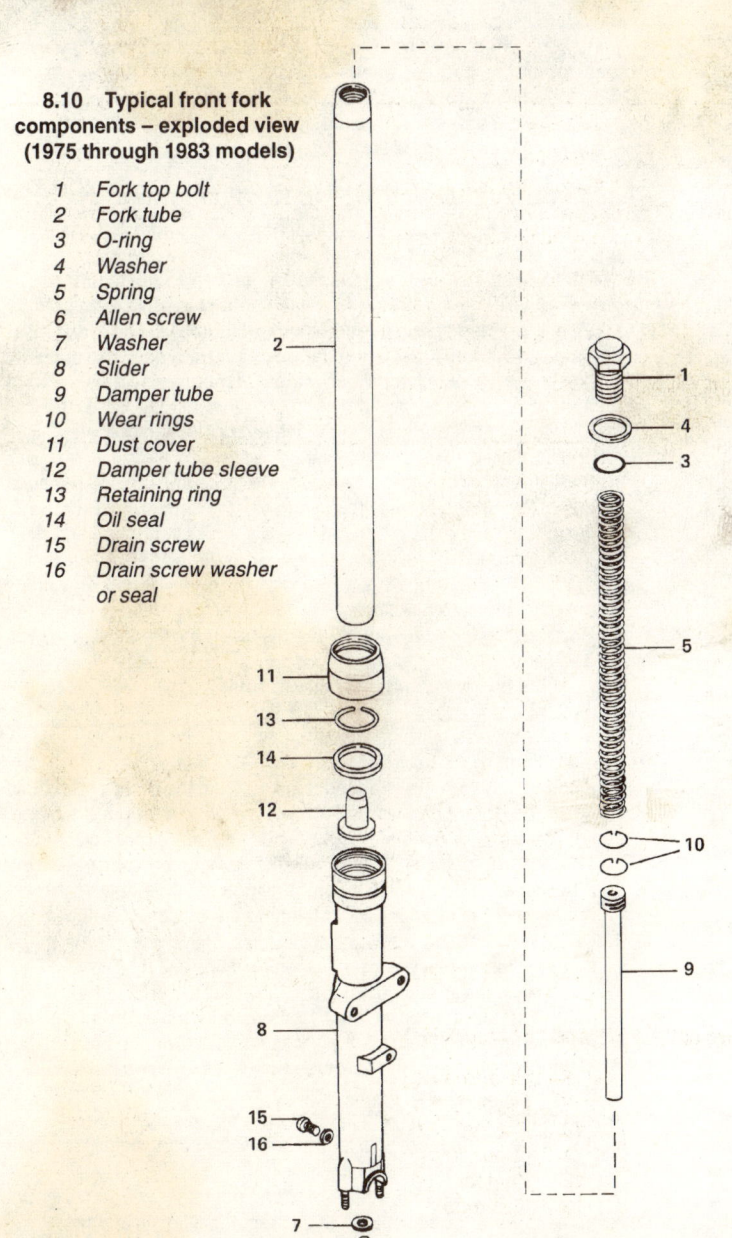

8.10 Typical front fork components – exploded view (1975 through 1983 models)

1. Fork top bolt
2. Fork tube
3. O-ring
4. Washer
5. Spring
6. Allen screw
7. Washer
8. Slider
9. Damper tube
10. Wear rings
11. Dust cover
12. Damper tube sleeve
13. Retaining ring
14. Oil seal
15. Drain screw
16. Drain screw washer or seal

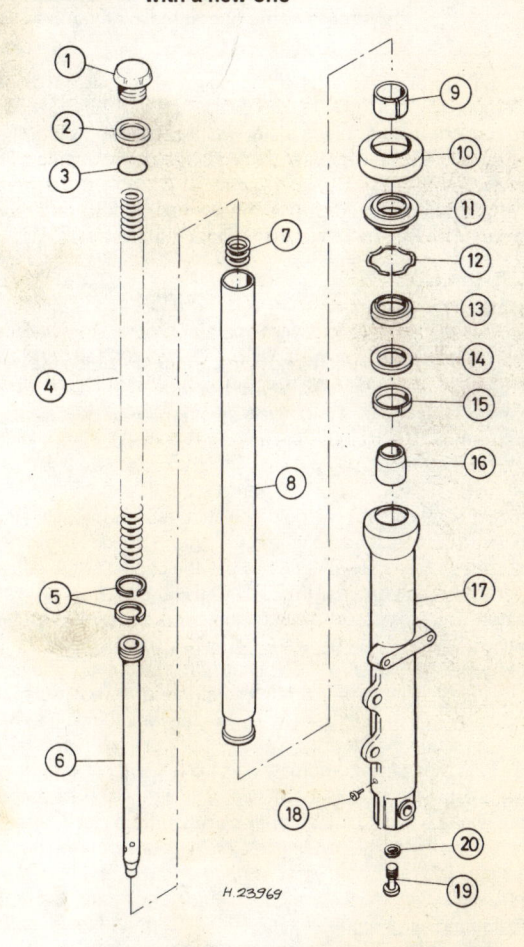

8.12 Typical front fork components – exploded view (1984-on 883 models)

1. Fork top bolt
2. Washer (where fitted)
3. O-ring
4. Fork spring
5. Wear rings
6. Damper tube
7. Rebound spring
8. Fork tube
9. Fork tube bushing
10. Dust cover
11. Dust seal
12. Retaining clip
13. Oil seal
14. Washer
15. Slider bushing
16. Damper tube sleeve
17. Slider
18. Oil drain screw
19. Allen screw
20. Washer

8.17 After you've unscrewed the fork cap, loosen the locknut, then remove the special tool or washer

13 Using a slide-hammer action, pull the fork tube out of the slider. This will cause the fork tube bushing (which has a larger OD than the ID of the slider bushing) to displace the oil seal, washer and slider bushing as the two components are separated. Slide these components off the top of the fork tube.
14 Invert the fork tube and tip out the damper tube and rebound spring. Invert the slider and tip out the damper tube sleeve.

Disassembly (1995-on XL 1200S)

Refer to illustration 8.17

15 Stand the fork upright and hold the inner fork tube with one hand. Unscrew the fork cap bolt from the inner tube, then lower the tube into the outer tube. The upper part of the spring will now be exposed.
16 Remove the stopper ring from the groove at the top of the fork, then remove the rebound adjuster and adjuster plate from the fork cap bolt.
17 To undo the fork cap bolt from the damper rod, you'll need to loosen the hex portion of the cap bolt. This requires compressing the spring and holding it compressed, so there will be room to place wrenches on the cap bolt hex and the flats at the bottom of the cap bolt (the flats are in the rebound adjuster portion of the cap bolt assembly). The spring is held in the compressed position by a slotted plate which is slipped under the locknut (see illustration). There's a special Harley tool available to compress the spring (HD-41549). If you don't have the special tool you can fabricate a substitute; just be sure it grips the spring securely, so it won't slip out. You'll probably need an assistant to slip the holder into position while you hold the spring compressed.
18 With the spring compressed, place a socket or wrench on the hex at the top of the cap bolt. Place an open-end wrench on the flats at the bottom of the cap bolt. Hold the open-end wrench (don't let it turn) and unscrew the hex portion of the cap bolt. **Note:** *Don't unscrew the rebound adjuster (the part you're holding with the open-end wrench) from the damper rod unless the damper or the rebound adjuster needs to be replaced.*
19 Remove the upper spacer, spring collar and lower spacer from the top of the spring. Take the spring out of the fork.
20 Place the fork in a vise, clamped just tight enough to keep the fork tube from turning. Unscrew the Allen bolt from the bottom of the fork leg, then pull the damper out of the fork tube.
21 Refer to Steps 12 and 13 above to complete disassembly of the fork.

Inspection

22 On models fitted with fork bushings, wear will taken by the bushings themselves, otherwise it will be present on the working surfaces of the fork tube and slider. Wear can be felt as shuddering when the front brake is applied and the increased amount of play can be detected by pulling and pushing on the handlebars when the front brake is applied hard (do not confuse this with steering head play though).

23 On models through 1974 replacing the bushings is a difficult operation, as the new bushings have to be reamed to size after installation; it is advised that this task is left to a Harley-Davidson dealer.
24 On 1984-on models the slider bushings is displaced during dismantling and a new one installed relatively easily. Only remove the fork tube bushing if it requires replacement – pry it apart at its split sufficiently to ease it off the end of the fork tube and install the new one by easing it over the end of the fork tube.
25 New oil seals must be fitted in the slider every time the fork is dismantled. A wire retaining clip secures each seal in the top of the slider. Take note of any washers present and install them in their original position. Apply fork oil to the oil seal lips to prevent damage during fork reassembly.
26 If fork damping action decreases, the damper assembly (pre-1975 models) should be examined carefully and the damper pistons replaced. On 1975-on models replace the damper tube wear rings if damping problems are evident and check that all holes in the damper tube are clear.
27 Before reassembling the forks, check the sliding surfaces of the fork tubes. If they're scuffed, scored or badly worn, they should be replaced, in conjunction with the fork bushings (if equipped).
28 It's not possible to straighten forks that have been bent in an accident, even if a repair service is available; there's no way of knowing whether the parts concerned have been overstressed. The tubes should always be checked for straightness by rolling them on a flat surface.
29 The fork yokes are also liable to twist or distort in an accident. As in the case of the fork tubes, they should be replaced and not repaired, especially since they are far more difficult to align correctly.

Reassembly (except 1995-on XL1200S models)

30 Reassembly is the reverse of disassembly. Be sure to install new O-rings and seals and coat them with fork oil before assembly.
31 Refer to Section 7 and reinstall the forks (don't forget to add oil).
32 When installing the 1984-on fork, take note of the following points.
  a) Fit the rebound spring over the damper tube and slide the damper tube into the fork tube; fit the damper tube sleeve over its exposed end and then install the slider without disturbing the damper components. Use a slim rod passed down through the top of the fork tube to hold the damper rod still while its retaining bolt is installed and tightened.
  b) Slide the slider bushing down over the fork tube and use a tubular drift of exactly the correct inside and outside diameter to drive it into its housing in the slider (an old bushing works well as an adaptor). Install the washer and drive the new oil seal into position in a similar manner. Retain all components with the retaining clip and press the dust seal into position.
33 Reassembly is the reverse of the disassembly steps, with the following additions:
  a) Coat the bushings and oil seal with Harley Type E fork oil before assembly.
  b) Install the guide bushing with its slit to one side of the fork, not facing front or rear.
  c) If you unscrewed the rebound adjuster from the damper rod, hold down the detent ball with a thumb and turn the rebound adjuster knob all the way counterclockwise, then turn it 13 clicks clockwise. Thread the locknut all the way onto the damper rod, then thread the rebound adjuster all the way on with fingers only. Tighten the locknut against the adjuster without letting the adjuster rotate.
  d) Use a new sealing washer on the Allen bolt and tighten it to the torque listed in this Chapter's Specifications.
  e) Pour half of the specified amount of fork oil into the fork, then slowly pump the damper rod at least 10 times and leave it compressed.
  f) Add the remaining fork oil, then measure the distance from the top of the oil to the top of the fork tube with a stiff tape measure. Add or remove oil to obtain the level listed in this Chapter's Specifications.
  g) Install the fork spring with its tight coils downward.
  h) Install the flat spacers above and below the spring collar with their sharp edges toward the collar.
  i) Tighten the fork cap onto the rebound adjuster, then into the fork tube, to the torques listed in this Chapter's Specifications.
  k) Use new O-rings on the rebound adjuster and coat them with fork oil.

Chapter 5 Frame and suspension

10.4a Pull the passenger strap forward (if equipped) to release the seat on later models

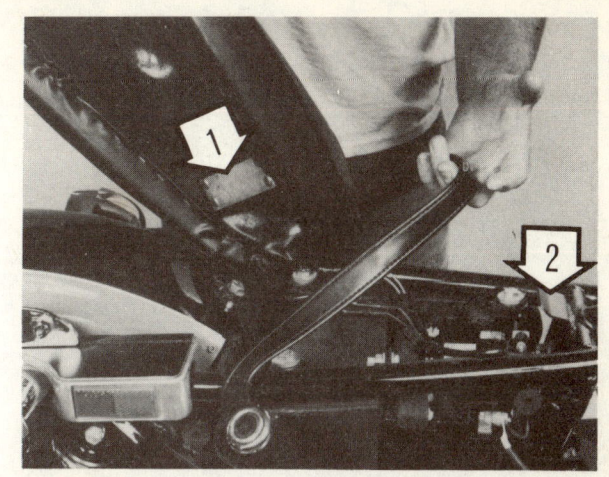

10.4b On later models, the bracket under the seat (1) slides under the protrusion on the frame (2)

9 Steering head bearings – maintenance

1 If the steering head bearing check (Chapter 1) reveals excessive play in the steering head bearings, the entire front end must be disassembled and the bearings and races (called cups/cones on pre-1975 models) replaced with new ones.
2 Refer to Chapter 6 and remove the front wheel.
3 Remove the forks from the steering head yokes as described in Section 7.
4 Remove the two bolts that secure the small cowl and warning light console to the upper fork yoke. The cowl and headlight can be suspended without disconnecting the wires.
5 Loosen the pinch bolt through the rear of the upper fork yoke and remove the large nut or bolt and washer from the top of the steering head.
6 Pre-1975 models have uncaged ball bearings in the steering head, so precautions should be taken to catch the bearings as they fall out. As the upper fork yoke is being removed, the steering head bearings on pre-1975 models will begin to fall out. Usually only the lower bearings will be disturbed and the bearings in the top cup usually stay in place.
7 Support the lower yoke while the upper yoke is removed, to prevent it from falling out of the steering head.
8 Clean and examine the cups and cones (pre-1975 models) or the races and bearings (1975 and later models). They should have a polished appearance with no dents or tracks if they are going to be reused.
9 If replacement is necessary, the bearings and races (cups/cones on early models) should be replaced as matched sets. The cups or races are a tight fit in the steering head and will have to be driven out. The lower head bearing cone is a tight fit on the base of the steering stem, but can usually be removed with the careful use of a chisel.
10 Clean and examine the bearings, which should also be highly polished with no signs of wear, pitting or surface cracks. If any of the ball bearings (used on pre-1975 models) are unusable, replace the entire set. Again, if the bearings are replaced, the bearing cups or races should also be replaced as a matched set.
11 When reassembling the steering head bearings, pack the cups or races and the bearing cones with grease. This will simplify the assembly of the ball bearing steering head used on pre-1975 models. Note that on pre-1975 models, the bearing cup shouldn't be full of bearings. there should be room to insert an extra ball bearing, a necessary arrangement to prevent the bearings from skidding on each other and wearing out prematurely. There is a total of 28 ball bearings – 14 bearings in each cup.
12 Install the forks as described in Section 7 and the front wheel as described in Chapter 6.
13 Adjustment of the steering head bearings is critical. They should be tightened sufficiently to eliminate all play, but they must not be over-tightened. Note that it is possible to place a load of several tons on the bearings by over-tightening and yet still be able to turn the handlebars. Adjustment is done by tightening the nut or bolt at the top of the steering head, immediately above the upper fork yoke. When the adjustment is correct, there should be no play in the bearings and, when the front wheel is raised off the ground, a light tap on the end of the handlebar should cause the forks to swing to the full lock position.
14 Over-tightened bearings in the steering head will cause the machine to roll at slow speeds, while loose bearings will cause front fork shudder when the front brake is applied.

10 Seat – removal and installation

Refer to illustrations 10.4a and 10.4b

1 On pre-1979 models, there's no need to remove the seat during overhaul, because there are no electrical or other components under it.
2 The seat is attached with a bolt at the front, passing through lugs on the seat and the frame. There's also a bolt securing the rear of the seat directly to the rear fender.
3 1979 and later models must have the seat removed to gain access to the battery and other electrical components, as well as the return line to the oil tank.
4 Remove the bolt that secures the rear of the seat to the rear fender. If equipped, slide the passenger strap toward the front of the seat **(see illustration)**. Lift the rear of the seat and withdraw the seat to the rear. The front of the seat is held in place with a bracket that slides over a protrusion on the frame **(see illustration)**.
5 Installation of the seat is the reverse of removal.

Chapter 6 Wheels, brakes and tires

Contents

General information 1	Rear drum brake – inspection and brake shoe replacement 12
Wheels – inspection and repair 2	Rear disc brake – inspection and brake pad replacement 13
Front wheel – removal and installation 3	Rear disc brake caliper – removal, overhaul and installation 14
Front drum brake – inspection and brake shoe replacement 4	Rear disc brake master cylinder – removal, overhaul
Front disc brake – inspection and brake pad replacement 5	and installation 15
Front brake disc – removal and installation 6	Rear brake pedal – check and adjustment 16
Front disc brake caliper – removal and installation 7	Rear wheel sprocket – removal and installation 17
Front disc brake caliper – overhaul 8	Rear brake disc – removal and installation 18
Front disc brake master cylinder – removal, overhaul	Tires – removal and installation 19
and installation 9	Tubes – repair 20
Disc brakes – bleeding 10	Wheel bearings – repack 21
Rear wheel – removal and installation 11	

Specifications

Wheels

Size (diameter)	**Front**	**Rear**
1970 through 1979	19 in	18 in (standard) 16 in (optional)
1980-on	19 in	16 in
Wheel bearing end play		
1970 through mid-1991	0.004 to 0.018 inch (0.10 to 0.46 mm)	
Mid-1991 on	0.002 to 0.006 inch (0.05 to 0.15 mm)	
Rim runout (front and rear)		
Spoke wheel	1/32 in (0.79 mm)	
Cast wheel		
Lateral (axial)	3/64 in (1.19 mm)	
Radial	1/32 in (0.79 mm)	

Chapter 6 Wheels, brakes and tires

Brakes

Minimum brake shoe/pad lining thickness	See Chapter 1
Maximum brake disc runout (warpage)	0.008 in. (0.2 mm)
Minimum disc thickness*	
Front	
1973 through 1977	0.188 in (4.78 mm)
1978 through 1985	0.160 in (4.06 mm)
1986-on	0.180 in (4.57 mm)
Rear (all years)	0.205 in (5.21 mm)
Caliper piston travel	0.020 to 0.025 in (0.51 to 0.63 mm)
Rear brake pedal free play (later disc brake models only)	
1979 through early 1987	1/16 in (1.6 mm)
Late 1987-on	None

*Note: *The minimum thickness is normally stamped on the disc – it supercedes information included here.*

Tires

Tire pressures	See Chapter 1
Minimum tread depth	See Chapter 1

Torque specifications

	Ft-lbs (unless otherwise indicated)	Nm
Axle nut		
Front		
1970 through 1987	50	68
1988-on	50 to 55	68 to 75
Rear	60 to 65	81 to 88
Axle pinch bolt		
1970 through 1972	11	15
1988-on	21 to 27	28 to 37
Brake caliper mounting bolts		
Front		
1973 through 1978	35	47
1979 through 1983	80 to 90 in-lbs	9 to 10
1984-on	25 to 30	34 to 41
Rear		
1979 through 1985	155 to 190 in-lbs	18 to 21
1986 through early 1987	13 to 16	18 to 22
Late 1987-on	15 to 20	20 to 27
Brake hose-to-caliper banjo fitting bolts		
With copper washers	30 to 35	41 to 47
With steel and rubber washers	17 to 22	23 to 30
Front disc mounting bolts/screws		
1973	35	47
1974 through 1983		
Spoke wheel	16 to 19	22 to 26
Cast wheel	14 to 16	19 to 22
1984 through 1990 (both types)	16 to 18	22 to 24
1991-on (both types)	16 to 24	22 to 33
Brake disc-to-rear hub bolts		
Through 1991	23 to 27	31 to 37
1992-on	30 to 45	41 to 61
Rear brake master cylinder mounting bolts		
Through 1990	13 to 16	18 to 22
1991-on	155 to 190 in-lbs	18 to 21
Brake drum-to-rear hub bolts	25	34
Rear sprocket mounting bolts		
1970 through 1983	45 to 50	61 to 68
1984 and 1985		
Spoke wheel	45 to 50	61 to 68
Cast wheel	50 to 55	68 to 75
1986 through 1990	50 to 55	68 to 75
1991-on spoke wheel	45 to 55	61 to 75
1991 cast wheel	65 to 70	88 to 95
1992 cast wheel	45 to 55	61 to 75
1993-on cast wheel	55 to 65	75 to 88

Chapter 6 Wheels, brakes and tires

1 General information

Depending on the model, either wire spoke wheels or cast alloy wheels are standard equipment. Wire wheels require frequent inspection and maintenance, while cast alloy wheels are virtually maintenance-free.

The brakes are different types, depending on year and model. Early models (through 1972) were equipped with drum brakes on both wheels. Models produced from 1973 through 1977 had a single disc brake at the front and a drum brake at the rear. Models produced in 1978 had dual front discs and a drum rear brake. Machines from 1979-on had either single or dual discs at the front and a single disc at the rear. **Warning:** *Dust created by the brake system contains asbestos, which is harmful to your health. Never blow it out with compressed air and don't inhale any of it. An approved filtering mask should be worn when working on the brakes. Do not, under any circumstances, use petroleum-based solvents to clean brake parts. Use brake system cleaner or denatured alcohol only!*

Tires are conventional, requiring an inner tube with spoke wheels. Later models with cast wheels have tubeless tires.

2 Wheels – inspection and repair

Spoke (wire) wheels

1 Wire wheels should be inspected frequently to ensure the wheel runs true and to prevent potential damage from loose or broken spokes.
2 Clean the wheels thoroughly to remove mud and dirt, then make a general check of the wheels and spokes as described in Chapter 1.
3 Raise the motorcycle so the front wheel is off the ground and support it securely with blocks. Because the frame is very narrow under the engine, be sure to support the motorcycle so it can't fall over sideways. Attach a dial indicator to the fork slider and position the stem against the side of the rim. Spin the wheel slowly and check the side-to-side (axial) runout of the rim. In order to accurately check the radial runout with the dial indicator, the wheel would have to be removed from the machine and the tire removed from the wheel. With the axle clamped in a vise, the wheel can be rotated to check the runout.
4 An easier, though slightly less accurate, method is to attach a stiff wire pointer to the fork slider and position the end a fraction of an inch from the wheel (where the wheel and tire join). If the wheel is true, the distance from the pointer to the rim will be constant as the wheel is rotated. Repeat the procedure to check the rear wheel.
5 A wheel that wobbles from side-to-side can be trued by loosening the spokes that lead to the hub from the high side of the rim and tightening the spokes that lead to the hub from the low side. This in effect will pull the bulge out of the rim. Always tighten/loosen spokes in small increments to avoid distorting the rim and make sure all the spokes are uniformly tight (see Chapter 1).
6 An out-of-round wheel can be trued by loosening the spokes (both sides of the hub) that lead to the low area, and tightening the spokes that lead to the high or bulged-out area.
7 Generally, the wheels will probably have a combination of side-to-side wobble and out-of-roundness. Keep in mind that tightening and loosening spokes will affect wheel runout in both directions. Wheel truing requires patience and practice to develop any degree of skill, so it's best left to a dealer service department or motorcycle repair shop.
8 If the inspection reveals a bent, cracked, or otherwise damaged rim, the entire wheel will have to be rebuilt using a new rim and spokes. This is a complicated task requiring experience and skills beyond those of the average home mechanic and should be done by a dealer service department or motorcycle repair shop.

Cast alloy wheels

9 The cast alloy wheels should be visually inspected for cracks, flat spots on the rim and other damage. Check the axial and radial runout as described previously for wire wheels.
10 If damage is evident, or if runout in either direction is excessive, the wheel will have to be replaced with a new one. Never attempt to repair a damaged cast wheel.

3 Front wheel – removal and installation

Refer to illustrations 3.7 and 3.8
Caution: *Do not operate the brake lever after the front wheel is removed – the piston in the caliper might be forced out. If the piston is forced out of the bore, the caliper will have to be completely disassembled and rebuilt.*

Removal

1 Raise the front wheel off the ground and support the machine securely on blocks. Be sure it's stable from side-to-side. The frame under the engine is very narrow, so you may have to support the sides of the motorcycle. **Note:** *Pay close attention to any spacers used on the front axle. Be sure to reinstall them in their original location(s).*

1970 through 1972

2 Disconnect the front brake cable from the front wheel by removing the clevis pin through the brake operating arm. Remove the brake plate anchor bolt and lock washer.
3 Unscrew the nut from the end of the axle.
4 Loosen the pinch bolts (at the bottom of each fork leg) that secure the axle to the forks.
5 Tap the axle out and remove the front wheel.

1973 only

6 Remove the axle nut and washer from the front axle.
7 Loosen the nuts securing the axle retaining cap to the bottom of the right fork slider **(see illustration)**.
8 Tap the axle out and lower the front wheel. Pull the speedometer drive out of the hub **(see illustration)**. Leave the speedometer drive unit connected to the speedometer cable and tie them up out of the way.

1974 through 1977

9 Remove the brake caliper mounting bolt, washers and locknut.
10 Remove the axle nut, washer and lock washer.
11 Loosen the bolts securing the axle retaining caps to the bottom of each fork slider.
12 Tap the end of the axle to loosen it, then pull it out of the forks and front hub.
13 Lower the front wheel until the speedometer drive unit can be disengaged from the front hub. Tie the speedometer drive unit and cable out of the way.
14 Remove the front wheel.

1978 through 1987

15 Detach the front brake caliper(s) from the fork leg(s) and tie them up, out of the way.
16 Remove the nut, washer and lock washer from the axle.
17 Loosen the slider cap nuts at the bottom of each fork leg. **Note:** *On 1986 and 1987 models, detach the cable from the speedometer, then turn it counterclockwise to separate it from the speedometer drive at the wheel.*
18 Tap the end of the axle to loosen it, then pull it out of the front wheel.
19 Lower the wheel from the forks and remove the speedometer drive unit from the hub. Remove the front wheel.

1988-on

20 Remove the axle nut and washers.
21 Loosen the pinch bolt and pull the axle out.
22 Lower the wheel until the brake disc clears the caliper, then remove the wheel and separate it from the speedometer drive unit.

Installation

23 Installation of the front wheel is the reverse of the removal procedure. Note the following important points:
   a) On disc brake models, be sure the speedometer drive unit engages properly in the hub.
   b) On 1973 models, position the brake pads so the brake disc fits between them during installation.

Chapter 6 Wheels, brakes and tires

3.7 Loosen the slider cap nuts before attempting to remove the axle

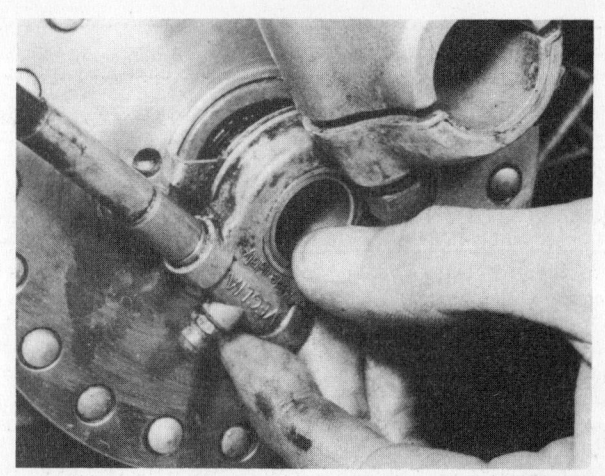

3.8 Lower the front wheel and disengage the speedometer drive unit from the wheel hub

c) Tighten the axle nut to the specified torque before tightening the slider cap nuts or pinch bolt.
d) **Caution:** *The calipers on some later models are secured with locknuts. When the locknuts are removed, they're destroyed and must be replaced with new ones!*
e) Check the front wheel bearing end play and compare it to the Specifications in this Chapter. If end play doesn't fall within the specified limits it will be necessary to install a different length hub bearing spacer on models through mid-1991, or a different thickness spacer shim on models from mid-1991-on. Refer to Chapter 1, Section 28 for access to spacer or shim and to a Harley-Davidson dealer for the appropriate replacement part.

4 Front drum brake – inspection and brake shoe replacement

Refer to illustration 4.4

1 Drum brakes don't usually require frequent maintenance, but they should be checked periodically to ensure proper operation. If the linkage is properly adjusted, the brake shoes aren't contaminated or worn out and the return springs and cables are in good condition, the brakes should work fine.
2 Check the cable ends to make sure they aren't frayed and check the lever pivot for binding and excessive play. As a general rule, the cable should be adjusted so the brake shoes don't drag when the lever is released and the lever doesn't touch the handlebar when the brake is applied.
3 If the lever doesn't operate smoothly, lubricate the cable, the cable ends and the pivot (refer to Chapter 1). If the brakes still don't operate smoothly, the problem is in the shoe actuating mechanism.
4 If the brake shoe wear check (refer to Chapter 1) indicates the shoes are near the wear limit, refer to Section 3 and remove the front wheel. Measure the thickness of the brake shoe lining **(see illustration)** and compare it to the Specifications in Chapter 1. If the shoes have worn beyond the allowable limits, or if they're worn unevenly, they must be replaced with new ones.
5 If the linings are acceptable as far as thickness is concerned, check them for glazing, high spots and hard areas. A light touch-up with a file or emery paper will restore them to usable condition. If the linings are extremely glazed, they have probably been dragging. Be sure to properly adjust the lever free play to prevent further glazing.
6 Occasionally the linings may become contaminated with grease from the wheel bearings or brake cam. If this happens, and it's not too severe, cleaning the shoes with a brake system solvent (available at auto parts stores) may restore them. Better yet, replace the shoes with new ones – the cost is minimal.
7 To remove the shoes from the backing plate, remove the pivot stud bolt, the operating shaft nut and the operating lever **(see illustration 4.4)**. Tap on the operating shaft and pull out the shaft, the pivot stud, the brake shoes and the return springs as an assembly.

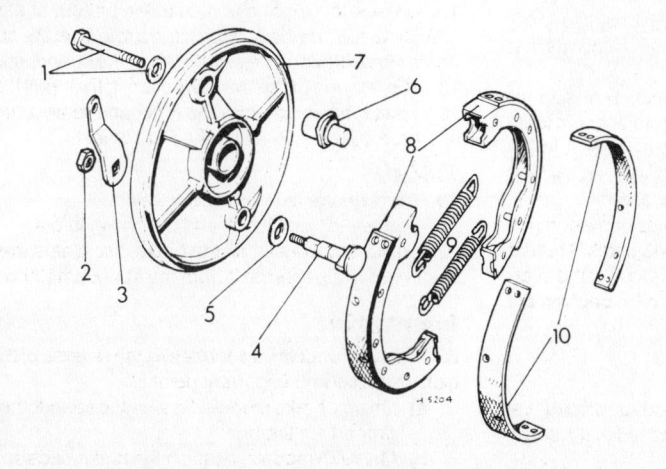

4.4 Front drum brake components – exploded view (typical)

1. Pivot stud bolt and washer
2. Operating shaft nut
3. Operating lever
4. Operating shaft
5. Operating shaft washer
6. Brake shoe pivot stud
7. Backing plate
8. Brake shoe
9. Return spring
10. Brake shoe lining

8 Remove the return springs from the shoes and check the springs for cracks and distortion. Replace them with new ones if defects are noted.
9 Clean the backing plate with solvent to remove brake dust and dirt. Also, clean the operating shaft and pivot stud. If compressed air is available, use it to dry the parts thoroughly.
10 Check the operating shaft and the hole in the backing plate for excessive wear. Slide the operating shaft back into the hole and make sure it turns smoothly without binding. If excessive side play is evident, the backing plate should be replaced with a new one. Check the shoe contact areas of the cam for wear also.
11 Before installing the new shoes, file a taper on their leading edges. Install the return springs, then apply a thin coat of high-temperature grease to the shoe contact areas of the cam and pivot stud.
12 Slip the brake shoe assembly into position in the backing plate, then install the pivot stud bolt and washer and the operating lever and shaft nut.
13 Clean the brake drum out with a wet rag. Don't use solvent in the brake drum because the rubber seals in the hub will be damaged by it.
Caution: *Don't blow the brake drum out with compressed air – the brake dust may contain asbestos, which can damage your lungs if inhaled.*
14 Check the drum for rough spots, rust and evidence of excessive wear. If the outer edge of the drum has a pronounced ridge, excessive wear has occurred. Measure the diameter of the drum at several places to determine if it's worn out-of-round. Excessive wear and out-of-roundness indicate the need for a new hub/drum. Slight rough spots and roughness can be removed with fine emery paper. Use one of the brake shoes as a sanding block so low spots aren't created in the drum.
15 Insert the brake shoe assembly into the brake drum and install the front wheel as described in Section 3.
16 Check the adjustment of the brake as described in Step 2 of this Section. If the brake needs adjusting, loosen the locknut on the adjusting sleeve, then turn the adjusting sleeve nut until the lever moves freely for about one-quarter of its full movement before the brakes begin to drag. Tighten the locknut against the adjusting sleeve.
17 If the adjustment is correct, but the brakes drag, the brake shoes must be centered in the brake drum. Loosen the pivot stud bolt and the axle nut, then spin the front wheel. While the wheel is spinning, apply the brake and tighten the pivot stud bolt and the axle nut. Recheck the adjustment.

5 Front disc brake – inspection and brake pad replacement

Inspection

Refer to illustration 5.4

1 Carefully examine the master cylinder, the hoses and the caliper for evidence of brake fluid leakage. Pay particular attention to the hoses. If they're cracked, abraded, or otherwise damaged, replace them with new ones. If leaks are evident at the master cylinder or caliper, they should be rebuilt by referring to the appropriate Sections in this Chapter.
2 Check the lever for proper operation. It should feel firm and return to its original position when released. If it feels spongy, or if lever travel is excessive, the system may have air trapped in it. Refer to Chapter 1 and bleed the brakes.
3 Check the brake pads for excessive wear by referring to Chapter 1.
4 Examine the brake disc for cracks and evidence of scoring. Measure the thickness of the disc. If its worn beyond the allowable limit **(see illustration)**, it must be replaced with a new one.
5 If the brake lever pulsates when the brake is applied during operation of the machine, the disc may be warped. Attach a dial indicator set up to the fork slider and check the disc runout. If the runout is greater than specified, replace the disc with a new one. If a dial indicator isn't available, a dealer service department or motorcycle repair shop can make this check for you.
6 If the brake pads are worn out or contaminated with brake fluid or dirt, they must be replaced with new ones. Failure to replace the pads when necessary will result in damage to the disc and severe loss of stopping power.

Pad replacement

Caution: *Do not operate the brake lever while the caliper is apart – the piston will be forced out of the caliper. Always replace all pads in the front brake caliper(s) at the same time; never replace only one pad or the pads in only one caliper on dual disc models.*

Caution: *Ensure that only the correct parts are fitted when replacing pads or discs. Modifications to the pad and disc material from 1992-on prevent the mixing of early and late model components.*

1973

Refer to illustrations 5.7 and 5.8

7 Remove the clamp securing the brake hose to the fork leg **(see illustration)**.
8 Unscrew the bolts holding the caliper together, then separate the outer half and the damper spring from the rest of the caliper **(see illustration)**.
9 Remove the mounting pin and the inner half of the caliper.
10 Disengage the brake pad mounting pins and detach the brake pads.

1974 through 1977

Refer to illustrations 5.11a, 5.11b and 5.11c

11 Remove the Allen-head bolts and locknuts holding the caliper together **(see illustration)**. Separate the two caliper halves **(see illustrations)**.
12 Remove the brake pads and check them for wear.

5.4 The minimum brake disc thickness is stamped near one of the mounting bolts

5.7 Remove the screw securing the brake hose to the fender or fork leg

Chapter 6 Wheels, brakes and tires

5.8 Front disc brake components – exploded view (1973)

1. Bolt
2. Washer
3. Outer caliper half
4. Damper spring
5. Mounting pin
6. Inner caliper half
7. Brake pad mounting pins
8. Brake Pads
9. Brake piston
9A. Brake piston assembly
10. Brake line
11. Piston boot
12. Snap-ring
13. Backing plate
14. Wave spring
15. Friction ring
16. O-ring
17. Bleeder valve
18. Bushing
19. Bushing
20. Bolt and lock washer
21. Brake disc
22. Brake disc spacer
23. Replacement mounting pin

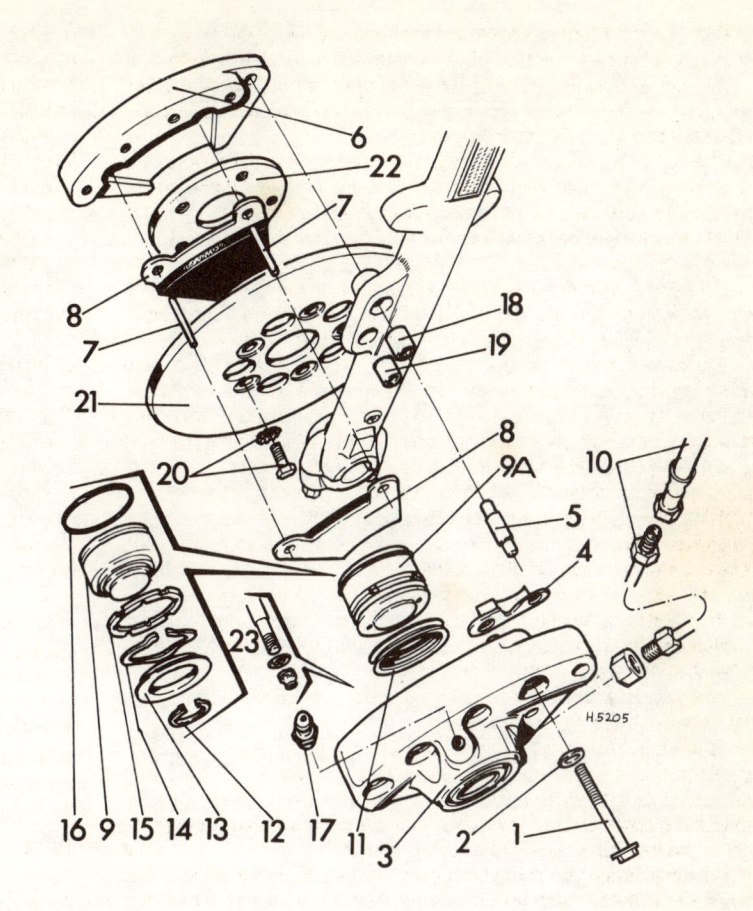

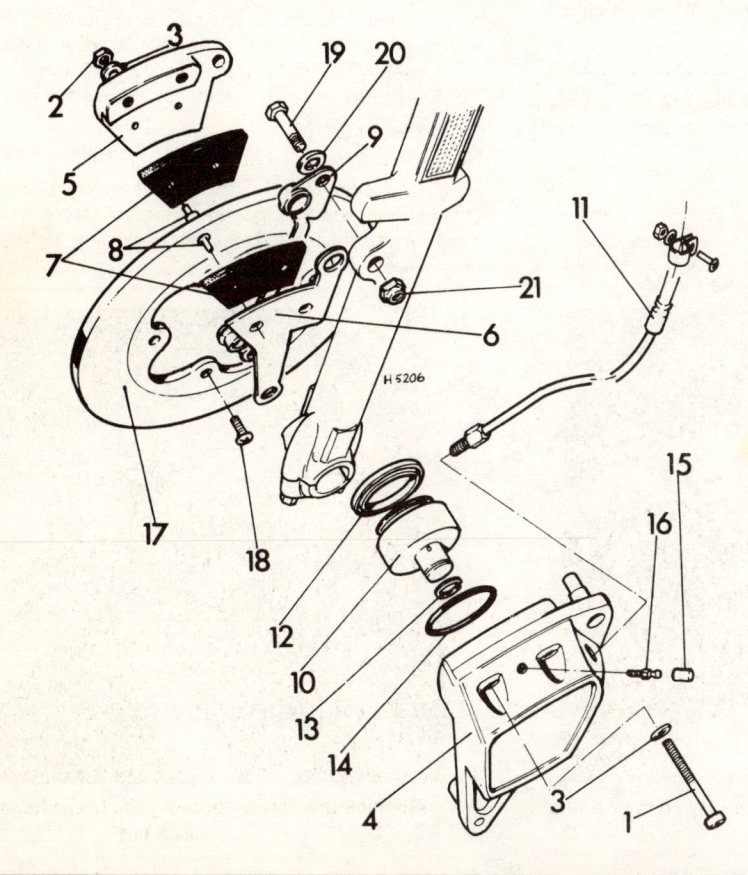

5.11a Front disc brake components – exploded view (1974 through 1977)

1. Allen-head bolt
2. Locknut
3. Washer
4. Outer caliper half
5. Inner caliper half
6. Backing plate
7. Brake pad
8. Rivet
9. Torque arm
10. Piston
11. Brake hose
12. Rubber boot
13. Friction ring
14. O-ring
15. Bleeder valve cap
16. Bleeder valve
17. Brake disc
18. Mounting bolt
19. Torque arm mounting bolt
20. Washer
21. Lockout

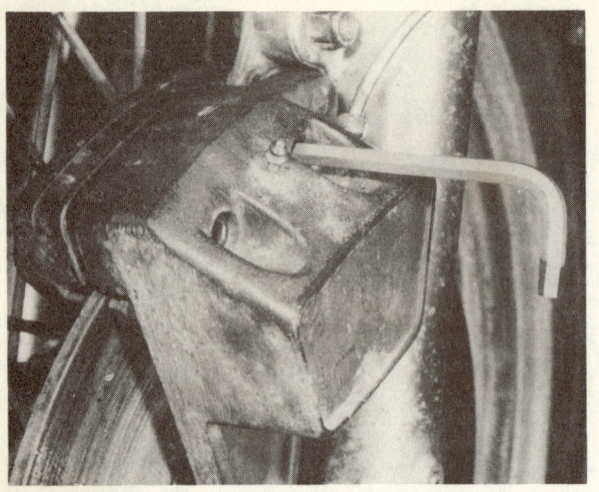

156 Chapter 6 Wheels, brakes and tires

5.11b Remove the Allen-head bolts, . . .

5.11c . . . and separate the caliper sections to get at the brake pads on 1973 models

1978 through 1983
Refer to illustrations 5.13, 5.15a and 5.15b
13 Using a socket, universal joint and extension, loosen the bolt securing the two halves of one of the calipers together **(see illustration)**. You have to work from the back side of the caliper to loosen the bolt.
14 Remove the Allen-head bolts and nuts attaching the caliper to the lower fork leg.

15 Detach the caliper from the forks and separate the two halves. Remove the brake pads from the guide pins **(see illustrations)**.
16 Repeat the procedure for the remaining caliper.

1984-on
Refer to illustrations 5.18 and 5.20
17 Loosen the pad retainer screw at the back (inner) side of the caliper.

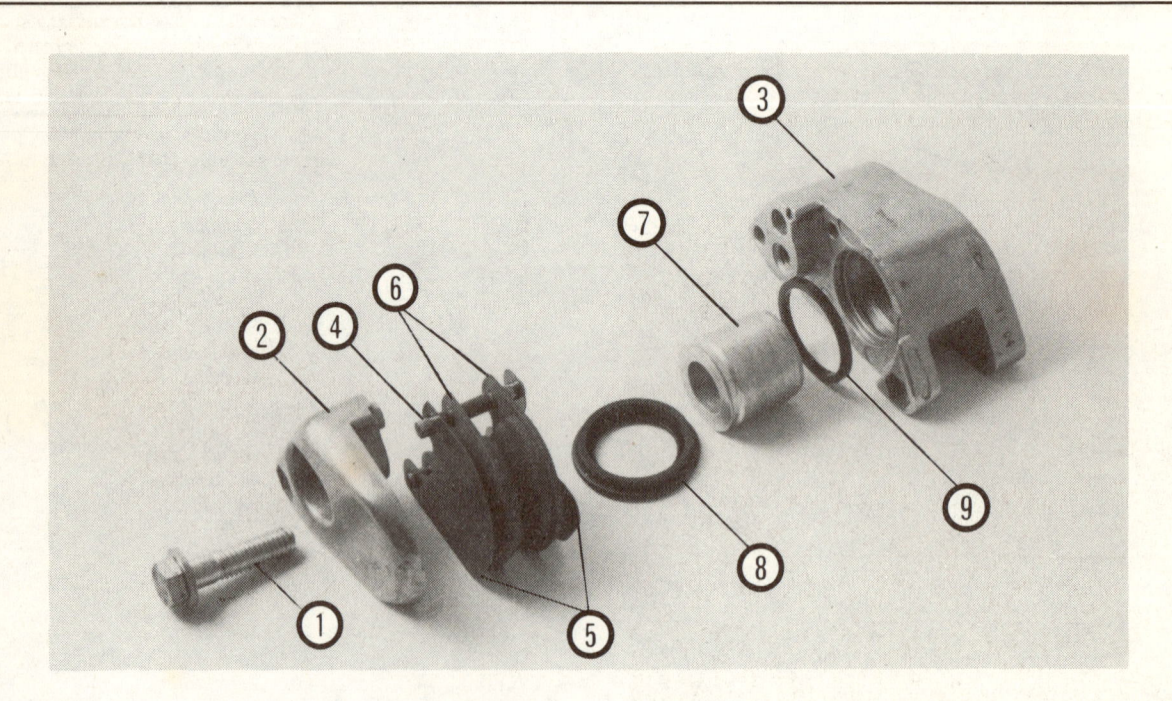

5.13 Front disc brake caliper components – exploded view (1978 through 1983)

1 Bolt (holds caliper sections together)	4 Pad guide pin (2)	7 Piston
2 Inner caliper half	5 Pad shims	8 Boot
3 Outer caliper half	6 Brake pads	9 Seal

Chapter 6 Wheels, brakes and tires

5.15a Remove the bolt holding the caliper sections together, . . .

5.15b . . . then lift off the inner caliper half and remove the pads and shims

18 Remove the upper mounting bolt and the lower mounting pin **(see illustration)**. Move the caliper to the rear and down slightly, away from the fork slider, then remove the outer pad, pad holder and spring clip from the caliper as an assembly. Pull out the bushing the upper mounting bolt threads into, then slide the caliper off the brake disc.
19 Note how the pad retainer is installed in the caliper, then remove the screw and detach the pad retainer and inner brake pad.
20 Note how the pad and spring clip are positioned in the holder **(see illustration)**. Push the pad out of the clip to remove it from the holder.

All models

Refer to illustrations 5.28a and 5.28b

21 While the caliper assembly is apart, check the movement of the piston in the outer caliper half (1973 through 1978 models only). Mount a dial indicator on the back of the outer caliper, so the plunger rests on the piston face. Apply the handlebar lever gently until the piston is extended and set the indicator to zero. Release the brake lever. If the piston isn't restricted, it should move from 0.020 to 0.025-inch.
22 Clean the disc surface with brake system cleaner, lacquer thinner or acetone. **Warning:** *Do not use petroleum-based solvents*.
23 The piston(s) in the caliper must be pressed into the bore as far as possible when new pads are installed.
24 Don't touch the faces of the brake pads when installing them. Make sure the pads are in the correct position and facing the right direction during reassembly.
25 Reassemble the calipers, with the new pads, in the reverse order of disassembly.
26 On 1974 through 1977 models, install new locknuts on the caliper.
27 On 1984 and later models, position the spring clip and the pad with the insulator backing material in the pad holder. The pad lining (friction face) and the spring clip loop must face away from the caliper piston **(see illustration 5.20)**.

5.18 On 1984 and later models, remove the upper mounting bolt and lower mounting pin to detach the caliper

1 Caliper mounting bolt
2 Caliper mounting pin
3 Brake hose-to-caliper banjo fitting bolt (12-point head)

5.20 This is what the outer brake pad and spring clip look like when correctly installed in the pad holder

Chapter 6 Wheels, brakes and tires

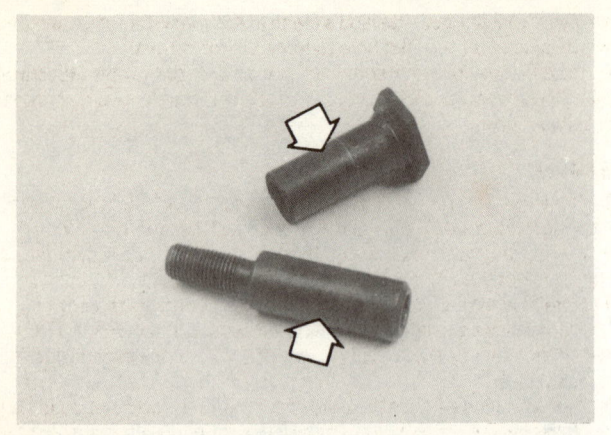

5.28a Clean them thoroughly, then apply high-temperature brake grease to the sliding surfaces (arrows) of the caliper pins/bushings before installation (components for 1984 and later models shown)

28 Lubricate the caliper pins with a smear of high-temperature grease **(see illustration)**. When installing the upper mounting bolt threaded bushing on 1993-on models, note that its flanged head must locate under the pad holder rivet so that one of its cutouts slots over the rivet body **(see illustration)**.

29 Check the brake fluid level in the master cylinder after the pads are installed. The level may be too high, requiring the excess fluid to be siphoned off.

6 Front brake disc – removal and installation

1 Remove the front wheel as described in Section 3.
2 Remove the bolts or screws and separate the disc from the hub. Some models may have Allen-head or Torx-head bolts which require a special tool for removal and installation.
3 Before installing the disc, be sure the threads on the bolts (and in the hub) are clean and undamaged. On later models, align the square notch on the disc rear surface with the corresponding cutout in the hub. It is recommended that new screws be used when installing the disc; unless the new screws are supplied with a patch of locking compound on them, apply a few drops of thread-locking compound yourself. Install the bolts and tighten them in a criss-cross pattern until the specified torque is reached.
Note: *New bolts containing the locking patch can be removed and installed up to three times, then they should be replaced.*

7 Front disc brake caliper – removal and installation

1973

1 The caliper is held in place by the four bolts that secure the two caliper halves together **(see illustration 5.8).**
2 Remove the clamp securing the brake hose to the fork leg, then remove the four caliper bolts.
3 The caliper can be removed by sliding the mounting pins out of the fork slider bushings after separating the outer caliper half from the inner half (see Section 5).

1974 through 1977

Refer to illustration 7.4

4 Remove the two Allen-head bolts and locknuts holding the caliper halves together **(see illustration 5.11a)**, then detach the caliper by pulling out on it until the pin is disengaged from the torque arm **(see illustration)**.

1978 through 1983

5 Loosen the bolt securing the two caliper halves together **(see illustration 5.15a)**. If both calipers are being removed, loosen the bolt in each caliper.

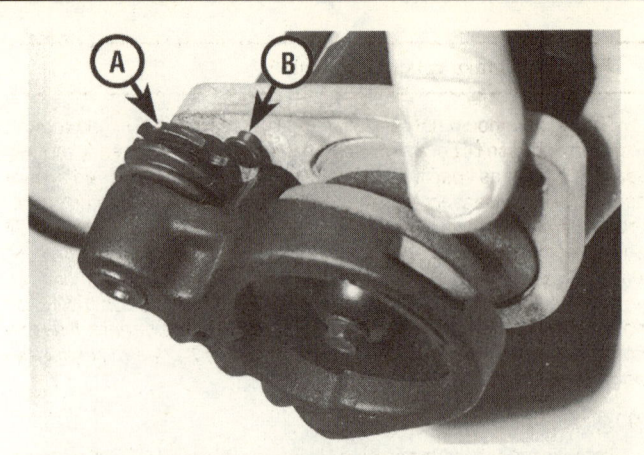

5.28b Caliper upper mounting bolt bushing flanged head (A) and rivet (B)

6 Remove the Allen-head bolts and nuts attaching the caliper(s) to the fork leg(s), then detach the caliper(s).

1984-on

7 Loosen the brake hose-to-caliper banjo fitting bolt, but don't unscrew it. A 12-point socket will be required to fit the bolt head.
8 Remove the upper mounting bolt and the lower caliper pin.
9 Detach the caliper.

All models

Caution: *While the caliper is apart or removed from the brake disc, do not operate the front brake lever – it will force the piston out of the caliper bore.*

10 Unscrew the brake line union or banjo fitting bolt at the caliper. Plug the end of the brake line and the opening in the caliper to prevent dirt from entering the hydraulic system.
11 Installation of the caliper is the reverse of the removal procedure. Apply Teflon tape to the threads of the brake line fitting before attaching the line to the caliper. On later models, discard the original washers used at the brake line banjo fitting and install new ones.
12 On 1974 through 1977 models, be sure the locating pin at the bottom of the outer caliper engages the backing plate for the brake pad as well as the torque arm **(see illustration 7.4).**
13 Coat the caliper pins (that allow the caliper to move back-and-forth) with high-temperature grease and tighten the mounting bolts to the specified torque.
14 After the calipers are installed, the brake system must be bled as described in Chapter 1.

7.4 On 1974 through 1977 models, pull the caliper straight out until the pin (arrow) clears the torque arm

Chapter 6 Wheels, brakes and tires

8 Front disc brake caliper – overhaul

1 The caliper should not be disassembled unless it leaks fluid around the piston or doesn't operate properly. If the piston travel on early models (Section 5) is not as specified, the piston will have to be removed and inspected. Before disassembling the caliper, read through the entire procedure and make sure you have the correct caliper rebuild kit. Also, you'll need some new, clean brake fluid of the recommended type and some clean rags. **Note:** *Disassembly, overhaul and reassembly of the brake caliper must be done in a spotlessly clean work area to avoid contamination and possible failure of the brake hydraulic system components. If such a work area isn't available, have the caliper rebuilt by a dealer service department or a motorcycle repair shop.*

1973

2 Remove the caliper as described in Section 7 but DO NOT disconnect the hydraulic brake line.
3 Remove the brake pads as described in Section 5.
4 Slowly pump the brake lever until the piston doesn't move any further.
5 Disconnect the brake line from the caliper and plug the line.
6 Pull the piston boot away from the groove in the piston, then remove the piston from the bore in the caliper.
7 Remove the snap-ring from the piston, then pull off the backing plate, the wave spring, the friction ring and the O-ring **(see illustration 5.8)**.
8 Unscrew the bleeder valve from the caliper.

1974 through 1977

9 Remove the caliper and the brake pads as described in Sections 5 and 7.
10 Disconnect the brake line from the caliper and plug it.
11 Remove the rubber boot, then, using two screwdrivers, pry the piston out of the caliper bore.
12 Check the friction ring at the end of the piston **(see illustration 5.11a)**. If it's damaged, remove it and install a new one.
13 Carefully pry the O-ring out of the caliper bore with a wood or plastic tool.
14 Unscrew the bleeder valve from the caliper.

1978 through 1983

Refer to illustration 8.16

15 Remove the caliper and brake pads as described in Sections 5 and 7. Disconnect the brake line from the caliper and plug it.
16 Carefully pry the piston out of the caliper, then remove the boot **(see illustration)**. If the piston cannot be pried out, place the caliper face down on a clean work surface. Position a clean towel under the piston and apply low pressure air to the inlet hole to force the piston out.
17 Carefully remove the seal from the caliper bore **(see illustration 5.13)**. Use a wood or plastic tool to remove the seal (to avoid scratching the caliper bore).

1984-on

18 Remove the caliper and brake pads as described in Sections 5 and 7. Disconnect the brake line from the caliper and discard the washers.
19 Pry out the boot retainer by inserting a small screwdriver into the notch at the bottom of the piston bore.
20 Note how it's installed, then remove the rubber piston boot.
21 Place the caliper face down on a clean work surface. Position a clean towel under the piston and apply low pressure air to the inlet hole to force the piston out.
22 Remove the seal from the caliper bore with a wood or plastic tool.
23 Note how they're installed, then remove the threaded bushing from the caliper and pull the pin boot out.
24 Remove the O-rings from the mounting bolt/pin holes.

All models

25 Clean all the brake components (except for the brake pads) with brake system solvent (available at auto parts stores), isopropyl alcohol or clean brake fluid. **Warning:** *Do not, under any circumstances, use petroleum-based solvents to clean brake parts. If compressed air is available, use it to dry the parts thoroughly.*
26 Check the caliper bore and the outside of the piston for scratches, nicks and score marks. If damage is evident, the caliper must be replaced with a new one.
27 Reassembly of the caliper is done in the reverse order of disassembly. Be sure to lubricate all of the components with clean brake fluid during reassembly and install new O-rings.
28 Install the brake pads as described in Section 5 and connect the brake line to the caliper (see Section 7).
29 Attach the caliper assembly to the forks and bleed the system as described in Chapter 1.

9 Front disc brake master cylinder – removal, overhaul and installation

Refer to illustrations 9.3 and 9.4

1 If the master cylinder is leaking fluid, or if the lever doesn't feel firm when the brake is applied, and bleeding the brakes doesn't help, master

8.16 On 1978 through 1983 models, the piston can be pried out of the caliper bore very carefully with two screwdrivers

9.3 On 1973 through 1981 models, detach the switch assembly (arrow) from the master cylinder and disconnect the brake light wires

9.4 Front disc brake master cylinder components – 1973 through 1981 models

1. Master cylinder housing
2. Master cylinder cover
3. Gasket
4. Cover screw
5. Brake line
6. Pivot pin snap-ring
7. Pivot pin
8. Brake lever
9. Brake lever pin
10. Plunger
11. Spring
12. Washer
13. Dust wiper
14. Snap-ring
15. Piston
16. O-ring
17. Piston cup
18. Spring cup
19. Spring
20. Union fitting

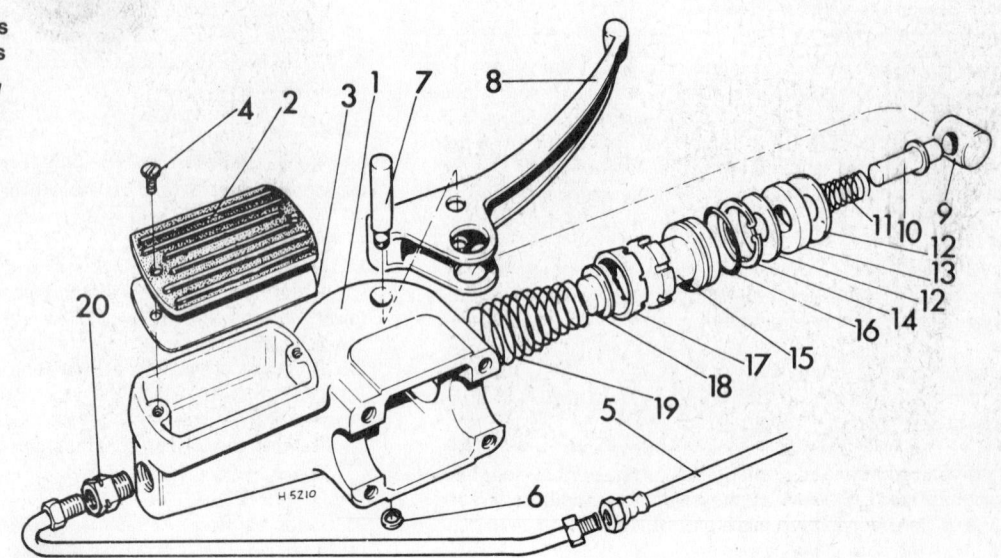

cylinder overhaul is recommended. Before disassembling the master cylinder, read through the entire procedure and make sure you have the correct rebuild kit. Also, you'll need some new, clean brake fluid of the recommended type, some clean rags and internal snap-ring pliers. **Note:** *Disassembly, overhaul, and reassembly of the brake master cylinder must be done in a spotlessly clean work area to avoid contamination and possible failure of the brake hydraulic system components. If such a work area isn't available, have the master cylinder rebuilt by a dealer service department or motorcycle repair shop.*

1973 through 1981

2 Turn the handlebars so the master cylinder is as level as possible and remove the cover and gasket. Disconnect the brake line from the master cylinder and catch the fluid in a container.
3 Remove the handlebar switch assembly and disconnect the brake light wires **(see illustration on previous page)**.
4 Remove the snap-ring and pivot pin so the handlebar brake lever can be removed. It'll pull out with the brake lever pin, plunger, spring, washers and dust wiper **(see illustration)**.
5 Remove the snap-ring from the master cylinder housing bore.
6 Pull out the piston, O-ring, piston cup, spring cup and piston return spring. Be careful when removing the piston so the bore of the master cylinder and the piston don't get scratched. If they do, the entire assembly must be replaced with a new one.

1982 through 1995

7 Open the bleeder valve on the front caliper and attach a piece of hose to the valve. Place the other end of the hose in a clean container and slowly pump the handlebar brake lever to drain the brake fluid.
8 Remove the bolt attaching the brake line to the master cylinder. Discard the washers on either side of the brake line.
9 Remove the cover and gasket from the master cylinder.
10 Remove the snap-ring securing the pivot pin, then lift out the pivot pin. Separate the brake lever and the reaction pin from the master cylinder.
11 Remove the master cylinder clamp and detach the assembly from the handlebars.
12 Pull the pushrod and switch out, followed by the dust boot, piston and O-ring, back-up disc (not used on 1983 and later models), cup, stop and piston return spring.
13 Remove the sight glass and grommet from the side of the master cylinder.

1996-on

14 Perform Steps 7 and 8 to drain the brake fluid from the master cylinder.
15 Make a wedge of cardboard 5/32-inch thick (wood or plastic will also work). Pull the brake lever and slip the wedge between the end of the lever and its bracket on the front side of the master cylinder (this protects the brake light switch plunger and boot).
16 Remove the two Torx screws and washers that secure the master cylinder clamp to the cylinder. Separate the clamp from the cylinder body and remove the master cylinder from the handlebar.
17 Wear eye protection and remove the lever snap-ring from the underside of the master cylinder. Withdraw the pivot pin, bushing and brake lever.
18 Pry the wiper seal out of the cylinder bore with a pointed tool, taking care not to scratch the bore.
19 Dump the piston cap, piston assembly and spring out of the bore. Note which way the primary cup is installed on the piston, then remove the primary cup and O-ring from their piston grooves.

All models

20 Clean all of the parts with brake cleaning solvent (available at auto parts stores), isopropyl alcohol or clean brake fluid. **Warning:** *Do not, under any circumstances, use petroleum-based solvent to clean brake parts.* If compressed air is available, use it to dry the parts thoroughly. Check the master cylinder bore for scratches, nicks and score marks. If damage is evident, the master cylinder must be replaced with a new one. Be sure the vent holes are open.
21 Before reassembling the master cylinder, soak the new rubber parts in clean brake fluid for 10 or 15 minutes. Lubricate the master cylinder bore with clean brake fluid, then carefully insert the piston and related parts in the reverse order of disassembly.
22 On 1982 and later models, install the sight glass and grommet in the side of the master cylinder.
23 If you're working on a 1995 or earlier model, apply a light coat of anti-seize compound to the reaction pin and insert it into the large hole in the brake lever. Connect the lever to the master cylinder, aligning the plunger (pushrod and switch on 1982 and earlier models) with the hole in the pin.
24 Install the pivot pin and secure it with the snap-ring.
25 On 1973 through 1981 models, attach the brake light wires and assemble the handlebar switch. On later models, clamp the master cylinder to the handlebars.

Chapter 6 Wheels, brakes and tires

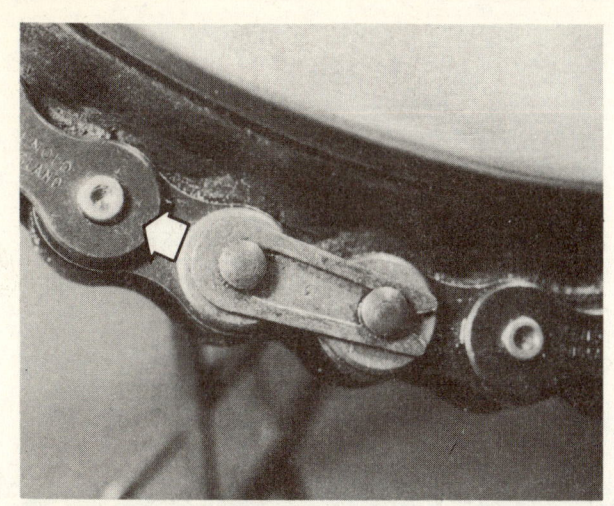

11.2 Position the master link on the rear sprocket to remove and install it – make sure the closed end of the spring clip points in the direction of chain travel (arrow)

11.3a On models with a drum rear brake, remove the brake adjusting nut . . .

11.3b . . . and the backing plate anchor bolt (if equipped)

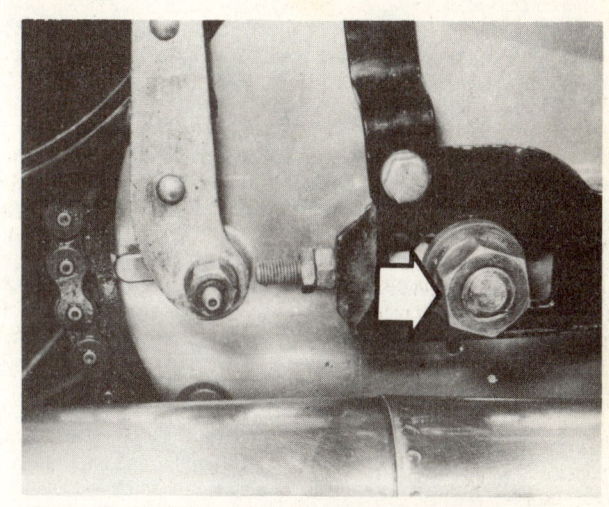

11.4 On 1970 through 1988 models, unscrew the axle nut (arrow) and pull out the axle

26 Make sure the relief port in the master cylinder is uncovered when the brake lever is released.
27 Connect the brake line to the master cylinder – use new washers on 1982 and later models.
28 Fill the master cylinder with the recommended brake fluid and bleed the system as described in Chapter 1.

10 Disc brakes – bleeding

Refer to Chapter 1, *Tune-up and routine maintenance,* for the disc brake bleeding procedure.

11 Rear wheel – removal and installation

Refer to illustrations 11.2, 11.3a, 11.3b, 11.4 and 11.8

Removal

1 Raise the rear of the motorcycle and support it securely on blocks so the tire is at least four inches off the ground (it must be stable from side-to-side so it won't tip over).

2 On chain drive models, rotate the rear wheel until the master link on the chain is positioned on the teeth of the rear sprocket **(see illustration)**. Remove the master link and disengage the chain from the rear wheel; lay the chain onto paper to prevent it picking up dirt from the ground. On belt drive models, first slacken the axle nut as described below, fully release the belt adjusters and push the wheel fully forward in the swingarm; this will create enough slack to allow the belt to be slipped off the sprocket.
3 On 1970 through 1978 models, unscrew the adjusting nut from the brake rod or cable and disconnect the rear brake operating lever **(see illustration)**. Remove the backing plate anchor bolt, if equipped **(see illustration)**.
4 On 1970 through 1988 models, remove the axle nut and any washers that are used **(see illustration)**. On 1989 and later models, remove and discard the cotter pin, then unscrew the nut and remove the washer.
5 Tap the end of the axle with a soft-face hammer to loosen it, then pull it out of the rear hub and swingarm. On drum brake models, an axle spacer must be removed from the left side.
6 Remove the wheel from the swingarm.

Installation

7 Apply a thin coat of anti-seize compound to the axle before installing the rear wheel.

Chapter 6 Wheels, brakes and tires

11.8 Install the spacer on the left side of the wheel before inserting the axle (drum brake models)

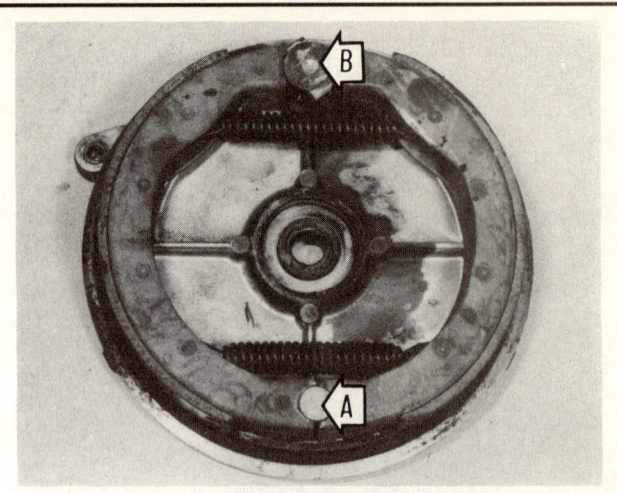

12.7 Location of the rear drum brake pivot stud (A) and operating shaft (B)

8   Position the wheel in the swingarm. On 1979 and later models, guide the brake disc into the caliper. Lift the wheel until the axle can be inserted through the swingarm and rear hub. Install all of the associated components with the axle in the reverse order of removal **(see illustration)**. Attach the brake rod or cable to the actuating lever on 1970 through 1978 models.
9   On chain drive models position the ends of the chain adjacent to each other on the rear sprocket and insert the master link with its closed end facing in the normal direction of wheel travel. On belt drive models, slip the belt over the sprocket.
10 Adjust the drive chain or belt as described in Chapter 1.
11 Tighten the axle nut to the specified torque. On 1989 and later models, install a new cotter pin. Bend one end up against the axle and the other end down against one of the nut flats.
12 On 1970 through 1978 models, adjust the rear brake as described in Section 12.
13 Check the rear wheel bearing end play and compare it to the Specifications in this Chapter. If end play doesn't fall within the specified limits it will be necessary to install a different length hub bearing spacer on models through mid-1991, or a different thickness spacer shim on models from mid-1991-on. Refer to Chapter 1, Section 28 for access to spacer or shim and to a Harley-Davidson dealer for the appropriate replacement part.

12 Rear drum brake – inspection and brake shoe replacement

Refer to illustration 12.7

Warning: *The dust created as the brake shoes wear may contain asbestos, which is harmful to your health. Never blow it out with compressed air and don't inhale any of it. An approved filtering mask should be worn when working on the brakes. Do not, under any circumstances, use petroleum-based solvents to clean brake parts. Use brake cleaner or denatured alcohol only!*

1   Drum brakes don't normally require frequent maintenance, but they should be checked periodically to ensure proper operation. If the linkage is properly adjusted, if the brake shoes aren't contaminated or worn out and if the return springs are in good condition, the brakes should work fine.
2   Check the brake pedal for proper operation. It shouldn't bind when depressed and should return completely when released.
3   If the brake doesn't operate properly, make sure nothing is interfering with the pedal or brake rod/cable and lubricate the pedal pivot. If the brakes still don't operate or return properly, the problem is in the shoe actuating mechanism.
4   If the brake shoe wear check (Chapter 1) indicates the shoes are near the wear limit, refer to Section 11 and remove the rear wheel. Measure the thickness of the brake shoe lining and compare it to the Specifications in Chapter 1. If the shoes have worn beyond the allowable limits, or if they're worn unevenly, they must be replaced with new ones.
5   If the linings are acceptable as far as thickness is concerned, check them for glazing, high spots and hard areas. A light touch-up with a file or emery paper will restore them to usable condition. If the linings are extremely glazed, they probably have been dragging. Be sure to adjust the pedal free play to prevent further glazing.
6   Occasionally the linings may become contaminated with grease from the wheel bearing or brake cam. If this happens, and it's not too severe, cleaning the shoes with brake system solvent (available at auto parts stores) may restore them. However, the best approach is to replace the shoes with new ones.
7   To remove the shoes from the backing plate, unscrew the operating lever retaining nut and pull off the lever and washer. Unscrew the pivot stud nut (anchor bolt on 1973 through 1978 models) and remove the washer **(see illustration)**. On 1970 through 1972 models a locating block is used; 1973 models have a spacer that must be removed next.
8   Gently tap the end of the operating shaft with a soft-face hammer and detach the brake shoes, springs, pivot stud and operating shaft/washer, as an assembly, from the brake backing plate.
9   Separate the springs from the brake shoes and check the springs for cracks and distortion. Replace the springs with new ones if they're cracked, bent or stretched.
10 Clean the backing plate to remove brake dust and dirt. Also, clean the operating shaft and pivot stud. If compressed air is available, use it to dry the parts thoroughly.
11 Check the operating shaft and pivot stud and the holes in the backing plate for excessive wear. Place the operating shaft in position in the backing plate and make sure it turns smoothly without binding. If excessive side play is evident, the operating shaft and/or backing plate will have to be replaced with new parts. Also check the shoe contact area on the operating shaft for wear.
12 Apply a thin coat of high-temperature grease to the operating shaft and pivot stud and attach the shoes to the pivot and shaft with one spring. Place the spring in the groove that's closest to the backing plate.
13 Place the washer over the operating shaft and install the assembly in the backing plate.
14 Install the locating block (or spacer), the nut and lock washer (anchor bolt and washer), the operating lever and the nut and washer that secure it to the backing plate.
15 Attach the remaining spring to the brake shoes.
16 Clean the brake drum out with a wet rag.
17 Check the drum for rough spots, rust and excessive wear (if the outer edge of the drum has a pronounced ridge, excessive wear has occurred). Measure the diameter of the drum at several places to determine if it's worn out-of-round. Excessive wear and out-of-roundness indicate the need for a new hub/drum. Slight rough spots can be removed with fine emery paper. Use one of the brake shoes as a sanding block so low spots

Chapter 6 Wheels, brakes and tires

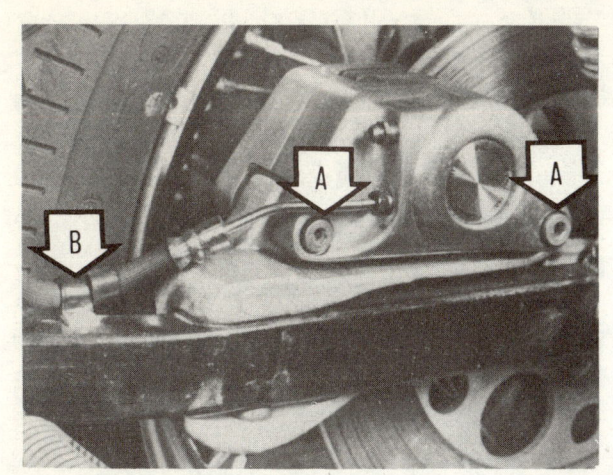

13.7 Remove the caliper mounting bolts (A) and the clamp (B) securing the brake line to the swingarm (1979 through 1981 models)

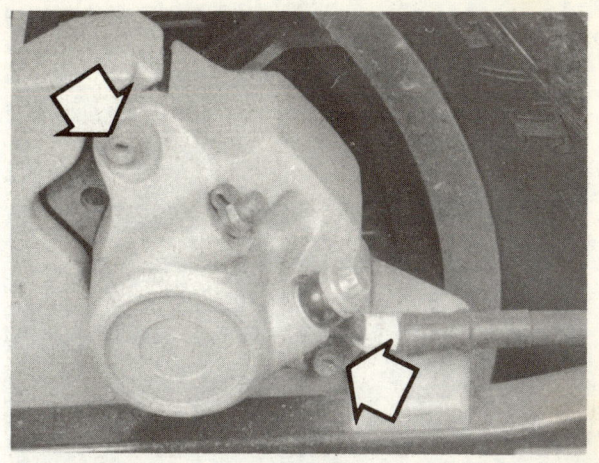

13.11 On late 1987 and later models, the rear brake caliper is attached to the bracket and slides on two mounting (pin) bolts

aren't created in the drum.
18 Insert the brake shoe/backing plate assembly into the brake drum and install the rear wheel as described in Section 11.
19 Insert the brake rod or cable through the operating lever and install the adjusting nut. Tighten the adjusting nut so the brake begins to make contact when the brake pedal is depressed 1-1/4 inches.
20 If the brake drags after the brake pedal is released, and the pedal is adjusted correctly, the brake shoes should be centered in the drum. Loosen the pivot stud nut or bolt and the axle nut. Spin the wheel and apply the brake. While the wheel is spinning, tighten the pivot stud nut or bolt and the axle nut. Check the brake pedal adjustment again.

13 Rear disc brake – inspection and brake pad replacement

Inspection

1 Carefully examine the master cylinder, the hoses and the caliper for evidence of brake fluid leakage. Pay particular attention to the hoses. If they're cracked, abraded, or otherwise damaged, replace them with new ones. If leaks are evident at the master cylinder or caliper, they should be rebuilt by referring to the appropriate Sections in this Chapter.
2 Check the pedal for proper operation. It should feel firm and return to its original position when released. If it feels spongy, or if pedal travel is excessive, the system may have air trapped in it. Refer to Chapter 1 and bleed the brakes.
3 Check the brake pads for excessive wear by referring to Chapter 1.
4 Examine the brake disc for cracks and score marks. Measure the thickness of the disc. If its worn beyond the specified limit, it must be replaced with a new one.
5 If the brake pedal pulsates when the brake is applied during operation of the machine, the disc may be warped. Attach a dial indicator to the swingarm and check the disc runout. If the runout is greater than specified, replace the disc with a new one. If a dial indicator isn't available, a dealer service department or motorcycle repair shop can make this check for you.
6 If the brake pads are worn out or contaminated with brake fluid or dirt, they must be replaced with new ones. Failure to replace the pads when necessary will result in damage to the disc and severe loss of stopping power. **Caution:** *Do not operate the brake pedal while the caliper is apart – the piston will be forced out of the caliper. Always replace both pads in the rear brake caliper at the same time; never replace only one pad.*

Pad removal

1979 through 1981
Refer to illustration 13.7

7 Loosen the large bolt on the back side of the caliper. Unscrew the two Allen-head bolts securing the caliper to the mounting bracket. Remove the clamp securing the brake line to the swingarm **(see illustration)**.
8 Detach the caliper from the mounting bracket and remove the large bolt from the back side. Separate the two halves of the caliper and pull the pads and plates off the pins.

1982 through early 1987
9 Remove the two Allen-head bolts securing the caliper to the bracket and detach the caliper.
10 Slide the brake pads out of position.

Late 1987-on
Refer to illustrations 13.11 and 13.12
Note: *On certain models you may have to remove the lower shock absorber mounting bolt and pivot the shock out of the way to remove the rear caliper mounting bolt.*

11 Remove the caliper mounting (pin) bolts **(see illustration)**, then pull the caliper up and off the brake disc and pads. It will probably be stubborn due to the piston clearing the pin on the back of the moving pad and it is advised that the caliper be rocked back and forth on the bracket to ease the piston back in its bore.
12 Carefully note how it's installed, then remove the wire retainer clip from the back side of the bracket **(see illustration)**. Slide the outer brake pad off the bracket.

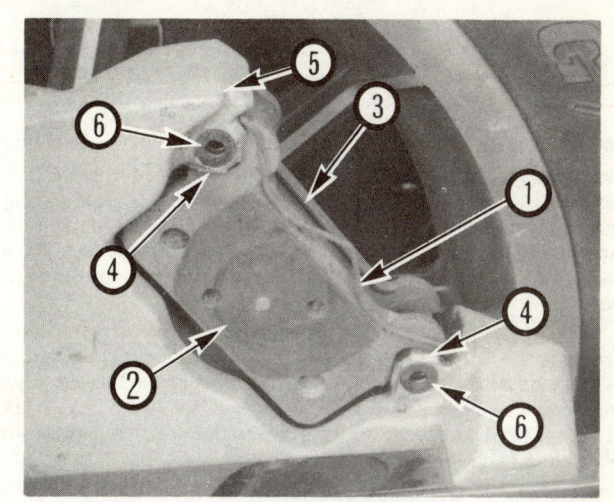

13.12 Brake pads and related components – late 1987 and later models

1	Wire retainer clip	4	Pad shim (2)
2	Outer brake pad	5	Bracket
3	Inner brake pad	6	Rubber bushing (2)

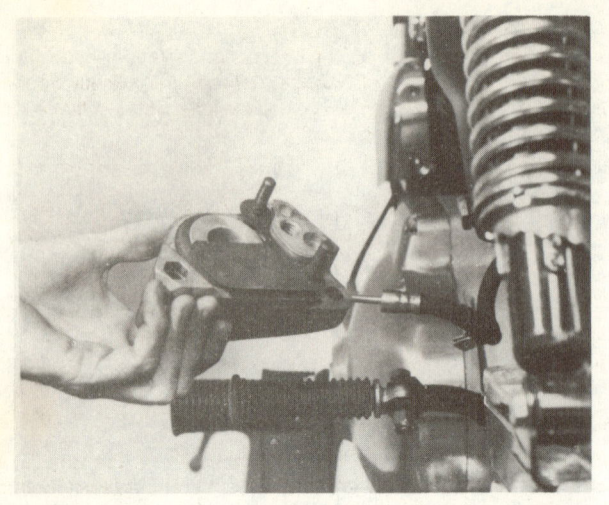

13.17 Slide the outer plate onto the guide pins, then . . .

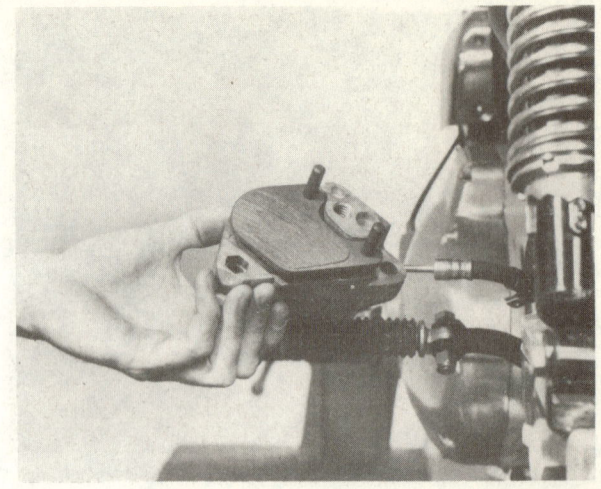

13.18a . . . place the outer brake pad over the guide pins, . . .

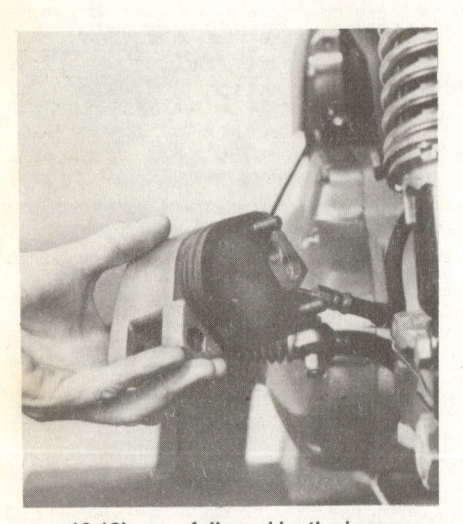

13.18b . . . followed by the inner brake pad . . .

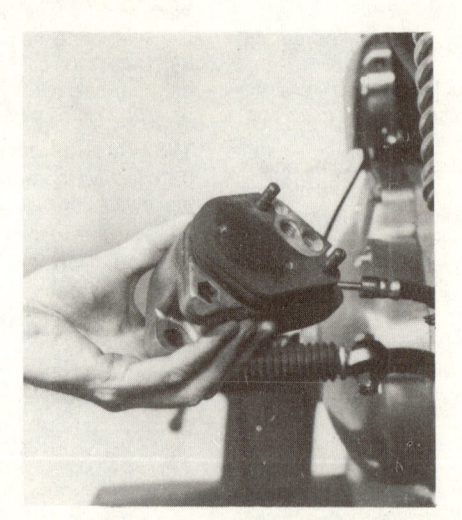

13.18c . . . and the inner plate

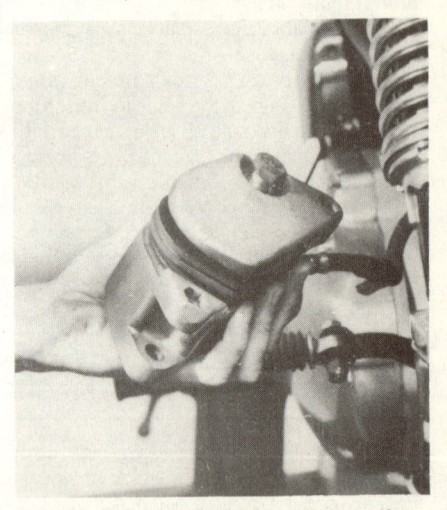

13.19 Connect the inner caliper to the outer caliper (don't forget to tighten the bolt after the caliper is installed)

13 Slide the inner pad off toward the wheel.
14 Detach the pad shims from the bracket.

Pad installation

15 Before installing the new pads, clean the brake disc with brake system cleaner (available at auto parts stores), lacquer thinner or acetone.
Warning: *Do not use petroleum-based solvents.*
16 Push the piston into the caliper bore as far as possible before installing the new brake pads.

1979 through 1981
Refer to illustrations 13.17, 13.18a, 13.18b, 13.18c and 13.19

17 Slide the outer plate over the pins on the outer caliper **(see illustration)**.
18 Install the new brake pads and the inner plate over the guide pins **(see illustrations)**.
19 Attach the inner caliper half to the outer half and install the bolt in the back side of the caliper finger-tight **(see illustration)**.
20 Position the caliper over the rear disc and attach it to the mounting bracket. The two Allen-head mounting bolts should be coated with anti-seize compound before installation. Use new locknuts on the mounting bolts. If new locknuts aren't available, coat the threads of the bolts with thread locking compound.

1982 through early 1987
21 Attach the pad spring to the top of the caliper with the long tab above the piston. The short tab must be hooked above the ridge on the caliper casting, opposite the piston, to hold the spring in place.
22 Position the pads on the bracket and attach the caliper to the bracket without turning the pins. The flat sides of the pin heads should be parallel with the opening in the bracket.
23 Install the caliper mounting bolts and tighten them to the torque listed in this Chapter's Specifications.

Late 1987-on
Caution: *Ensure only the correct parts are fitted when replacing pads, discs or shims. Modifications to the pad and disc material from 1992-on and redesign of the shims and pad shape from mid-1991 prevent the mixing of early and late components.*

24 On all models through mid-1991 position the pad shims on the mounting bracket with the tabs seated in the holes **(see illustration 13.12)**. From mid-1991 fit the shims so that their looped ends are positioned outwards (towards the piston) and hold them in place while the pads and wire retainer are fitted **(see illustration)**.
25 Slide the inner pad onto the shims from the wheel side. Slide the outer pad on from the outside.
26 Insert the wire retainer clip ends into the mounting bracket holes and position the clip over the outer brake pad **(see illustration 13.12)**.

Chapter 6 Wheels, brakes and tires

13.24 Position pad shims (arrows) with looped ends outward – mid-1991 on

Caution: *Make sure the pads are still riding on the shims after the retainer clip is installed.*

27 Make sure the mounting (pin) bolts are clean so the caliper can move freely. Carefully lower the caliper over the pads and disc and align the holes, then install the mounting bolts.
28 Tighten the mounting bolts to the torque listed in this Chapter's Specifications. Install the lower shock mounting bolt.

All models

29 Check the brake fluid level in the master cylinder. After installing new brake pads, the level may be too high. If so, carefully siphon off the excess fluid.

---

14 Rear disc brake caliper – removal, overhaul and installation

1 If the caliper is leaking fluid around the piston, it should be removed and overhauled to restore braking performance. Before disassembling the caliper, read through the entire procedure and make sure you have the correct caliper rebuild kit. Also, you'll need some new, clean brake fluid of the recommended type and some clean rags. **Note:** *Disassembly, overhaul and reassembly of the brake caliper must be done in a spotlessly clean work area to avoid contamination and possible failure of the brake hydraulic system components. If such a work area isn't available, have the caliper rebuilt by a dealer service department or a motorcycle repair shop.*
2 Remove the caliper as described in Section 13.

1979 through 1981

3 Disconnect the brake line from the caliper and plug the end. Do not lose the hose seat.
4 Separate the two halves of the caliper and remove the brake pads.
5 Carefully pry the piston out of the caliper, then remove the rubber boot.
6 To avoid scratching the bore, use a wood or plastic tool to remove the seal.

1982 through early 1987

7 Remove the brake pads from the caliper.
8 Detach the rubber boots from the upper and lower pins, then remove the pins from the caliper.
9 Note how the pad spring is installed on the caliper body, then remove it.
10 Pry the retaining wire out of the caliper and remove the piston boot.
11 Try to withdraw the piston and seal from the caliper bore. If it can't easily be removed, pump the brake pedal slowly until the piston is forced out as far as possible. Detach the brake line from the caliper and place the caliper on a workbench with the piston facing down. Place a clean rag under the piston and apply low air pressure to the brake line inlet hole (a bicycle tire pump should work for this). The air pressure should be enough to force the piston out of the bore, but it may be necessary to tap lightly around the caliper with a soft-face hammer at the same time. Remove the seal from the bore with a wood or plastic tool.

Late 1987-on

12 Detach the retaining wire and remove the rubber boot from the caliper.
13 Slowly pump the brake pedal until the piston doesn't move any further.
14 Disconnect the brake line from the caliper and plug the line.
15 Remove the piston and seal from the caliper. You may have to force the piston out of the caliper with air pressure (see Step 11).
16 Remove the rubber bushings from the mounting bracket bores. If they're worn or damaged, install new ones.

All models

17 Clean all of the brake components (except for the brake pads) with brake cleaning solvent (available at auto parts stores), isopropyl alcohol or clean brake fluid. **Warning:** *Do not, under any circumstances, use a petroleum-based solvent to clean brake parts. If compressed air is available, use it to dry the parts.*
18 Check the caliper bore and the outside of the piston for scratches, nicks and score marks. If damage is evident, the caliper must be replaced with a new one. **Warning:** *Don't attempt to rebore or hone the caliper.*
19 Before reassembling the caliper, soak the new rubber parts in clean brake fluid for 10 or 15 minutes. Lubricate the bore of the caliper with clean brake fluid, then install the seal in the bore. **Warning:** *Always install a new seal – do not reuse the old one if it was removed from the bore.*
20 Carefully insert the piston as far as possible into the bore. Install the boot and retaining wire (1982 through early 1987 models).
21 On 1979 through 1981 models, begin reassembly as described in Section 13.
22 On 1982 through early 1987 models, attach the pad spring to the caliper with the long tab extending above the piston. The short tab should be hooked above the ridge on the caliper casting to hold the spring in place. Lubricate the pins and pin bores with silicone grease and install the pins. The pin with the nylon sleeve fits in the top hole. Turn the pins so the flat edges on the heads are parallel with the opening in the bracket.
23 Reattach the brake hose and complete reassembly as described in Section 13. **Note:** *If the brake hose is attached to the caliper with a banjo fitting and bolt, use new washers when installing the bolt. The replacement washers must be the same type as the originals (some are zinc-plated copper, while others are steel with a rubber O-ring). Be sure to tighten the banjo fitting bolt to the correct torque – it's different, depending on the type of washers used.*
24 Fill the master cylinder with the recommended brake fluid and bleed the system as described in Chapter 1.

---

15 Rear disc brake master cylinder – removal, overhaul and installation

1 If the master cylinder is leaking fluid, or if the pedal doesn't feel firm when the brake is applied – and bleeding the brakes doesn't help – master cylinder overhaul is recommended. Before disassembling it, read through the entire procedure and make sure you have the correct rebuild kit. Also, you'll need some new, clean brake fluid of the recommended type, some clean rags and internal snap-ring pliers. **Note:** *Disassembly, overhaul and reassembly of the master cylinder must be done in a spotlessly clean work area to avoid contamination and possible failure of the brake hydraulic system components. If such a work area isn't available, have the master cylinder rebuilt by a dealer service department or motorcycle repair shop.*

Removal

2 Disconnect the master cylinder-to-caliper brake line from the master cylinder and plug the end. Use a flare-nut wrench, if possible, to avoid rounding off the fitting. Late 1987 and later models have a banjo fitting and bolt, so a flare-nut wrench isn't necessary. **Caution:** *If brake fluid is spilled on a painted surface, wipe it off immediately or the paint will be damaged.*

Chapter 6 Wheels, brakes and tires

15.8 The master cylinder is attached to the engine with two bolts

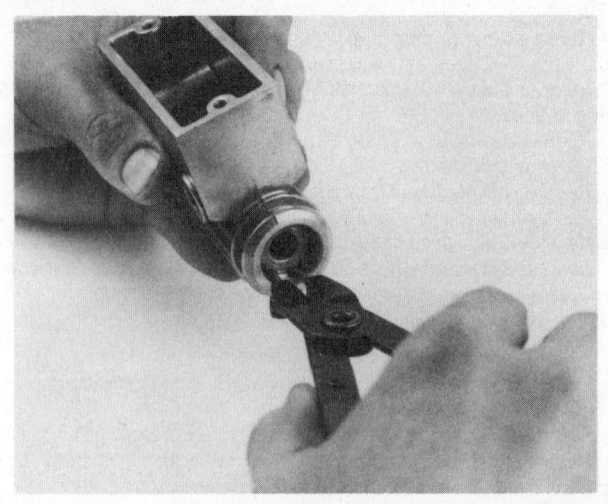

15.11a Remove the snap-ring that secures the piston assembly in the master cylinder bore

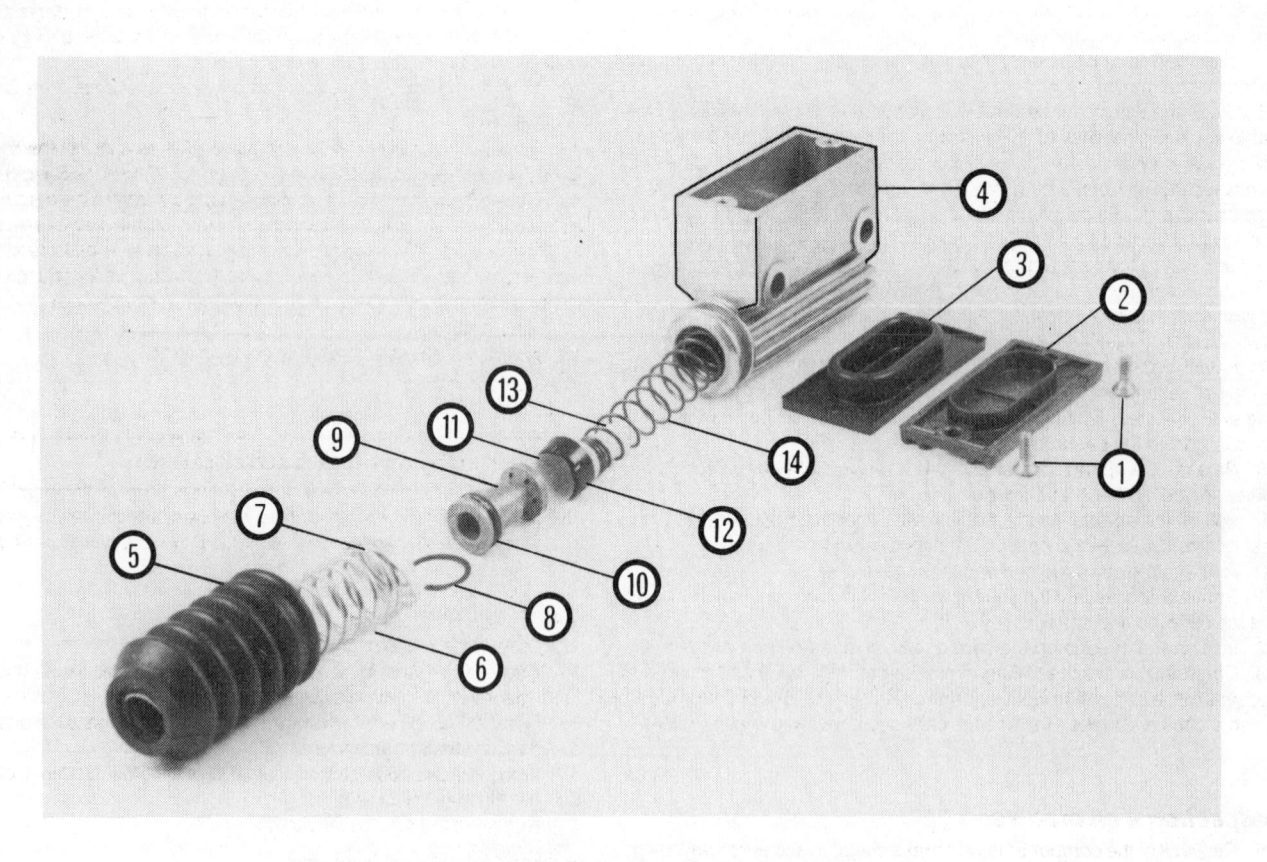

15.11b Rear disc brake master cylinder components – exploded view
(1979 through 1981 shown – 1982-on slightly different)

1	Screws	6	Spring	11	Wafer (thru 1981 only)
2	Cover	7	Cupped washer (2)	12	Piston cup
3	Gasket	8	Snap-ring	13	Spring seat
4	Master cylinder body	9	Piston	14	Spring
5	Dust boot	10	O-ring		

Chapter 6 Wheels, brakes and tires

1979
3 Remove the snap-ring from the clevis pin, then remove the clevis pin and the pushrod from the outer pivot arm.
4 Remove the two mounting bolts and lock washers, then detach the master cylinder.
5 Remove the cover and gasket from the top of the master cylinder and drain the brake fluid out.
6 Remove the dust boot from the end of the master cylinder.
7 Remove the internal snap-ring from the master cylinder bore, followed by the piston assembly, wafer, piston cup, spring seat and spring. Remove the O-ring from the piston.

1980 through early 1987
Refer to illustrations 15.8, 15.11a and 15.11b

8 Unscrew the two master cylinder mounting bolts and pull the master cylinder off the pushrod **(see illustration)**.
9 Remove the cupped washers, the spring and the dust boot.
10 Remove the cover and the gasket from the top of the master cylinder and drain the brake fluid out.
11 Remove the internal snap-ring from the master cylinder bore, followed by the piston assembly, wafer (1980 and 1981 models only), piston cup, spring seat and spring. Remove the O-ring from the piston **(see illustrations)**.

Late 1987-on
12 Back-off the adjuster locknut on the threaded brake pedal rod.
13 Remove the two Allen-head bolts and detach the master cylinder from the sprocket cover on the engine. Free the steel brake line from its clamp on the rear sprocket cover bolt.
14 Turn the internal master cylinder pushrod (that the pedal rod threads into) until the master cylinder is free.
15 Clean the exterior of the master cylinder, then thread the brake line banjo bolt back into the fitting. The bolt will protect the sealing surface in the master cylinder cartridge body during disassembly. **Note:** *To avoid confusion during reassembly, pay close attention to how the parts fit together as the master cylinder is disassembled – lay them out in the proper order and make a simple sketch if necessary.*
16 Don't allow the cartridge body to be contaminated with dirt or grease and don't disassemble it – it contains the piston and related components, which aren't sold separately. If the piston seal is leaking, install a new cartridge body.
17 Hold the master cylinder upright on a clean workbench with the banjo fitting bolt resting on the bench, then press down on the large washer to compress the return spring. With the spring compressed, remove the snap-ring from the pushrod groove.
18 Release the spring and remove the washer, rubber boot, spring retainer (inside the boot) and spring.
19 Pull off the adapter (early models only), then unscrew the large nut that retains the cartridge body in the master cylinder.
20 Remove the cartridge body/pushrod assembly.
21 Remove and discard the snap-ring inside the cartridge body, then pull out the pushrod and spacer.
22 Inspect the threads on the cartridge body, pushrod and banjo bolt.
23 Carefully remove the O-rings from the cartridge body and clean the grooves with a soft cotton cloth saturated with alcohol. Check the grooves for scratches and nicks. Replace the cartridge body if damage is evident.

Inspection – all models
24 Clean all of the components with brake cleaning solvent (available at auto parts stores), isopropyl alcohol or clean brake fluid. **Warning:** *Do not, under any circumstances, use petroleum-based solvents to clean brake parts.* If compressed air is available, use it to dry the parts.
25 Check the master cylinder bore for scratches, nicks and score marks. If damage is evident, the master cylinder must be replaced with a new one.
26 Make sure the vent holes in the master cylinder are open.
27 Before reassembly, soak the new rubber seals in clean brake fluid for 10 or 15 minutes. Lubricate the master cylinder bore with clean brake fluid.

Installation

1979 through early 1987
28 Attach the spring seat to the spring and insert them into the master cylinder bore.
29 Slide the piston cup into the bore, over the spring seat. Place the wafer against the piston cup (1979 through 1981 models only).
30 Install a new O-ring on the piston and push the piston into the bore. Secure it with the snap-ring.
31 Insert the pushrod carefully through the dust boot, into the piston assembly.
32 Attach the pushrod and clevis pin to the outer pivot arm and secure the clevis pin with the snap-ring (1979 models only).

Late 1987-on
33 Install the pushrod and spacer in the cartridge bore and secure them with the snap-ring. Make sure the snap-ring is seated in the groove in the cartridge bore after installing the pushrod. The pushrod must rotate freely.
34 Install new O-rings in the cartridge grooves and lubricate them with clean brake fluid.
35 Insert the cartridge into the master cylinder. The notch in the threaded end of the cartridge body must engage the lug on the inside of the master cylinder.
36 Carefully grip the master cylinder in a vise equipped with soft jaws. The pushrod end must face up.
37 On early models install the adapter over the pushrod (large diameter end in the master cylinder).
38 Install the spring, retainer and boot over the pushrod. The small end of the boot must seat against the retainer and the vent hole must be positioned so it's at the bottom of the boot when the master cylinder is installed.
39 Slip the washer over the pushrod and compress the spring, then install a new snap-ring in the pushrod groove.
40 Attach the master cylinder to the threaded pedal rod (see Step 14).

All models
41 Position the master cylinder on the engine and install the mounting bolts. Tighten the bolts to the torque listed in this Chapter's Specifications.
42 Connect the brake line to the master cylinder and fill the reservoir with brake fluid. If a banjo fitting is used, be sure to install new washers of the correct type (DO NOT interchange zinc-plated copper washers with steel/rubber washers) and tighten the banjo bolt to the torque listed in this Chapter's Specifications – the torque is different, depending on the type of washers used.
43 Bleed the brakes as described in Chapter 1.
44 Adjust the brake pedal.

16 Rear brake pedal – check and adjustment

Note: *The specified amount of free play between the brake pedal and master cylinder is very important. Without it, the rear disc brake may drag or fail to operate correctly.*

1979
1 Loosen the brake pedal stop bolt locknut.
2 Depress the pedal by hand until resistance is felt as the pushrod contacts the master cylinder piston.
3 Note how far the end of the outer pivot arm (attached to the pushrod in the master cylinder) has moved.
4 Turn the brake pedal stop bolt as needed to obtain 1/16-inch of movement at the outer pivot arm.
5 Retighten the stop bolt locknut.

1980 through early 1987
6 Loosen the locknuts on the brake pedal stop bolt and adjust the brake pedal until the desired height is obtained, then retighten the locknuts.
7 Depress the pedal by hand until slight resistance is felt as the pushrod contacts the master cylinder piston.
8 Measure the gap between the pedal and stop bolt – it should be 1/16-inch. If it isn't, loosen the locknut and turn the pushrod (that contacts the master cylinder piston) until it is, then retighten the locknut.

Chapter 6 Wheels, brakes and tires

Late 1987-on

9 With the motorcycle on a level surface, loosen the adjuster locknut on the threaded brake pedal rod.
10 Turn the internal master cylinder pushrod (that the pedal rod threads into) until the brake pedal is parallel with the floor.
11 Retighten the adjuster locknut.

17 Rear wheel sprocket – removal and installation

1 Remove the rear wheel as described in Section 11.

Drum brake models

Note: *Special tools are required to install a new sprocket on drum brake models. Read the procedure before beginning removal – you may decide to have a Harley-Davidson dealer do the job.*

2 The rear wheel sprocket is riveted to the brake drum. If examination indicates some of the teeth are chipped, broken or hooked, the brake drum must be removed from the wheel. It's attached to the wheel by five bolts.
3 Remove the bolts securing the brake drum to the hub.
4 the sprocket is attached to the brake drum with rivets. Using a sharp chisel, cut the heads off the rivets and dowel pins.
5 If the rivet holes aren't worn or elongated, the new rivets can be installed in the same holes. However, if they're worn or elongated, a new set of holes should be drilled midway between the existing holes and the dowel pin holes. A 9/64-inch drill bit is required to bore the holes.
6 New dowel pin holes should be drilled also (use a 3/16-inch drill bit). The dowel pins must be a tight fit. Use the new sprocket as a template.
7 Drill a rivet hole, then place the sprocket in position on the drum. Insert a rivet through the hole but don't head the rivet. Drill another hole across from the first rivet and insert another rivet. Again, don't head the rivet.
8 Drill the remaining holes, then drill the four dowel pin holes.
9 Remove the sprocket from the brake drum and deburr the newly drilled holes.
10 Position the drum and sprocket on the center support of the riveting jig (special tool no. 95600-33A).
11 The dowel pins are installed first, followed by the rivets. The dowel pins and the rivets must be installed through the brake drum side.
12 Using a hollow driver, seat the dowel pin and rivet simultaneously, driving the sprocket and hub flange together.
13 Using a concave punch, flare the end of the dowel pin. Head the end of the rivet until it extends 3/64-inch above the face of the sprocket.
14 Repeat Steps 11 through 13, seating the rivets and dowel pins on the opposite side of the hub until all of them are in place. This will prevent distortion of the sprocket as the rivets are installed.
15 Install the brake drum on the hub and reinstall the rear wheel as described in Section 11.

Disc brake models

16 The sprocket is attached to the rear wheel hub with five bolts.
17 On chain drive models, if examination of the sprocket indicates the teeth are chipped, broken or hooked, the sprocket should be replaced with a new one. If the rear wheel sprocket is worn, the final drive sprocket on the transmission is probably worn also. It's a good idea to replace both sprockets at the same time, as well as the chain. This will alleviate rapid wear that results from mixing old and new parts.
18 On belt drive models, examine the sprocket teeth and flange for signs of damage. The sprocket will not wear to the same extent as chain drive models, it being more likely that the belt itself will require replacement.
19 Remove the bolts securing the sprocket to the hub. Lift the sprocket off and place the new one in position.
20 Install the sprocket mounting bolts and washers (having applied a drop of thread-locking compound to the bolt threads) and tighten the bolts to the torque listed in this Chapter's Specifications.
21 Install the rear wheel as described in Section 11.

18 Rear brake disc – removal and installation

1 Remove the rear wheel as described in Section 11.
2 The brake disc is attached to the rear hub with five bolts. Remove the bolts and detach the brake disc from the hub.
3 Before installing the disc, be sure the threads on the bolts and in the hub are clean and undamaged. It is recommended that new screws be used when installing the disc; unless the new screws are supplied with a patch of locking compound on them, apply a few drops of thread-locking compound yourself. Install the bolts and tighten them in a criss-cross pattern until the specified torque is reached. **Note:** *New bolts containing the locking patch can be removed and installed up to three times, then they should be replaced.*

19 Tires – removal and installation

Note: *This procedure does not apply to tubeless tires!*

1 To properly remove and install tires, you'll need at least two motorcycle tire irons, some tire mounting lubricant (available at motorcycle dealers and accessory stores) or a solution of water and liquid soap, some talcum powder and a tire pressure gauge.
2 Begin by removing the wheel from the machine. If the tire is going to be reused, mark it next to the valve stem or wheel balance weight.
3 Deflate the tire by removing the valve stem core. When it's completely deflated, push the tire bead away from the rim on both sides. In some extreme cases, and on all 16-inch wheels, this can only be accomplished with a bead breaking tool, but usually it can be done with tire irons. Riding on a deflated tire to break the tire bead is not recommended – damage to the rim and tire will result.
4 Dismounting a tire is easier when the tire is warm, so an indoor tire change is recommended in cold climates. The rubber gets very stiff and is difficult to manipulate when cold.
5 Place the wheel on a thick pad or old blanket. This will help keep the wheel and tire from slipping around. On disc brake equipped wheels, place the disc down or remove it from the hub.
6 Once the bead is completely off the rim, lubricate the inside edge of the rim and the tire bead with a solution of water and liquid soap or tire mounting lubricant. Remove the locknut and push the tire valve stem through the rim.
7 Insert one of the tire irons under the bead at the valve stem and lever the bead up over the rim. This should be fairly easy. Take care not to pinch the tube as this is done. When removing tires from cast wheels, cushion the rim with a shop towel or some other type of pad to avoid gouging the rim. If it's difficult to pry the bead up, make sure the rest of the bead opposite the valve stem is in the dropped center section of the rim.
8 Hold the tire iron down with the bead over the rim, then move about three or four inches to either side and insert the second tire iron. Be careful not to cut or slice the bead or the tire may split when inflated. Also, don't catch or pinch the inner tube as the second tire iron is levered over (this is why tire irons are recommended over screwdrivers or other implements).
9 With a small section of the bead up over the rim, one of the levers can be removed and reinserted a few inches around the rim until about one-quarter of the tire bead is above the rim edge. Make sure the rest of the bead is in the dropped center of the rim. At this point, the bead can usually be pulled up over the rim by hand.
10 Once all of the first bead is over the rim, the inner tube can be withdrawn from the tire and rim. Push in on the valve stem, lift up on the tire next to the stem, reach inside the tire and carefully pull out the tube. It usually isn't necessary to completely remove the tire from the rim to repair the inner tube. It's sometimes recommended though because checking for foreign objects in the tire is difficult while it's still partially mounted on the rim.
11 To remove the tire completely, make sure the bead is broken all the way around on the remaining edge, then stand the tire and wheel up on the tread and grab the wheel with one hand. Push the tire down over the same edge of the rim while pulling the rim away from the tire. If the bead is correctly positioned in the dropped center of the rim, the tire should roll off and separate from the rim very easily. If tire irons are used to work the last bead over the rim, the outer edge of the rim may be marred. If a tire iron is necessary, be sure to pad the rim as described earlier.

TIRE CHANGING SEQUENCE - TUBED TIRES

A. Deflate tire. After pushing tire beads away from rim flanges push tire bead into well of rim at point opposite valve. Insert tire lever next to valve and work bead over edge of rim.

B. Use two levers to work bead over edge of rim. Note use of rim protectors

C. Remove inner tube from tire

D. When first bead is clear, remove tire as shown

E. To install, partially inflate inner tube and insert in tire

F. Work first bead over rim and feed valve through hole in rim. Partially screw on retaining nut to hold valve in place.

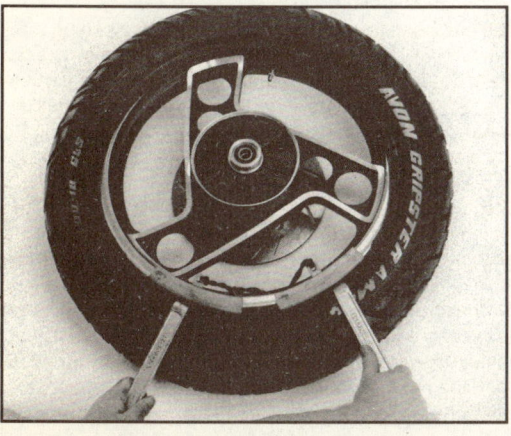

G. Check that inner tube is positioned correctly and work second bead over rim using tire levers. Start at a point opposite valve.

H. Work final area of bead over rim while pushing valve inwards to ensure that inner tube is not trapped.

12   Refer to Section 20 for inner tube repair procedures.
13   Mounting the tire is basically the reverse of removal. Some tires have a balance mark and/or directional arrows molded into the tire sidewall. Look for these marks so the tire can be installed properly. The dot should be aligned with the valve stem.
14   If the tire wasn't removed completely to repair or replace the inner tube, the tube should be inflated just enough to make it round. Sprinkle it with talcum powder, which acts as a dry lubricant, then carefully lift up the tire edge and install the tube with the valve stem next to the hole in the rim. Once the tube is in place, push the valve stem through the rim and start the locknut on the stem.
15   Lubricate the tire bead, then push it over the rim edge and into the dropped center section opposite the inner tube valve stem. Work around each side of the rim, carefully pushing the bead over it. The last section may have to levered on with tire irons. If so, be careful not to pinch the inner tube as this is done.
16   Once the bead is over the rim edge, check to see if the inner tube valve stem is pointing to the center of the hub. If it's angled slightly in either direction, rotate the tire on the rim to straighten it out. Run the locknut the rest of the way down the stem but don't tighten it completely.
17   Inflate the tube to approximately 50 psi and check to make sure the guidelines on the tire sidewalls are the same distance from the rim around the entire circumference of the tire.
18   After the tire bead is correctly seated on the rim, allow the tire to deflate. Replace the valve core and inflate the tube to the recommended pressure, then tighten the valve stem locknut securely and install the cap.

20   Tubes – repair

1   Inner tube repair requires a patching kit that's usually available from motorcycle dealers, accessory stores and auto parts stores. Be sure to follow the directions supplied with the kit to ensure a safe, long-lasting repair.
2   If the inner tube has been patched previously, or if there's a tear or large hole, it should be replaced with a new one. Sudden deflation can cause loss of control and an accident, particularly if it occurs on the front wheel.
3   To repair a tube, remove it from the tire, then inflate and immerse it in a tub or sink full of water to pinpoint the leak. Mark the position of the leak, then deflate the tube, Dry it off and thoroughly clean the area around the puncture.
4   Most tire patching kits have a buffer to rough up the area around the hole for proper adhesion of the patch. Roughen an area slightly larger than the patch, then apply a thin coat of the patching cement to the roughened area. Allow the cement to dry until tacky, then apply the patch.
5   You may have to remove a protective covering from the top surface of the patch after its been attached to the tube. Keep in mind that inner tubes made from synthetic rubber may require a special type of patch and adhesive to achieve a satisfactory bond.
6   Before replacing the tube, check the inside of the tire to make sure the object that caused the puncture isn't lying inside it. Also check the outside of the tire, particularly the tread area, to make sure nothing is projecting through the tire that may cause another puncture. Check the rim for sharp edges and damage. On wire spoke wheels, make sure the rubber rim band is in good condition and properly installed before inserting the tube.

21   Wheel bearings – repack

Refer to Chapter 1 for wheel bearing inspection and maintenance procedures.

Chapter 7 Electrical system

Contents

General information 1	Starter motor – disassembly, inspection and reassembly 13
Electrical troubleshooting – general information 2	Starter solenoid – check 14
Charging system check – general information 3	Handlebar switches – removal and installation 15
Battery – check and maintenance 4	Ignition and light switch – removal and installation 16
Battery – charging 5	Horn – adjustment 17
Battery specific gravity – check 6	Brake light switches – adjustment 18
Headlight bulb – replacement 7	Warning light bulbs – replacement 19
Taillight and turn signal bulbs – replacement 8	Evaporative emission control system – solenoid test (1992-on
Generator/alternator – check 9	California models) 20
Generator/alternator – removal, overhaul and installation 10	Turn signal relay/cancel unit – location 21
Regulator – general information 11	Wiring diagrams – general information 22
Starter motor – removal and installation 12	

Specifications

Battery
Voltage	12-volts
Electrolyte specific gravity (at 80-degrees F)	
100-percent charge	1.250 to 1.270
75-percent charge	1.220 to 1.240
50-percent charge	1.190 to 1.210
25-percent charge	1.160 to 1.180
Ground connection	Negative

Generator
Type	Two brush
Minimum brush length	1/2 in (12.7 mm)

Starter motor brush length (minimum)
1970 through 1980	
Prestolite	1/4 in (6.35 mm)
Hitachi	0.438 in (11.13 mm)
1981-on	0.354 in (8.99 mm)

Bulbs – 1970 through 1983
Headlight	
1970 through 1978	45/35 W
1979-on	50/35 W
Brake/taillight	
1970 through 1978	32/4 cp
1979 through 1983	32/3 cp
Generator warning light	4 cp
Oil pressure warning light	4 cp
High beam indicator light	2 cp
Speedometer/tachometer lights	2 cp
Turn signal lights	32 cp

Bulbs – 1984-on .... Consult a Harley-Davidson dealer

Chapter 7 Electrical system

Torque specifications

	Ft-lbs (unless otherwise indicated)	Nm
Starter through bolts		
1970 through 1978	20 to 25 in-lbs	2.3 to 2.8
1979 and 1980	60 to 80 in-lbs	6.8 to 9
Starter mounting bolts (1981-on)	13 to 20	18 to 27
Battery cable-to-starter terminal (1979 and 1980)	65 to 80 in-lbs	7.3 to 9
Alternator stator Torx screws	30 to 40 in-lbs	3.4 to 4.5
Alternator rotor bolts (1991-on)	90 to 110 in-lbs	10 to 12

1 General information

All models covered in this manual are equipped with a 12-volt electrical system. The charging system on early models (through early 1984) is made up of a two-pole, two-brush DC generator, driven by the timing gears. The output of the generator is controlled by a voltage regulator to keep the battery charged and to meDet the requirements of the motorcycle. Since the generator output is DC, there's no need for a rectifier.

On late 1984 and later models, the charging system consists of an alternator and a rectifier/regulator. On models through 1990 the alternator stator is bolted to the transmission access cover, behind the clutch on the left-hand side of the engine, and the rotor is mounted on the rear of the clutch outer drum. From 1991 the alternator stator is mounted on the left crankcase at the front, and the rotor is bolted to the rear of the primary drive sprocket. The rectifier/regulator is attached to the frame downtubes in front of the engine, where it's cooled by airflow as the machine is moving.

A large capacity battery is installed on models with an electric starter. A solenoid relay provides power to the starter motor directly from the battery. The solenoid is controlled by a switch on the handlebars.

Keep in mind that electrical parts, once purchased, can't normally be returned. To avoid unnecessary expense, make very sure the defective component has been positively identified before buying a replacement part.

Caution: *When working on the electrical system, the battery should be disconnected to avoid accidentally causing a short circuit in the system. Always disconnect the negative cable first, followed by the positive cable.*

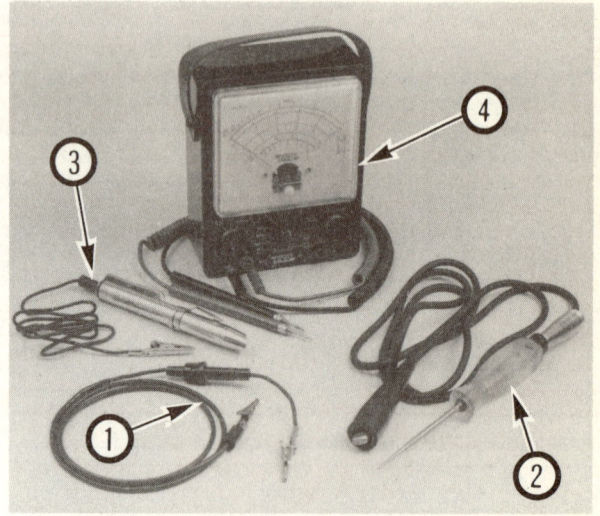

2.5 **Basic electrical troubleshooting can be done with an assortment of tools from a simple jumper wire to a multimeter (volt/ohm/ammeter)**

1	Jumper wire (with in-line fuse)	3	Continuity tester
2	12-volt test light	4	Multimeter

2 Electrical troubleshooting – general information

Refer to illustration 2.5

A typical electrical circuit consists of an electrical component, any switches, relays, motors, fuses or circuit breakers related to that component and the wiring and connectors that link the component to both the battery and the frame. To help pinpoint electrical circuit problems, wiring diagrams are included at the end of the manual.

Before tackling any troublesome electrical circuit, first study the appropriate wiring diagram to get a complete understanding of what makes up the circuit. Trouble spots, for instance, can often be narrowed down by noting if other components related to the circuit are operating properly. If several components or circuits fail at one time, chances are the problem is in a fuse or ground connection, because several circuits are often routed through the same ones.

Electrical problems usually stem from simple causes, such as loose or corroded connections, a blown fuse or a bad relay. Visually check the condition of all fuses, wires and connections in a problem circuit before troubleshooting it.

If test instruments are going to be utilized, use the diagram to plan ahead of time where to make the connections in order to accurately pinpoint the trouble spot.

The basic items needed for electrical troubleshooting include a 12-volt test light, a voltmeter, a self-powered continuity tester (which includes a bulb or buzzer, battery and set of test leads), and a jumper wire, preferably with a fuse incorporated, which can be used to bypass electrical components **(see illustration)**.

Voltage checks

A voltage check should be done if a circuit isn't functioning properly. Connect the test light alligator clip to the negative battery terminal. Use the test light probe to contact the connector wire terminal(s) in the circuit being tested, starting with the one nearest to the battery or fuse. If the test bulb lights, voltage is present, which means the part of the circuit between the connector and battery is okay. Continue checking the rest of the circuit in the same manner. When you reach a point where no voltage is indicated, the problem is between that point and the last test point with voltage. Most of the time the problem can be traced to a loose connection.

Note: *Keep in mind that some circuits receive voltage only when the ignition key is in the On position.*

Finding a short

One method of finding a short in a circuit is to remove the fuse and connect a test light or voltmeter in its place to the fuse terminals. The circuit should be off (no voltage present in the circuit). Move the wiring harness from side-to-side while watching the test light. If the bulb lights, there's a short to ground somewhere in that area, probably where the insulation has rubbed through. The same test can be performed on each component in the circuit, even a switch.

Ground check

Perform a ground check to see if a component is properly grounded. Disconnect the battery and connect one lead of a self-powered test light, known as a continuity tester, to a known good ground. Connect the other lead to the wire or ground connection being tested. If the bulb lights, the ground is good. If the bulb doesn't light, the ground is bad.

Chapter 7 Electrical system

Continuity check

A continuity check is done to determine if there are any breaks in a circuit – if it's capable of passing electricity properly. With the circuit off (no power in it), a self-powered continuity tester can be used to check it. Connect the test leads to both ends of the circuit (or to the "hot" wire and a good ground). If the test light comes on, the circuit is passing current properly. If the light doesn't come on, there's a break (called an open) somewhere. The same procedure can be used to test a switch, by connecting the continuity tester leads to the switch terminals. With the switch on, the test light should come on.

Finding an open circuit

When looking for possible open circuits, it's almost impossible to locate them by sight because oxidation and terminal misalignment are hidden by the connectors. Sometimes, wiggling a connector in the wiring harness will correct the open circuit condition temporarily. Remember this when an open circuit is indicated when troubleshooting a circuit. Intermittent problems may also be caused by oxidized or loose connections.

Electrical troubleshooting is simple if you keep in mind that all electrical circuits are basically electricity running from the battery, through the wires, switches, relays, fuses, etc. to each electrical component (light bulb, motor, etc.) and to ground, then back to the battery. All electrical problems are an interruption in the flow of electricity to and from the battery.

3 Charging system check – general information

1 If the battery isn't being recharged, the charging system should be checked first, followed by testing of the individual components (the generator/regulator or alternator/regulator/rectifier). Before beginning the checks, make sure the battery is fully charged and all system connections are clean and tight (particularly the battery cables).
2 Checking the output of the charging system and the operation of the components in the system requires special electrical test equipment. A voltmeter and ammeter or a multimeter are the absolute minimum tools required. In addition, an ohmmeter is generally required for checking the remainder of the electrical system.
3 When making the checks, follow the procedures carefully to prevent incorrect connections and short circuits – irreparable damage to electrical system components may result if a short circuit occurs. Because of the special tools and expertise required, checking the electrical system normally should be left to a dealer service department or a reputable motorcycle repair shop.

4 Battery – check and maintenance

1 The battery on models through 1996 has removable filler caps that allow the addition of water to the electrolyte. Later models are equipped with a sealed, maintenance-free battery. Never remove the cap strip or attempt to add water to a sealed battery.
2 Most battery damage is caused by heat, vibration and/or low electrolyte level. Keep the battery securely mounted and make sure the charging system is functioning correctly. On models through 1996, check the electrolyte level frequently. Refer to Chapter 1 for the electrolyte level checking procedure.
3 On maintenance free batteries, condition is indicated by the battery's open circuit voltage. To check this, disconnect the battery negative cable, then the positive cable. Connect the positive terminal of a voltmeter to the battery positive terminal and the voltmeter's negative terminal to the negative terminal of the battery. Readings are as follows:
  a) 13 volts – 100 percent charged
  b) 12.8 volts – 75 percent charged
  c) 12.5 volts – 50 percent charged
  d) 12.2 volts – 25 percent charged

4 On models through 1996, check around the base of the battery for sediment, which is the result of sulfation caused by low electrolyte levels. These deposits will cause internal short circuits, which can quickly discharge the battery. On all models, look for cracks in the case. Replace the battery if either of these conditions are found.
5 Check the battery terminals and the cable ends for tightness and corrosion. If corrosion is evident, remove the cables from the battery and clean the terminals and cable ends with a wire brush or a knife and emery cloth. Reconnect the cables and apply a thin coat of petroleum jelly to the connections to slow further corrosion.
6 The battery case should be kept clean to prevent current leakage, which can discharge the battery over a period of time (especially when it sits unused). Wash the outside of the case with a solution of baking soda and water. *Caution: Do not get any baking soda solution in the battery cells. Rinse the battery thoroughly, then dry it.*
7 If acid has been spilled on the frame or battery box, neutralize it with the baking soda and water solution, dry it thoroughly, then touch up any damaged paint. Make sure the battery vent tube is directed away from the frame and the final drive chain or belt and isn't kinked or pinched.
8 If the motorcycle sits unused for long periods of time, refer to Section 5 and charge the battery approximately once every month.

5 Battery – charging

1 If the machine sits idle for extended periods of time or if the charging system malfunctions, the battery can be charged from an external source.
2 To charge the battery properly, you will need a charger of the correct rating. If you're working on a fillable battery, you'll also need a hydrometer, a clean rag and a syringe for adding distilled water to the battery cells.
3 The maximum charging rate for any battery is 1/10 of the rated amp/hour capacity. For example, the maximum charging rate for a 22 amp/hour battery would be 2.2 amps; the maximum rate for the 1997 maintenance free battery, which is rated at 18 amp/hours, is 1.8 amps. If the battery is charged at a higher rate, it could be damaged.
4 Don't allow the battery to be subjected to a so-called quick charge (high rate of charge over a short period of time) unless you're prepared to buy a new battery.
5 When charging the battery, always remove it from the machine before hooking it to the charger. If you're working on a fillable battery, be sure to check the electrolyte level and add distilled water to any cells that are low before you start charging the battery.
6 If you're working on a fillable battery, loosen the cell caps and cover the top of the battery with a clean rag. Hook up the battery charger leads (positive to battery positive and negative to battery negative). Then – and only then – plug in the battery charger. **Warning:** *Remember, the hydrogen gas escaping from a battery cell is explosive, so keep open flames and sparks well away from the area. Also, the electrolyte is extremely corrosive and will damage anything it comes in contact with.*
7 If you're working on a fillable battery, allow the battery to charge until the specific gravity is as specified. The charger must be unplugged and disconnected from the battery when making specific gravity checks.
8 If you're working on a maintenance free battery, charge at a constant 1.8 amps for the following periods of time, depending on the open circuit voltage reading obtained in Section 4:
  a) 12.8 volts – three to five hours
  b) 12.5 volts – four to seven hours
  c) 12.2 volts – ten hours
If a tapering charger or trickle charger is used, you'll need to charge the battery for a longer period of time.
9 If the battery gets warm to the touch or gases excessively, the charging rate is too high. Either disconnect the charger and let the battery cool down or lower the charging rate to prevent damage to the battery.
10 If one or more of the cells do not show an increase in specific gravity after a long slow charge (fillable batteries only) or if the battery as a whole doesn't seem to want to take a charge, it's time for a new battery.
11 When the battery is fully charged, unplug the charger first, then disconnect the charger leads from the battery. Install the cell caps (fillable batteries only) and wipe any electrolyte off the outside of the battery case.

6 Battery specific gravity – check

Caution: *Be extremely careful when handling or working around the battery. The electrolyte is very caustic and an explosive gas is given off when the battery is being charged.*

1 To check and replenish the battery electrolyte, you'll have to remove the battery from the motorcycle.

1970 through 1979

2 Loosen the nuts securing the battery retaining strap and release the strap from the bottom of the battery box.
3 Tilt the side cover to clear the frame and release it.
4 Disconnect the battery cables (negative cable first) and lift the battery out of the motorcycle. Note how the vent hose is routed.

1980-on

Refer to illustration 6.6

5 Remove the seat as described in Chapter 5 and disconnect the battery cables from the battery (negative cable first).
6 Remove the side cover from the right side of the motorcycle **(see illustration)**.
7 Remove the battery retaining strap and carefully slide the battery out the right side of the motorcycle. The battery may be stuck to the rubber anti-vibration pad it rests on. If so, simply rock it from side-to-side until the seal is broken. Note how the vent hose is routed.

All models

8 Check the electrolyte level in each battery cell. If any of the cells are low, fill them to the proper level with distilled water. **Note:** *Do not use tap water (except in an emergency) and do not overfill the battery. The cell holes are quite small, so it might help to use a plastic squeeze bottle with a small spout to add the water.*
9 Next, check the specific gravity of the electrolyte in each cell with a small hydrometer made especially for motorcycle batteries. They are available at most dealer parts departments and motorcycle accessory shops.
10 Remove the caps, draw some electrolyte from the first cell into the hydrometer and note the specific gravity. Compare the reading to the Specifications. Return the electrolyte to the appropriate cell and repeat the check for the remaining cells. When the check is complete, rinse the hydrometer thoroughly with clean water.
11 If the specific gravity of the electrolyte in each cell is as specified, the battery is in good condition and is apparently being charged by the machine's charging system.
12 If the specific gravity is low, the battery isn't fully charged. This may be due to corroded battery terminals, a dirty battery case, a malfunctioning charging system, or loose or corroded wiring connections. On the other hand, it may be because the battery is worn out, especially if the machine is old, or that infrequent use of the motorcycle prevents normal charging from taking place. If the specific gravity of any two cells varies by more than 50 points, replace the battery with a new one.
13 Be sure to correct any problems and charge the battery if necessary before reinstalling it in the machine. Refer to Sections 4 and 5 for additional battery maintenance and charging procedures.
14 Install the battery by reversing the removal sequence. Be very careful not to pinch or otherwise restrict the battery vent tube, as the battery may build up enough internal pressure during normal charging system operation to explode.

7 Headlight bulb – replacement

Refer to illustrations 7.2, 7.3 and 7.4

1 Models through 1991 have a sealed beam headlight, and 1992-on models have a quartz bulb headlight. The sealed beam unit comprises the reflector, glass and bulb as a single unit, whereas the quartz bulb can be replaced separately from the reflector unit.
2 To gain access to the bulb, remove the screw through the chrome plated clamp that surrounds the headlight **(see illustration)**. Take off the clamp, then pry the headlight unit away from the rubber mount.
3 Lift the headlight out of the housing until the wire harness connector can be unplugged from the rear **(see illustration)**.
4 Where a quartz bulb is fitted, release the wire clip from its slot and hinge the clip backward. Holding the bulb by its wire connector terminals, withdraw it from the reflector **(see illustration)**. **Caution:** *When fitting a new bulb, do not touch its glass envelope – bulb life will be shortened by skin contact.*
5 Attach the wire connector to the rear of the new headlight and position the headlight in the housing.
6 Install the chrome plated clamp and tighten the screw securely.
7 Adjustment should not have been altered while changing the headlight. If necessary, refer to Chapter 1 for the headlight adjustment procedure.

8 Taillight and turn signal bulbs – replacement

Refer to illustrations 8.1a and 8.1b

1 Remove the screws securing the plastic lens cover to the taillight or the turn signal and detach the cover **(see illustrations)**.
2 Push in on the bulb and simultaneously turn it counterclockwise.

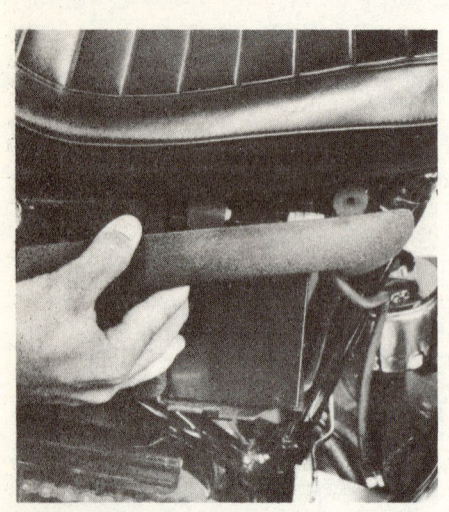

6.6 On 1980 and later models, remove the right side cover to gain access to the battery

7.2 Loosen the clamp screw to remove the headlight

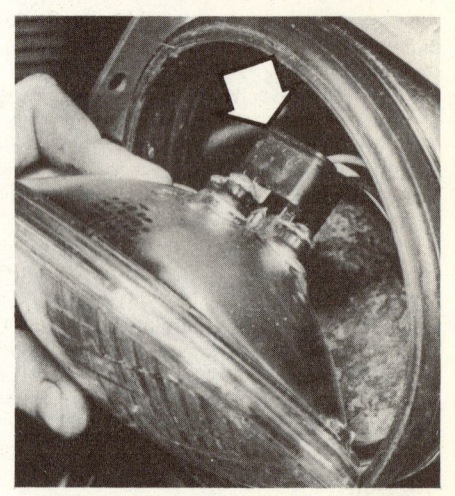

7.3 Unplug the wire harness connector from the rear of the headlight

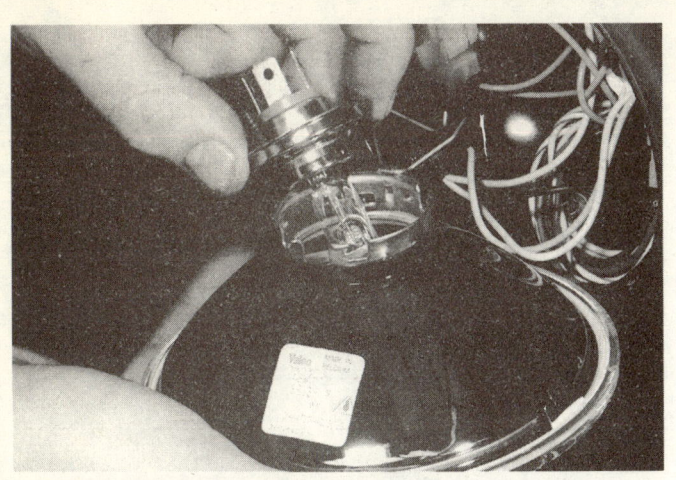

Chapter 7 Electrical system

7.4 Removing the quartz headlight bulb from the reflector – 1992-on models

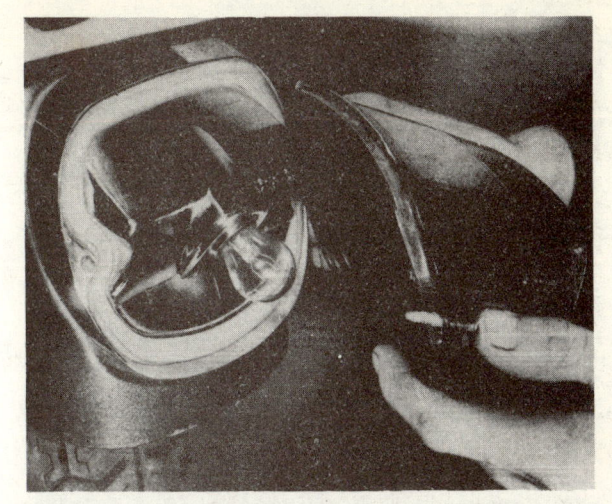

8.1a Remove the plastic lens for access to the taillight/brake light bulb

3 Replace the bulb with a new one of the same type by pushing it into the socket and turning it clockwise. **Note:** *The taillight bulb is a two filament bulb. The pins at the base of the bulb are offset so the bulb can only be inserted into the socket one way. If the bulb will not go into the socket, pull it out and rotate it 180-degrees, then reinsert it into the socket.*

4 Place the lens in position on the housing. Be sure the rubber seal is in good condition and makes contact all around the perimeter of the lense. Secure the lens with the two mounting screws.

9 Generator/alternator – check

1 The generator or alternator can be tested without removing it from the motorcycle. **Note:** *For all tests to be accurate, the battery must be in good condition and fully charged.*

Generator

1970 through 1978

Refer to illustration 9.2

2 On 1970 through 1977 models, test the circuit to the generator warning light to be sure it isn't grounded. Remove the wire or wires from the regulator terminal marked D or GEN and position them so they don't make contact with any part of the machine **(see illustration).** Turn the ignition on and note whether the warning light glows. If it does, there's a short circuit somewhere in the wiring, which is probably the cause of the generator malfunction.

3 **Caution:** *Never ground either the F terminal (XL models) or the BT terminal (XLCH models) before the wires to the terminal are disconnected. Failure to observe this precaution will result in permanent damage to the regulator.* If the light doesn't glow, or if the test is being made on a 1978 model, remove the wire from the F or field terminal of the generator. Connect a short jumper wire to the terminal, so it's grounded to some convenient part of the frame. A good ground connection is important. Disconnect the wire from terminal A of the generator and attach the positive lead from an ammeter to the vacant terminal.

4 Start the engine and run it at approximately 2000 rpm. Momentarily connect the negative lead of the ammeter to the positive terminal of the battery or to the terminal marked BAT on the regulator. If a reading of 10-amps or more is obtained, the generator is operating properly. If a somewhat lower reading or no reading at all is obtained, the generator should be removed from the engine for additional checks.

1979-on

5 Disconnect the wires from terminals A and F of the generator. Connect the positive lead of a voltmeter, adjusted to 10-volts on the DC scale, to terminal A. Connect the negative lead of the voltmeter to ground.

8.1b Turn signals have bayonet-type bulbs (push in and turn them counterclockwise for removal)

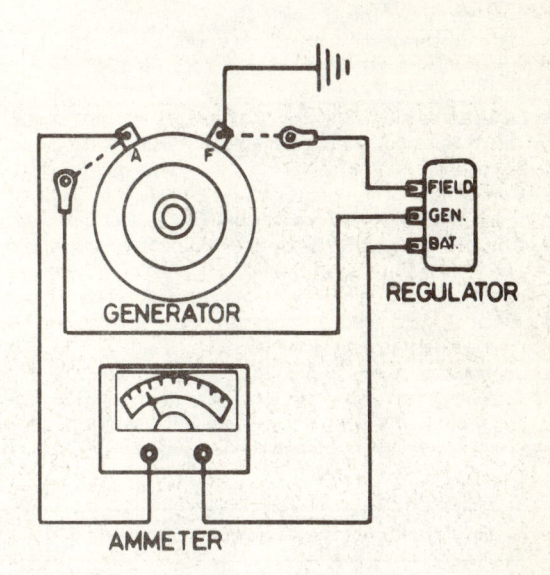

9.2 Hook up the ammeter as shown here to check the generator on 1970 through 1978 models

10.1 Label the wires end terminals, then detach the two wires from the generator

10.2a Check the brushes for wear and damage

6 Start the engine and run it at 2000 rpm. The voltmeter should read a minimum of 2.0-to-2.5 volts DC. If the meter registers the correct output, the generator is in good condition.

7 If zero or very little voltage is registered on the voltmeter, polarize the generator and test the output again. To polarize the generator while it's still installed on the engine, connect one end of a jumper wire to the terminal on the generator armature (A) then momentarily touch the other end of the jumper wire to the positive terminal of the battery.

8 If the voltmeter still registers zero or very low voltage, the generator will have to be disassembled and repaired.

9 With the voltmeter connected to the generator and the engine running at 2000 rpm, momentarily (not more than 10 seconds) connect a jumper wire to terminal F. The voltage should be 25-to-30 volts DC. If not, the generator will have to be disassembled and repaired.

Alternator

10 Checking of the alternator output requires access to test equipment not normally available to the home mechanic, plus a degree of skill to determine the alternator's condition from the results. It is recommended that the motorcycle be taken to a Harley-Davidson dealer service department for testing on the approved load tester.

11 It is, however, possible to perform a continuity check of the alternator windings as described in Steps 12 and 13. No test details are supplied by the manufacturer for the regulator/rectifier unit; if failure is suspected it can only be tested by the substitution of a new unit. Note that many regulator faults are due to poor ground contact at the unit mounting.

12 Disconnect the alternator at the wiring harness connector, just to the rear of the left side of the engine (models through 1990) or just down from the regulator/rectifier (1991–on models). Connect an ohmmeter (selector switch on the R x 1 scale) between one of the stator pins in the engine side of the connector and a good ground on the engine – there should be no continuity, i.e. infinite resistance. Repeat this test between the other pin in the connector and ground. Any reading other than infinite resistance indicates a grounded stator, which must be replaced.

13 Connect the ohmmeter between both pins on the engine side of the wire harness connector. Very low resistance should be indicated (0.2-to-0.4 ohm). If the resistance is much higher, or no meter needle movement occurs, the stator must be replaced.

10 Generator/alternator – removal, overhaul and installation

Generator removal

Refer to illustration 10.1

1 After disconnecting the two wires **(see illustration)**, the generator can be removed from the engine by removing the two long screws that pass through the timing cover and secure the generator in position. Be careful when the generator is being lifted from the engine, so the oil slinger on the end of the driveshaft will clear the idler gear.

Generator overhaul

1970 through 1981

Refer to illustrations 10.2a and 10.2b

2 Remove the brush strap from the generator and examine the brushes for wear, broken wires or a tendency to stick in the holders **(see illustrations)**. The brushes must be replaced with new ones when the longest side measures 1/2-inch or less.

3 The brush holder mounting plate can be removed from the generator by unscrewing the commutator end cover nuts and washers and the two long through bolts that hold the generator together. You'll have to disconnect the two black brush wires and the positive brush cable.

4 Remove the brushes from the holders, then make sure they slide in the holders without binding. If necessary, the brush holders can be cleaned with solvent to remove accumulated carbon dust. Clean the commutator at the same time; don't use harsh abrasives such as emery cloth.

1982-on

5 Remove the two long through bolts and detach the rear cover from the generator. **Note:** *There are a number of thrust washers between the end cover and the rear bearing. Be sure none of them are lost because they must be reinstalled during reassembly of the generator.*

6 Separate the brush holder mounting plate and the body assembly from the front cover. Inspect the brushes for wear and broken wires. Measure the length of the brushes. If the length of the longest side is 1/2-inch or less, the brushes should be replaced as a set.

7 Check the brush springs for wear. The springs should exert a constant even pressure on the brushes. Replace the springs, if necessary, with new ones.

All models

Refer to illustration 10.9

8 If no reason has been found for malfunctioning of the generator to this point, take it to a Harley-Davidson dealer service department or an auto electric shop. The more detailed inspection required demands equipment and expertise the home mechanic isn't likely to have.

9 Reassemble the generator in the reverse order of disassembly. Be sure to align the notch in the end cover with the locating pin in the main body **(see illustration)**.

Chapter 7 Electrical system

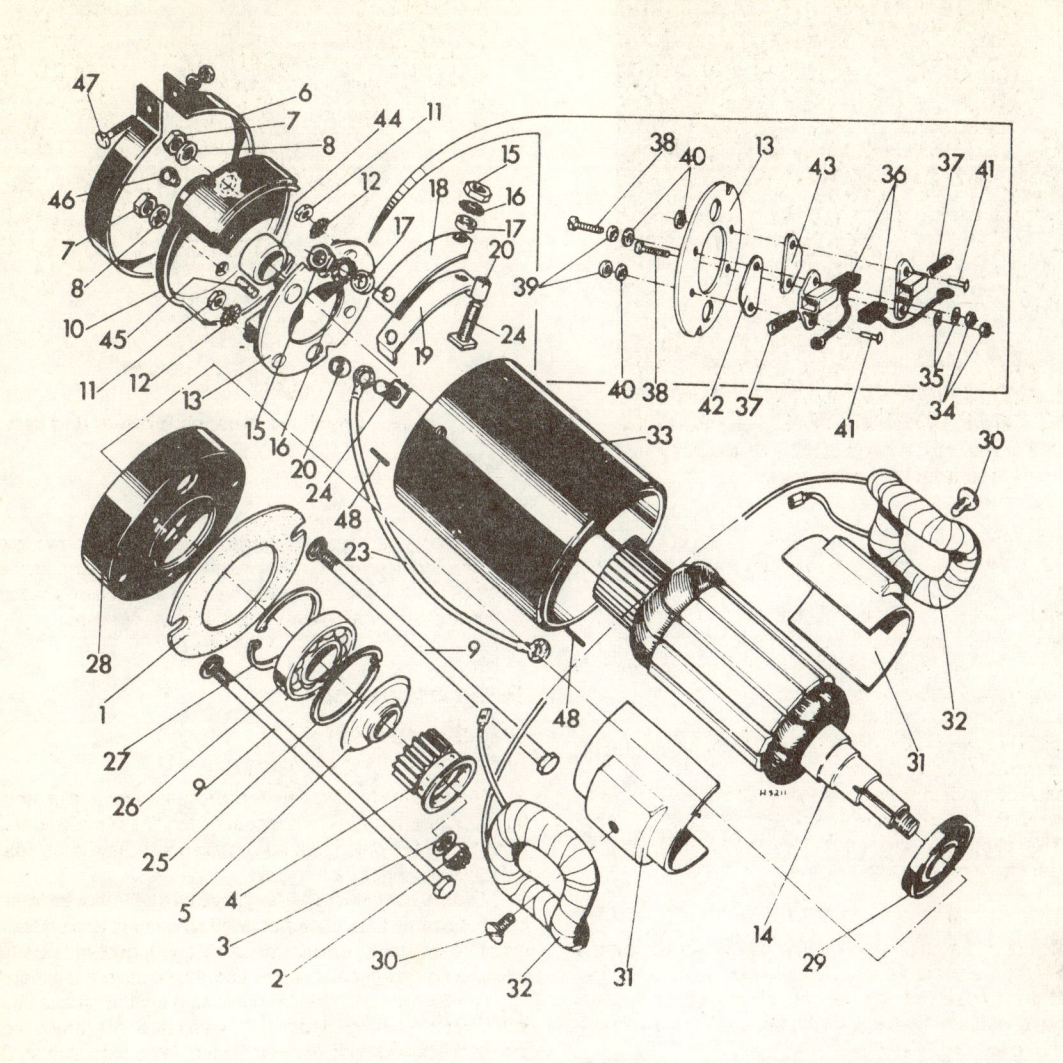

10.2b Exploded view of the generator components – 1970 through 1981 models

1	Gasket	17	Insulating washer	33	Main body
2	Driveshaft nut	18	Terminal insulator	34	Terminal screw nut
3	Washer	19	Terminal bolt clip	35	Lock washer
4	Drive gear	20	Terminal screw bushing	36	Brush
5	Oil thrower	21	Bracket insulator	37	Brush spring
6	Brush cover strap	22	Terminal screw	38	Screw
7	Nut	23	Positive brush cable	39	Washer
8	Washer	24	Terminal screw	40	Washer
9	Through bolts	25	Bearing retainer	41	Rivet
10	Commutator end cover	26	Armature bearing	42	Brush holder insulation
11	Nut	27	Bearing retainer	43	Brush holder spacer
12	Washer	28	Drive end plate	44	End cover bearing
13	Brush holder mounting plate	29	Armature oil seal	45	Oiling wick
14	Armature	30	Pole shoe screw	46	Commutator end cover oil cup
15	Terminal screw nut	31	Pole shoe	47	Brush cover screw, lock washer and nut
16	Lock washer	32	Field coil	48	End locating pin

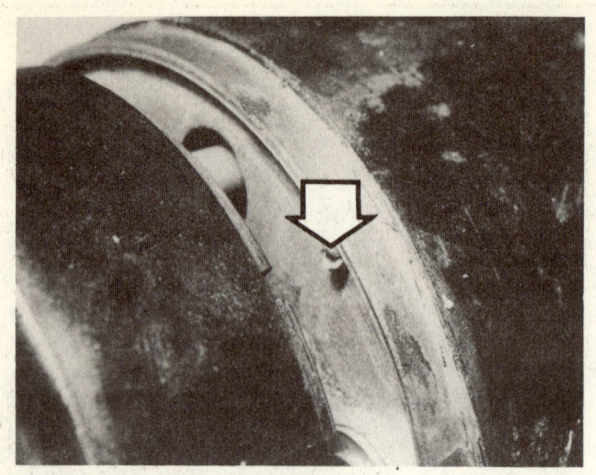

Chapter 7 Electrical system

10.9 Align the notch in the cover with the locating pin (arrow) during reassembly

10.17 Rotor is bolted to back of primary sprocket on 1991-on models

10.20 Stator is secured by four screws (1991-on location shown) – stator wiring must be clamped to the casing

Generator installation
10 Installation is the reverse of removal.
11 After the generator is assembled and installed, it must be polarized to be sure it will charge correctly. If the generator isn't polarized, permanent damage may occur to the generator and the regulator.
12 On 1970 through 1977 models, momentarily bridge the BAT and GEN terminals of the regulator with a jumper wire. This should be done after all of the electrical connections have been made, but before the engine has been started for the first time.
13 On 1978 and later models, the generator can be polarized on the motorcycle or on the workbench. While on the workbench, connect the positive battery cable to the armature terminal of the generator. Momentarily (not more than 10 seconds) connect the negative battery cable to the field terminal of the generator. If the generator is installed on the motorcycle, connect a jumper wire to the armature terminal of the generator and momentarily touch the other end of the jumper wire to the positive battery terminal.
14 The generator should be polarized any time a new one is installed, the wires are disconnected or after extended periods of non-use.

Alternator removal
Refer to illustration 10.17

15 Access to the alternator is gained by removing the chaincase cover as described in Chapter 2, Section 8, and withdrawing the engine sprocket, clutch and primary chain as a complete unit. **Note:** *You may notice slight drag from the rotor magnets as the stator and rotor separate.*
16 On models through 1990, to separate the rotor from the back of the clutch unit, remove the large snap ring. Unscrew the four Torx screws to release the stator from the transmission access cover and release its wiring.
17 On 1991-on models, remove the bolts holding the rotor to the engine sprocket and press the engine sprocket boss out of the rotor (take care to support both components while this is done) **(see illustration)**. The stator is retained to the left crankcase half by four Torx screws, but first disconnect the wiring at the connector just below the regulator/rectifier, free it from any ties and withdraw it from between the gearcase. **Note:** *Due to the difficulty in threading the wiring back through the casing on installation, it is advised that string or a length of spare wire be drawn temporarily into place as the wiring is removed.*

Alternator inspection
18 Clean all traces of dirt or metallic particles which have become attached to the rotor magnets. **Caution:** *Do not drop the rotor – damage to its magnetism will result.*
19 Clean the stator coils with contact cleaner and check for signs of damage. If the stator coil test in Section 9 has indicated a coil failure the stator must be replaced.

Alternator installation
Refer to illustration 10.20

20 Install the stator, first making sure its wiring is routed correctly and secured by any clamps provided **(see illustration)**. On 1991-on models, it will be necessary to insert the grommet in the crankcase, and route the wiring across the top of the case, then down through the gearcase (use of string or spare wire as a guide will make this easier) and tie it to the inner side to the frame tube; take care to position it well forward of the gearcase mounting lug to avoid contact with the drive belt/chain. On all models, use new Torx screws to secure the stator and tighten them evenly to the specified torque. These screws contain a locking patch, and can be used once only.
21 On models through 1990 the manufacturer advises that a new snap ring is used when installing the rotor on the clutch unit.
22 When reassembling the rotor and engine sprocket on 1991-on models, align the bolt holes and press the two components together, applying pressure to the rotor boss, not its periphery. Apply thread-locking compound to each bolt and tighten to the specified torque.
23 Refit the primary drive and clutch as described in Chapter 2, Sections 26 and 27, followed by the chaincase. Replenish the transmission oil supply.

11 Regulator – general information

1 The usual signs of a defective regulator include a battery that will not remain charged, the need to fill the battery more often than usual, or lights

Chapter 7 Electrical system

13.1a Remove the screws . . .

13.1b . . . and detach the cover from the starter motor

13.2 Disengage the spring to release the brush from the holder

that increase significantly in intensity as engine speed increases. If these signs pass unnoticed or ignored, the battery and the regulator itself will suffer permanent damage.

2   Sophisticated test equipment is required to check the regulator for proper operation. This is especially true of Bosch regulators, which are factory set and sealed.

3   If the regulator is suspected of being defective, take it to a dealer and have it checked. Normally the regulator doesn't require attention during routine maintenance, although Delco-Remy regulators may benefit from an occasional cleaning of the points. The cleaning and adjusting of the points should be left to a Harley-Davidson dealer service department or an auto electric shop.

12 Starter motor – removal and installation

Refer to the appropriate Sections in Chapter 2 for starter motor and solenoid removal and installation procedures.

13 Starter motor – disassembly, inspection and reassembly

1970 through 1980

Refer to illustrations 13.1a, 13.1b, 13.2, 13.3 and 13.6

1   Remove the end cover from the rear of the starter motor **(see illustrations)**.

2   Check the length of the brushes and compare the measurement to the Specifications **(see illustration)**. If necessary, replace the brushes with a new set. Make sure the brushes can move freely in the holders. If they aren't free to move, remove them and clean the holders with solvent to remove any carbon deposits that may be built-up. Clean the commutator with solvent at the same time. Do not use a harsh abrasive such as emery cloth on the commutator.

3   To replace the brushes on a Prestolite starter, remove the terminal and brush assembly from the main body shell **(see illustration)**. Install a new terminal and brush assembly. The other two brushes can be removed

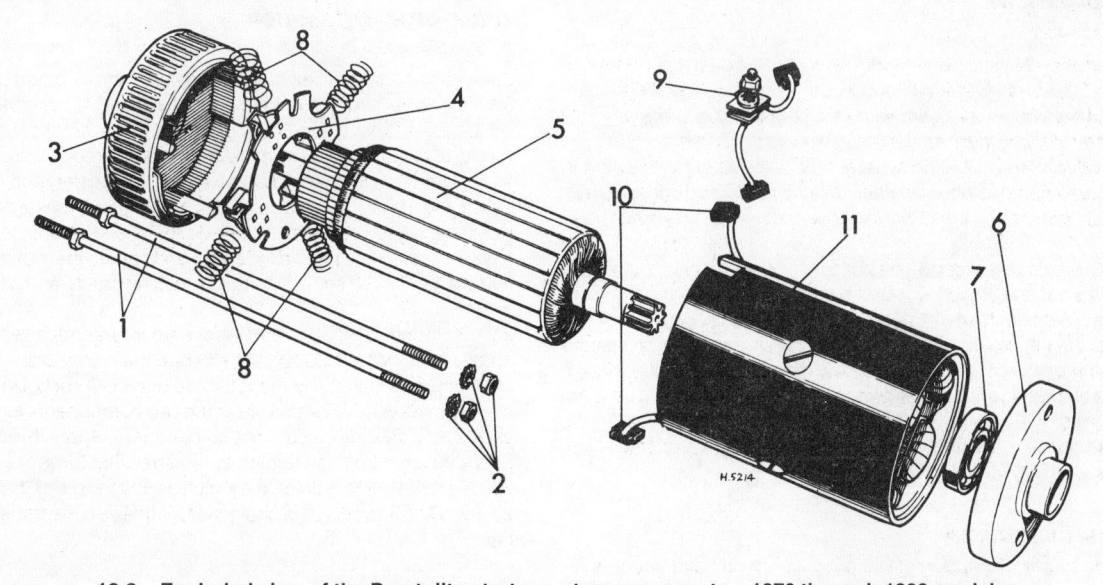

13.3 Exploded view of the Prestolite starter motor components – 1970 through 1980 models

1 Through bolt
2 Washer and lock washer
3 Commutator end cover
4 Brush plate and holder assembly
5 Armature
6 Drive end cover
7 Drive end ball bearing
8 Brush spring
9 Terminal and brush assembly
10 Ground brush
11 Main body shell and field coil assembly

Chapter 7 Electrical system

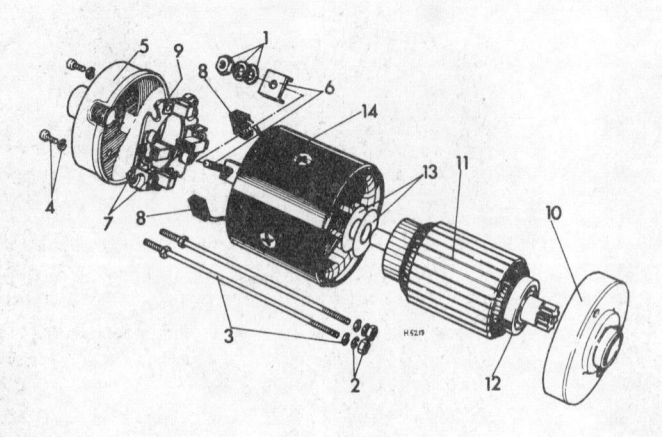

13.6 Exploded view of the Hitachi starter motor components – 1970 through 1978 models

1 Terminal nut, lock washer and washer
2 Through bolt nut and lock washer
3 Through bolt and lock washer
4 Rear cover screw and lock washer
5 Rear (commutator end) cover
6 Terminal and insulator
7 Negative brush
8 Positive brush
9 Brush holder assembly
10 Front (drive end) cover
11 Armature
12 Armature ball bearing
13 Thrust washer
14 Main body shell

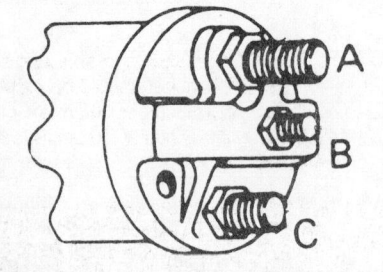

14.2 Solenoid terminal locations – 1970 through 1980 models

from the field coils by cutting the brush lead wire where it connects to the field coil lead.
4 File the old coil connection until the coil lead is thoroughly cleaned. Strip the insulation from the coil lead as far as necessary to make a new solder connection.
5 Solder the leads from the new brushes to the field coil lead with rosin flux. Be sure the new lead is in the same position as the original lead. Do not use too much heat or solder, otherwise the leads will become flow coated with solder and lose their flexibility.
6 To replace the brushes in Hitachi starters, the brush leads must be unsoldered from the brush holder **(see illustration)**. Solder the new brushes into position on the brush holder. Do not use too much heat or solder, otherwise the leads will become flow coated with solder and lose their flexibility.

1981-on

7 Release the field coil wire from the unit. Remove the two long bolts which pass from the end cover to the housing. Remove the two screws from the end cover and withdraw the cover, complete with large O-ring on 1200 model.
8 Release the springs from the ends of the brushes with a wire hook, then lift the brush assembly off the commutator. Slide the new brush assembly over the commutator and position the brushes in the holder. Use a wire hook to attach the springs to the ends of the brushes.
9 Inspect the brushes and commutator as described in Step 2.

All models

10 If the starter motor has been used a lot, the mica insulation between the individual copper segments of the commutator may need servicing. The commutator must be undercut 1/32-inch by an automotive electric shop or a Harley-Davidson dealer service department.
11 Other defects in the starter motor require professional attention or the installation of a new unit.

12 Be sure the brush leads do not contact the body of the motor before reinstalling the starter motor on the motorcycle.

14 Starter solenoid – check

1970 through 1980

Refer to illustration 14.2
1 Disconnect the negative cable from the battery, followed by the positive cable. Remove the wires from the solenoid.
2 Connect solenoid terminals A and C to a 12-volt battery; A to the positive terminal and C to the negative terminal **(see illustration)**. Attach a jumper wire to the positive terminal of the battery and touch the other end of the jumper wire to terminal C of the solenoid. The solenoid should make a clicking sound. If a click or heavy spark at the terminal doesn't occur, the solenoid is defective and must be replaced with a new one.

1981-on

3 Disconnect the wire from terminal C on the solenoid.
4 Connect terminal 50 (on the solenoid) to the positive terminal of a 12-volt battery. Connect the negative terminal of the battery to terminal C and to the body of the solenoid. The starter gear should pull-in forcefully. If not, the solenoid is defective and must be replaced with a new one.
5 Disconnect the negative cable from terminal C. The starter gear should remain in the pulled-in position. If not, the solenoid is defective and must be replaced with a new one.
6 Connect terminals C and 50 to the positive terminal of the 12-volt battery. Ground the negative terminal of the battery on the solenoid body. Disconnect the test lead from terminal 50. This should result in the starter gear returning to its original position. If the gear doesn't return to its original position, the solenoid is defective and must be replaced with a new one.
7 Replacement of the solenoid on 1981 and later models requires disassembly of the starter motor and should be left to a Harley-Davidson dealer service department.

15 Handlebar switches – removal and installation

Refer to illustration 15.1
1 Generally speaking, the handlebar switches give little trouble, but if necessary they can be removed by separating the sections that form a split clamp around the handlebars **(see illustration)**.
2 To prevent the possibility of a short circuit, disconnect the battery before removing the switches.
3 Most troubles are caused by dirty contacts, which can be cleaned with an aerosol contact cleaner specially formulated for this purpose.
4 Repair of the switches is usually impractical. In the event of damage or pronounced wear of some internal part, the switch should be replaced with a new one.

Chapter 7 Electrical system 181

15.1 Remove the four screws securing the switch to the handlebar

16.1 Unscrew the knurled ring around the ignition switch to release the switch from the bracket – models through 1991

16 Ignition and light switch – removal and installation

Warning: *Disconnect the battery negative cable before working on the switch.*

Refer to illustrations 16.1 and 16.2

1 The main switch that controls both the ignition system and the lights is attached to the top engine mounting bracket on the left side of the motorcycle. On models through 1991, the switch is attached to a plate by a knurled ring. Once the ring is unscrewed, the switch can be pressed through the plate and removed from the rear after its wires have been disconnected **(see illustration)**.

2 On 1992-on models unscrew the retaining ring from the front of the switch and remove the locknut from the stud on its underside **(see illustration)**. Lift the switch assembly and its seal away from the engine bracket and remove the switch cover. Disconnect the wires from the switch.

3 If the switch malfunctions, replace it with a new one – repair isn't possible. Note that if the switch is replaced, the ignition key will have to be replaced also.

4 The switch doesn't normally require attention and it should never be oiled. If the switch is oiled, there's a risk of oil reaching the electrical contacts and acting as an insulator.

17 Horn – adjustment

1 The horn is equipped with an adjusting screw on the back side so the volume can be varied.

2 To adjust the tone or volume, turn the screw 1/2-turn in either direction and check the sound. If the sound is weaker or lost altogether, turn the screw in the opposite direction. Continue adjusting until the desired note and volume are achieved.

3 If the horn malfunctions and can't be restored by adjustment, install a new one. Horn repair isn't possible, because the assembly is riveted together.

18 Brake light switches – adjustment

Rear brake light switch

Refer to illustrations 18.1 and 18.3

1 The switch on drum brake models is bolted to a small lug on the right side of the lower frame tube, close to the rear wheel. The plunger-type switch is actuated by a small right angle bracket that's clamped to the rear brake operating rod **(see illustration)**.

2 To adjust the point at which the switch operates and the brake light comes on, the screw through the bracket should be loosened and the bracket moved up-or-down the rod, as required. Moving it to the rear of the machine makes the light operate later, since the switch is in the off position while the plunger is depressed.

16.2 Release the retaining ring (A) and locknut (B) to free the switch – 1992-on models

18.1 Location of the brake light switch on rear drum brake models

Chapter 7 Electrical system

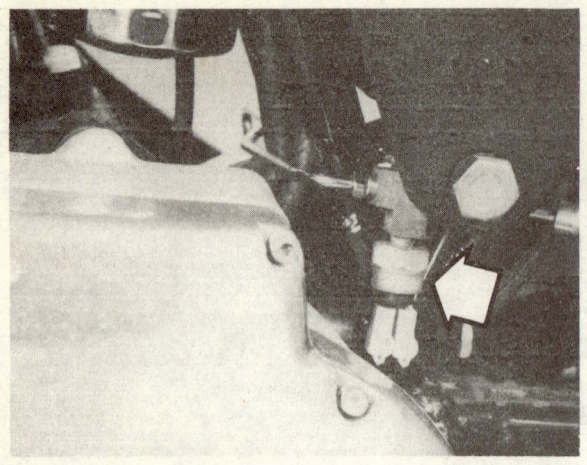

18.3 On rear disc brake models, the brake light switch is mounted below the oil tank

19.3 Remove the upper two Allen-head bolts to release the instrument bracket, then . . .

3 The brake light switch on rear disc brake models is located on the left side of the frame, just below the oil tank **(see illustration)**. The rear disc brake is self-compensating so adjustment of the switch isn't necessary or possible.

Front brake light switch

4 Early models with front drum brakes have the brake light switch incorporated in the brake cable. Later disc brake models have the brake light switch built into the master cylinder. Since the disc brakes are self-compensating, the switch doesn't require adjustment.

19 Warning light bulbs – replacement

Early models

Refer to illustrations 19.3 and 19.5

1 Two separate warning lights are mounted in a small display panel, just above the headlight. The left light is the generator charging light and the right light is the oil pressure warning light. Both lights should illuminate as soon as the ignition switch is turned on, but should go out soon after the engine is started.
2 Access to the bulbs is from the back of the panel, which the bulbs fit into.
3 Remove the upper two Allen-head bolts securing the handlebar clamp **(see illustration)**. Carefully loosen the lower two bolts until the instrument bracket can be slid out from under the clamp. Hold the handlebars so they don't slip out of position while the clamp is loose.
4 Pry the plug out of the top of the warning light panel and loosen the headlight adjusting nut.
5 Remove the two bolts securing the warning light panel and lift the panel off **(see illustration)**. Be careful – there are washers between the panel and mounting bracket which may be lost when the panel is removed.
6 Unplug the defective light and install a new bulb.
7 Place the panel in position with the washers under it and install the bolts.
8 Slide the instrument mounting bracket into place between the handlebar mounting clamps. Insert the two upper mounting bolts and begin to tighten them. Be sure the nuts at the bottom of the upper yoke that the bolts thread into are in position.
9 Place the handlebars at the desired level and tighten the mounting bolts securely.

Later models

Refer to illustration 19.11

10 Each warning light is a sealed assembly; bulb replacement is not possible.
11 To remove a defective light, first remove the headlamp reflector for access to the warning light wiring connector. Disconnect the relevant wiring and any ground connections. Pry the inner cover from the warning light display and withdraw the defective light unit from the housing or instrument bracket **(see illustration)**.
12 Installation is a reverse of the removal procedure.

19.5 . . . remove the bolts securing the warning light panel

19.11 Warning light pulls out of housing on 1993 model shown

Chapter 7 Electrical system

Speedometer and tachometer illumination

13 Refer to Chapter 5, Section 3 for details.

20 Evaporative emission control system – solenoid test (1992-on California models)

1 The solenoid is clamped to the rear of the air cleaner baseplate and operates a butterfly valve located in the bottom of the air cleaner housing, a mechanical linkage connects the two components. If operating normally, it will shut the valve when the engine is stopped (ignition switch in OFF position), open it when the starter circuit is operated via the pull-in winding, and keep it open while the engine is running via its hold-in winding.
2 To test the two windings trace the wiring up to the 4-pin connector and separate it at this point. Making the tests on the solenoid side of the connector, use an ohmmeter to measure the resistance between the pull-in winding wires (black/red and grey/black on early models; green and black on later models). A reading of 4 to 6 ohms should be obtained. Take another reading between the hold-in winding wires (white and black on early models; white/black and black on later models). A reading of 21 to 27 ohms should be obtained.
3 If either resistance reading is widely different from that specified, the solenoid is confirmed faulty and must be renewed.
4 If the windings prove sound, yet the butterfly valve still fails to operate normally make continuity checks along the supply and ground circuits to isolate the fault – it will most likely be due to a corroded connector or broken wire.

21 Turn signal relay/cancel unit – location

Models through 1990

1 The turn signal relay resides in the headlamp shell. Remove the headlamp sealed beam unit as described in Section 7 for access. If a fault occurs in the turn signal circuit which cannot be traced to the bulbs, handlebar switches or associated wiring, the relay must be renewed.
2 If installing a new relay check that it fits into its clip and will not contact the headlamp unit.

1991-on models

3 The self-cancelling circuit comprises the cancel relay (bolted to the rear of the ignition module bracket), the reed switch in the speedometer head and the handlebar switches. The relay automatically cancels the turn signals after receiving distance information from the reed switch, or on manual control via either handlebar switch. An internal circuit of the relay provides a hazard flasher function.
4 Various tests can be made on the relay if it is suspected of failure, but first check that the fault is not due to a blown bulb, poor ground connection, corrosion or damaged wiring – these are more likely than a failed relay.
5 If the fault only affects one side of the system, check for battery voltage (12Vdc) at the harness side of the relay block connector. Connect the positive voltmeter probe to the white/brown terminal (right-hand side) or white/violet terminal (left-hand side) and the negative probe to ground. With the ignition switch ON, press the appropriate turn signal switch. If battery voltage is shown on the meter the switch is proved good and the fault lies in the relay or lamp wiring.

6 If the complete system fails, including the hazard function, check for battery voltage at the orange wire terminal of the connector with the ignition switch ON. If battery voltage is shown, the relay is confirmed faulty, although make sure that its ground wire is sound. No battery voltage indicates a power supply problem; refer to the wiring diagram at the end of this Chapter and work back through the system to identify the cause.
7 If the turn signals work, but will not cancel, check the reed switch operation. Connect an ohmmeter between the white/green wire of the relay connector and ground. With the front wheel raised, have someone spin it while you observe the meter; it should vary from zero to infinity if the reed switch is functioning correctly.

22 Wiring diagrams – general information

Since it isn't possible to include all wiring diagrams for every year covered by this manual, the following diagrams are typical.

Prior to troubleshooting a circuit, check the fuse to make sure it's in good condition. Make sure the battery is fully charged and check the cable connections (Sections 4, 5 and 6).

When checking a circuit, make sure all connectors are clean, with no broken or loose terminals or wires. When unplugging a connector, do not pull on the wires. Pull only on the connector housings themselves.

Refer to the accompanying table for the wire color codes.

Wiring diagram color code key

B	BLACK
W	WHITE
O	ORANGE
R	RED
GN	GREEN
Y	YELLOW
V	VIOLET
BE	BLUE
BN	BROWN
GY	GRAY

184   Chapter 7   Electrical system

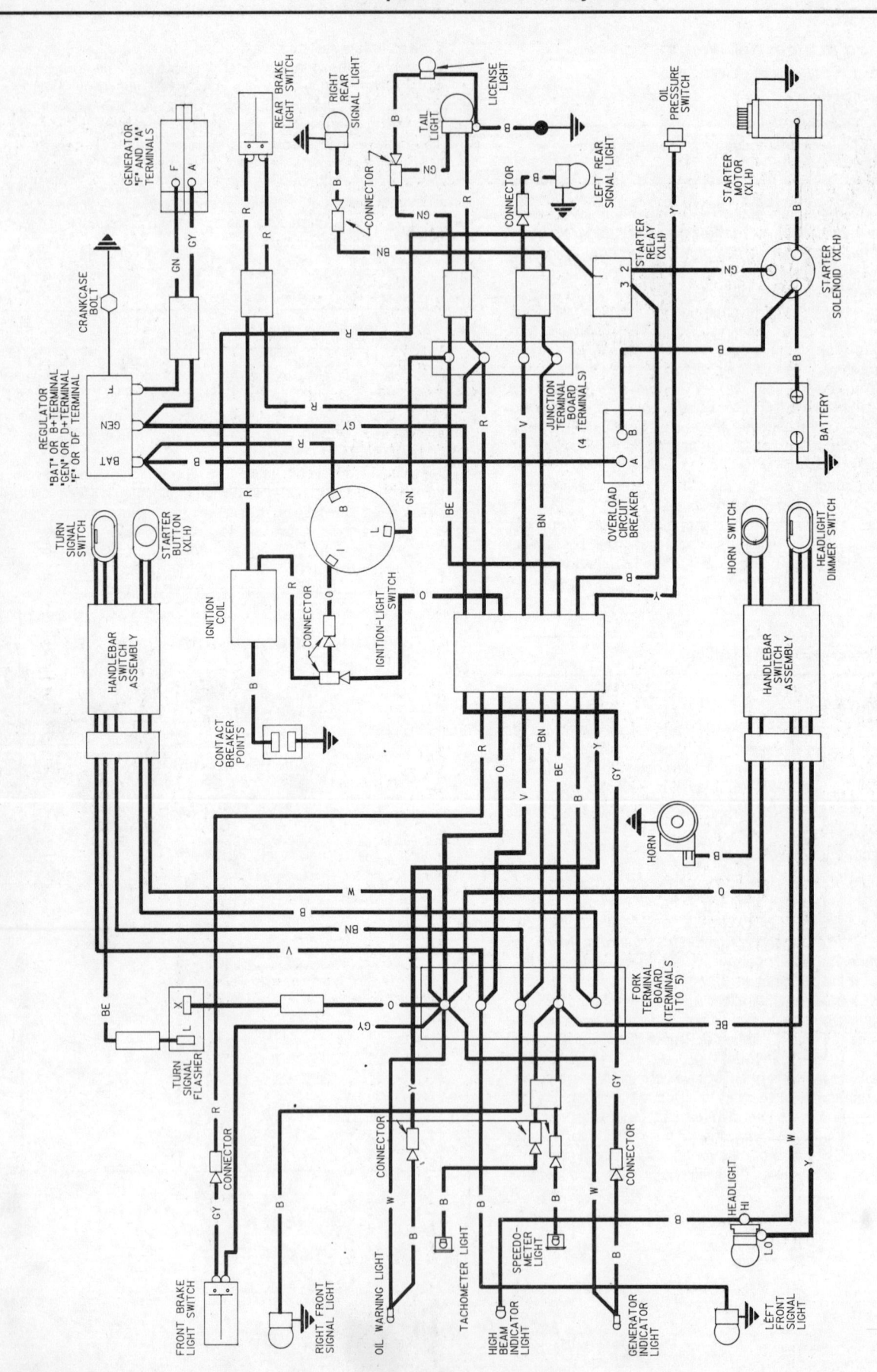

Wiring diagram (1970 through 1972) – typical

See page 183 for color key

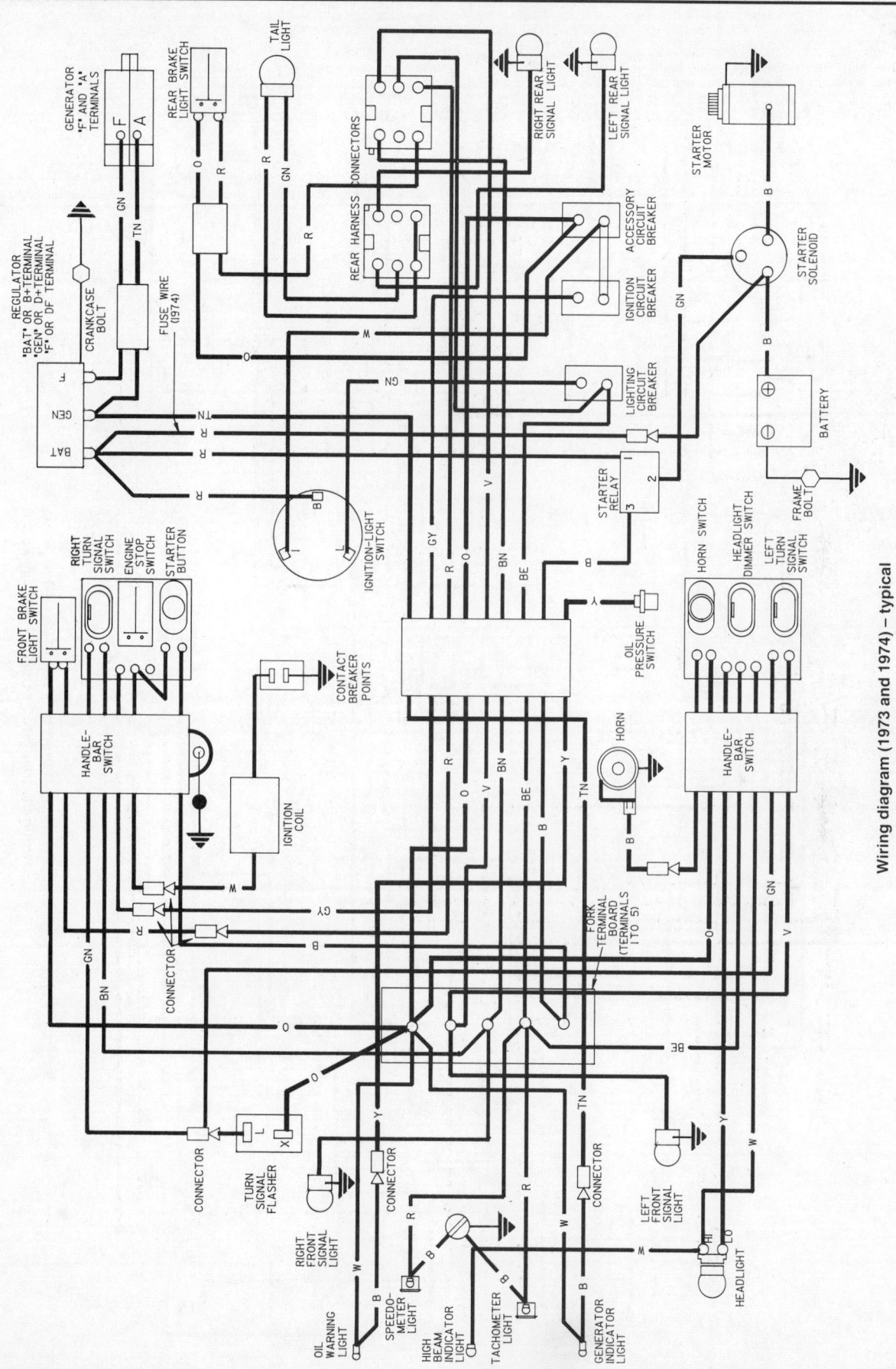

Chapter 7 Electrical system

Wiring diagram (1973 and 1974) – typical

See page 183 for color key

Chapter 7 Electrical system

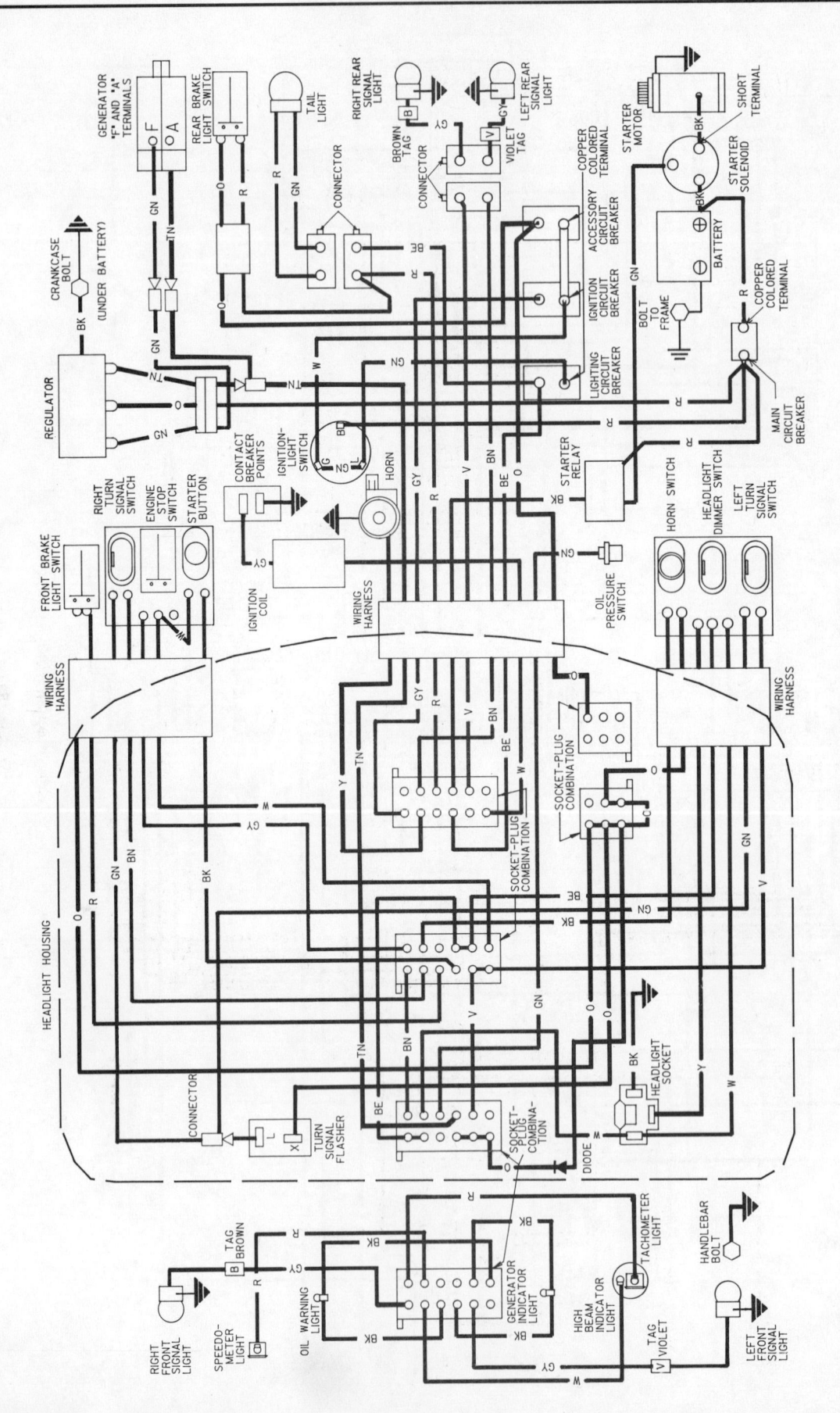

Wiring diagram (1975 through 1978) – typical

See page 183 for color key

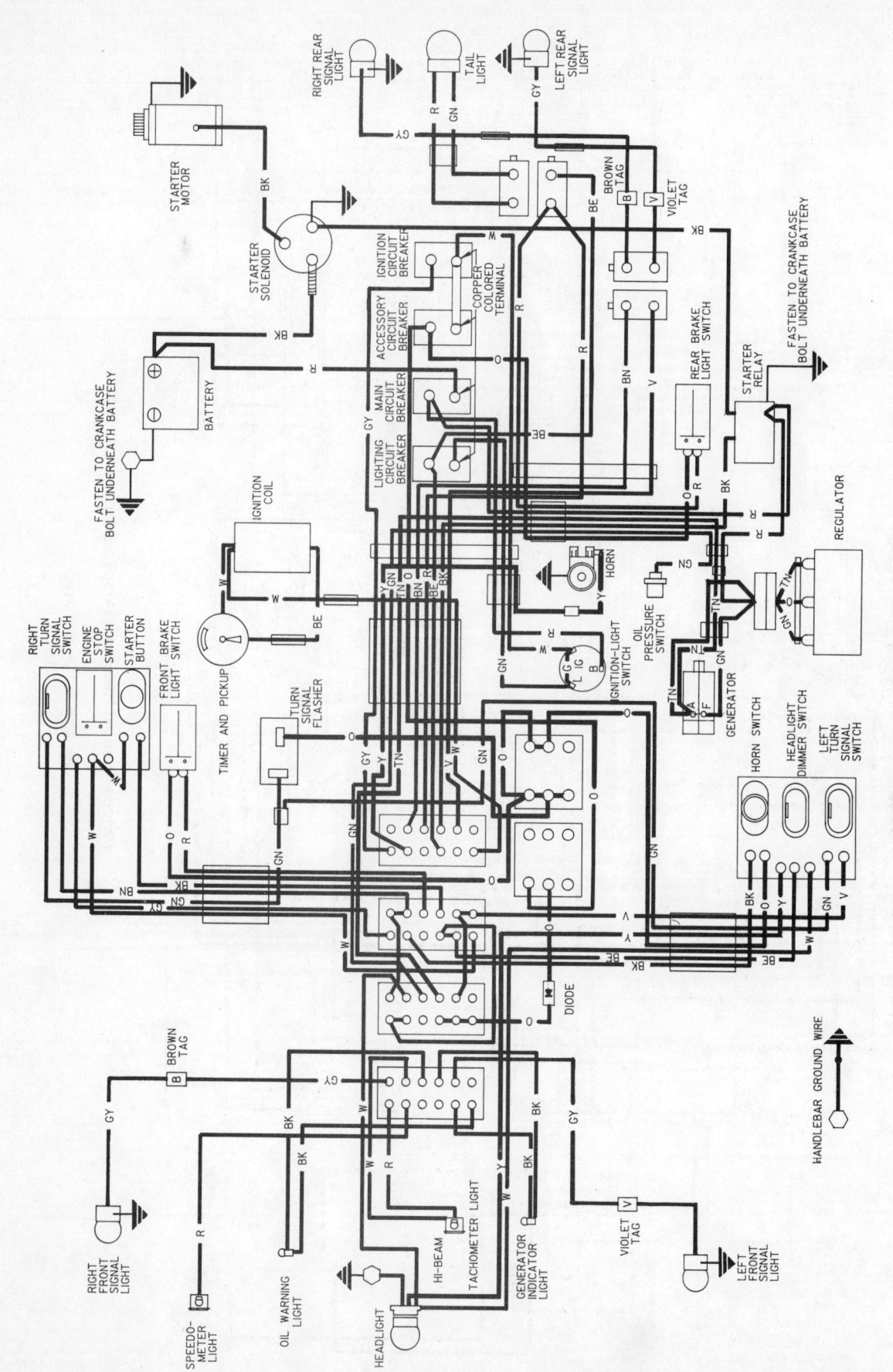

Chapter 7 Electrical system

Wiring diagram (1979) – typical

See page 183 for color key

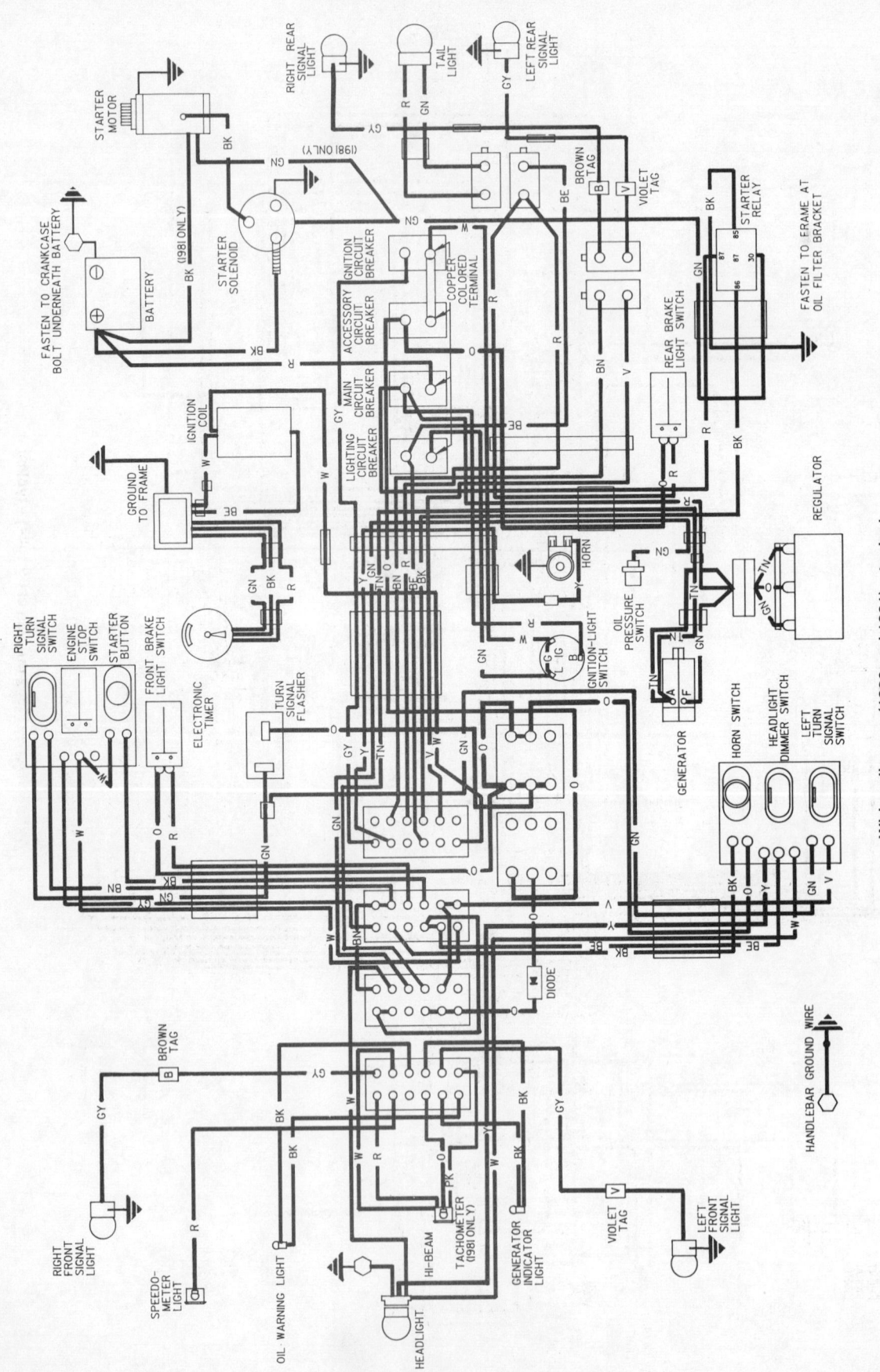

Chapter 7 Electrical system

Wiring diagram (1980 and 1981) – typical

See page 183 for color key

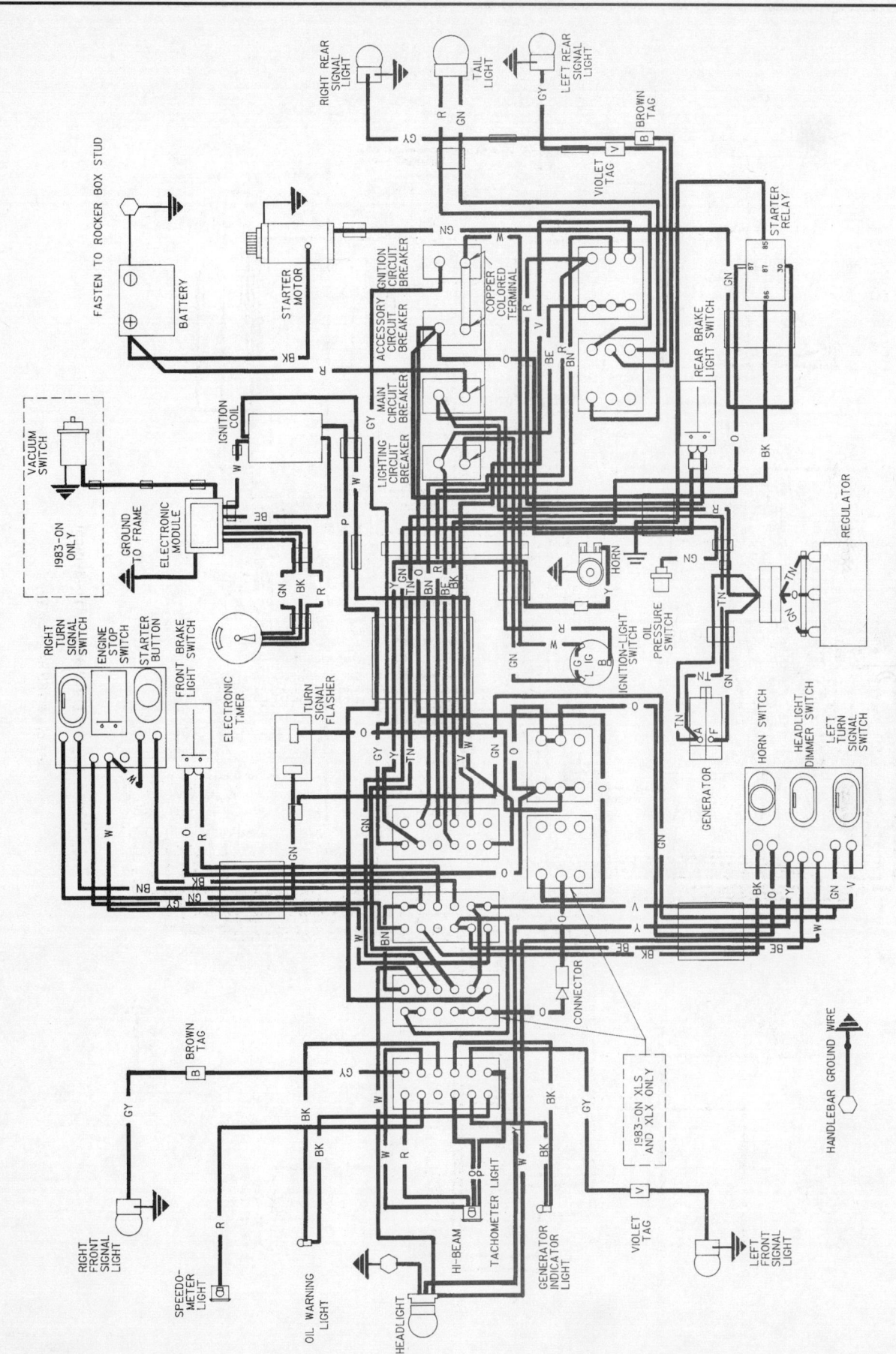

Chapter 7 Electrical system

Wiring diagram (1982 through early 1984) – typical

See page 183 for color key

Wiring diagram component key

1	Front stop light switch	24	Horn
2	Right turn signal switch	25	Ignition switch
3	Engine stop switch	26	Horn switch
4	Starter switch	27	Left turn signal switch
5	Turn signal relay	28	Dimmer switch
6	Ignition sensor	29	Right front turn signal/running light
7	Ignition module	30	Speedometer
8	VOES	31	High beam warning light
9	Ignition coil	32	Oil pressure warning light
10	Battery	33	Headlight
11	Voltage regulator	34	Turn signal warning light
12	Alternator stator	35	Neutral light
13	Rear cylinder spark plug	36	Tachometer (where fitted)
14	Front cylinder spark plug	37	Left front turn signal/running light
15	Starter motor	38	Handlebar ground wire
16	Rear right turn signal	39	Turn signal cancel unit
17	Tail/stop light	40	Ground connection to frame beneath brake caliper hose clamp
18	Rear left turn signal		
19	Fuse	41	Ground connection to engine mounting bracket
20	Rear stop light switch		
21	Starter relay	42	Right turn signal warning light
22	Neutral switch	43	Left turn signal warning light
23	Oil pressure switch	44	Emission control solenoid (California models)

Chapter 7 Electrical system

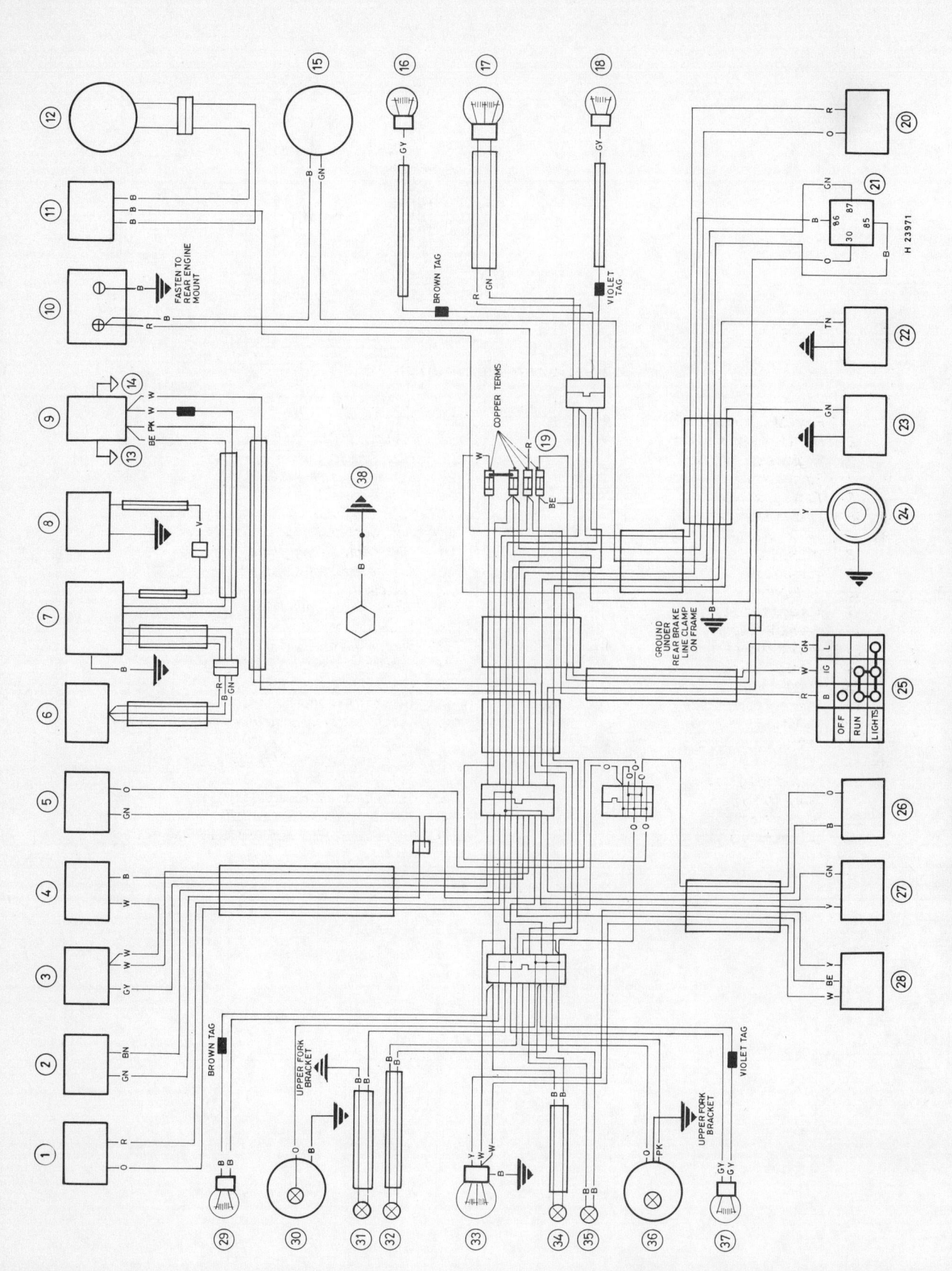

Wiring diagram (1986 through 1990 models) – late 1984 and 1985 models similar

See page 190 for key

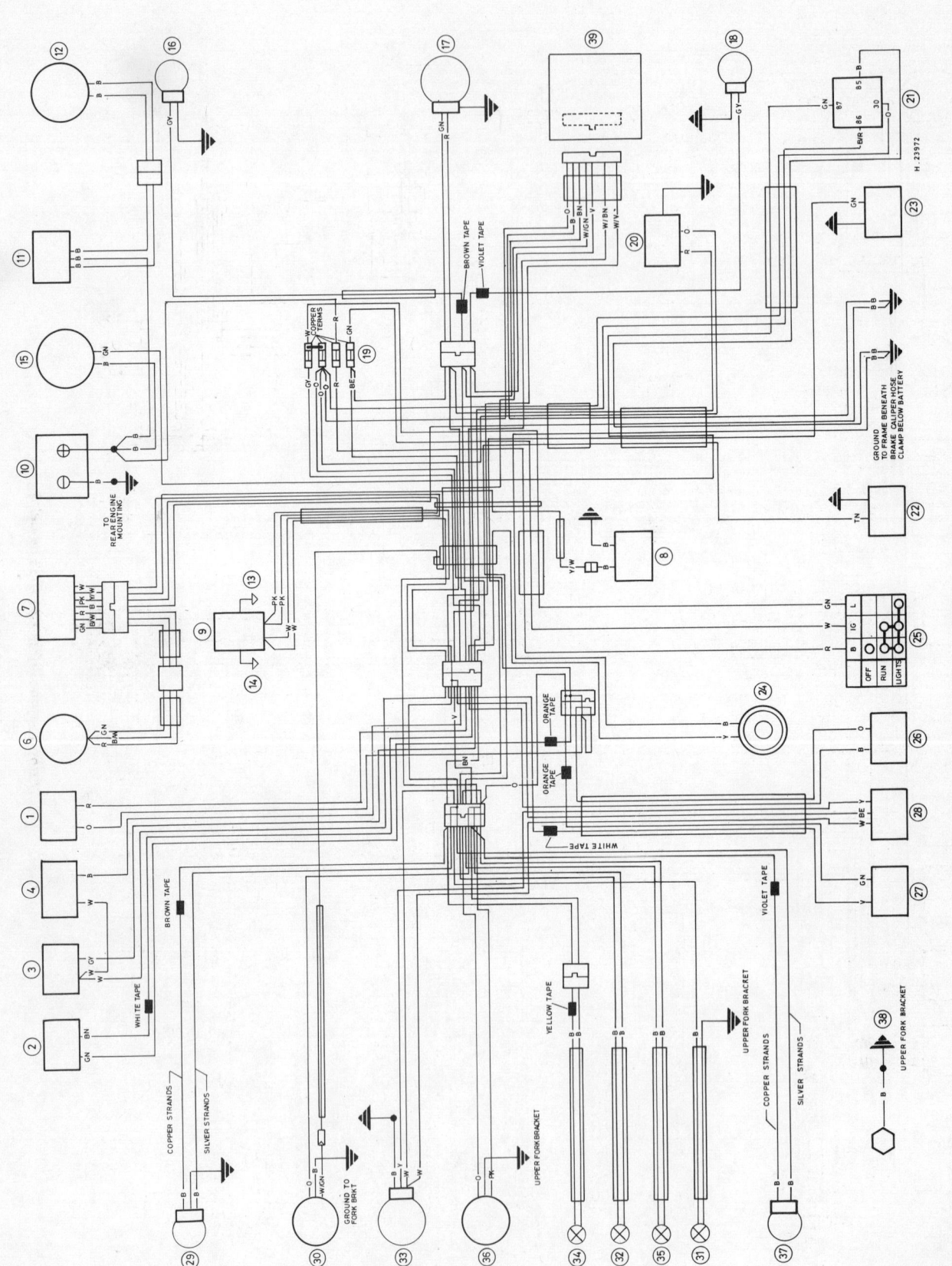

192   Chapter 7 Electrical system

Wiring diagram (1991 models)

See page 190 for key

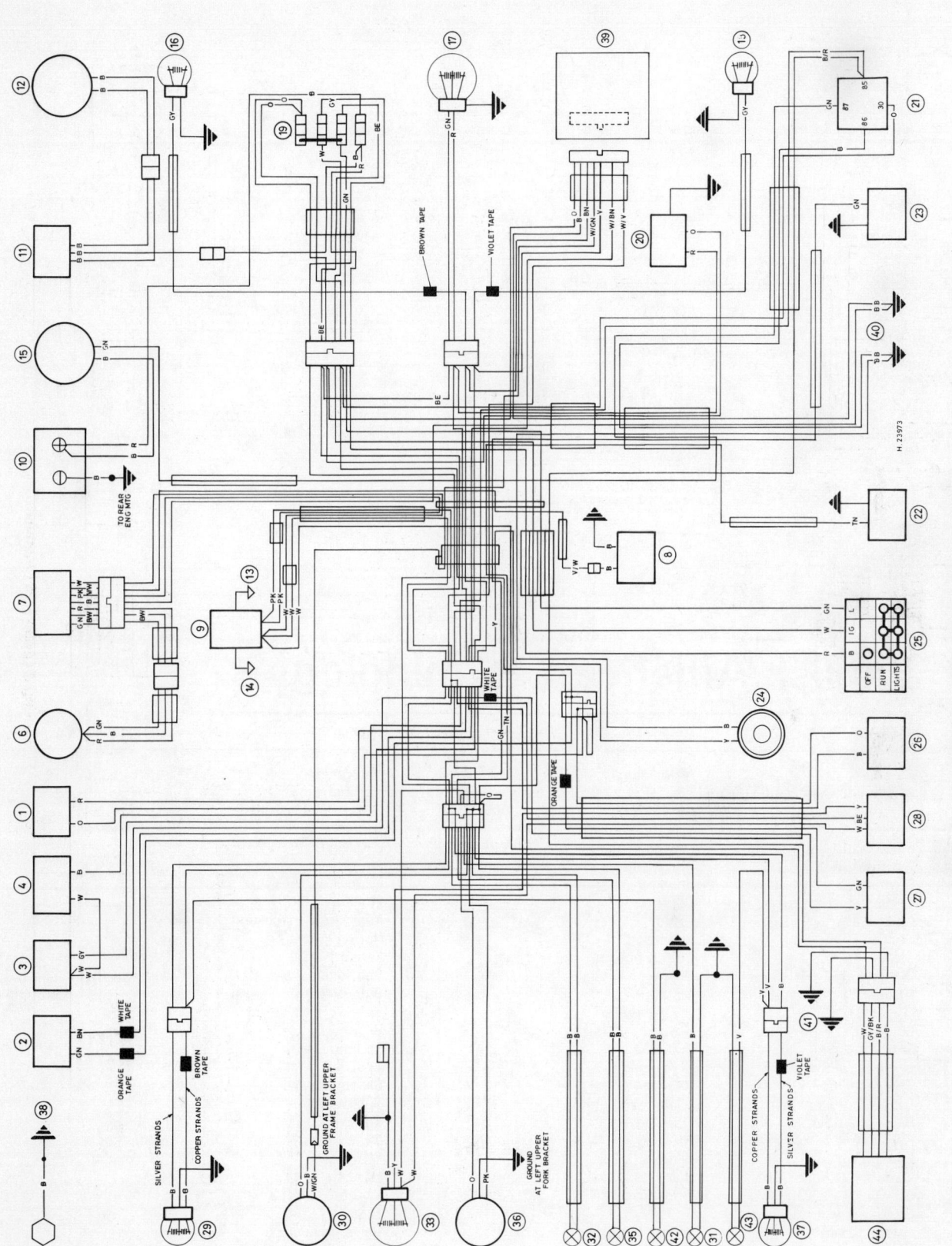

Chapter 7 Electrical system

Wiring diagram (1992 and 1993 models)

See page 190 for key

English/American terminology

English	American	English	American
Air filter	Air cleaner	Mudguard	Fender
Alignment (headlamp)	Aim	Number plate	License plate
Allen screw/key	Socket screw/wrench	Output or layshaft	Countershaft
Anticlockwise	Counterclockwise	Panniers	Side cases
Bottom/top gear	Low/high gear	Paraffin	Kerosene
Bottom/top yoke	Bottom/top triple clamp	Petrol	Gasoline
Bush	Bushing	Petrol/fuel tank	Gas tank
Carburettor	Carburetor	Pinking	Pinging
Catch	Latch	Rear suspension unit	Rear shock absorber
Circlip	Snap-ring	Rocker cover	Valve cover
Clutch drum	Clutch housing	Selector	Shifter
Dip switch	Dimmer switch	Self-locking pliers	Vise-grips
Disulphide	Disulfide	Side or parking lamp	Parking or auxiliary light
Dynamo	DC generator	Side or prop stand	Kickstand
Earth	Ground	Silencer	Muffler
End float	End play	Spanner	Wrench
Engineer's blue	Machinist's dye	Split pin	Cotter pin
Exhaust pipe	Header	Stanchion	Tube
Fault diagnosis	Troubleshooting	Sulphuric	Sulfuric
Float chamber	Float bowl	Sump	Oil pan
Footrest	Footpeg	Swing arm	Swingarm
Fuel/petrol tap	Petcock	Tab washer	Lock washer
Gaiter	Boot	Top box	Trunk
Gearbox	Transmission	Two/four stroke	Two/four cycle
Gearchange	Shift	Tyre	Tire
Gudgeon pin	Wrist/piston pin	Valve collar	Valve retainer
Indicator	Turn signal	Valve collets	Valve keepers
Inlet	Intake	Vice	Vise
Input shaft or mainshaft	Mainshaft	Wheel spindle	Axle
Kickstart	Kickstarter	White spirit	Stoddard solvent
Lower leg	Slider	Windscreen	Windshield

Index

A

About this manual – 6
Acknowledgements – 2
Adjustments
    Air gap – 133
    Carburetor – 36, 128
    Clutch – 34
    Drive belt – 49
    Drive chain – 32
    Drive chain oiler – 37
    Float level – 127, 128, 129
    Headlight aim – 48
    Horn – 181
    Idle speed – 36
    Ignition timing – 40
    Points – 39
    Primary drive chain – 43
    Spark plug gap – 39
Air cleaner
    Removal and installation – 129
    Servicing – 31
Air gap – 133
Alternator
    Check – 175
    Removal and installation – 178
Automatic chain oiler – 37

B

Basic maintenance techniques – 10
Barrels – cylinder
    Inspection – 85
    Installation – 107
    Removal – 62
Battery
    Charging – 173
    Check and maintenance – 173
    Electrolyte level check – 31
    Specific gravity – 174
Bearings
    Camshaft – 92
    Crankshaft and connecting rod – 84
    Steering head
        Check – 44
        Maintenance – 149
    Swingarm – 140
    Wheel – 45
Bleeding hydraulic brakes – 48

Brakes
    Bleeding – 48
    Fluid level check – 30
    Fluid replacement – 49
    Front cable lubrication – 35
    Front caliper
        Overhaul – 159
        Removal and installation – 158
    Front disc
        Inspection – 154
        Removal and installation – 158
    Front disc brake pads
        Check – 33
        Replacement – 154
    Front drum brake
        Shoe check – 33
        Shoe replacement – 153
    General check – 35
    General information – 152
    Master cylinder
        Front – 159
        Rear – 165
    Rear brake pedal check and adjustment – 167
    Rear caliper – 165
    Rear disc
        Inspection – 163
        Removal and installation – 168
    Rear disc brake pads
        Check – 33
        Replacement – 163
    Rear drum brake
        Shoe check – 33
        Shoe replacement – 162
    Shoe and pad inspection – 33
    Specifications – 27, 151
Bulbs
    Replacement – 174, 182
    Specifications – 172
Buying parts – 10

C

Cable lubrication – 35
Camshaft
    Inspection – 92
    Installation – 99
    Removal – 73
Carburetors
    Adjustments – 36, 128

Index

Disassembly, inspection and reassembly – 120
Float adjustment – 127, 128, 129
Overhaul – 119
Removal and installation – 119
Specifications – 116, 117
Chain – final drive
Automatic oiler maintenance and adjustment – 37
Check, adjustment and lubrication – 32
Chain – primary
Adjustment – 43
Installation – 101
Removal – 68
Chain oiler – 37
Charging system check – 173, 175
Clutch
Adjustment – 34
Cable lubrication – 35
Inspection – 95
Installation – 101
Removal – 68
Specifications – 54
Coil
Check, removal and installation – 133
Specifications – 132
Compression check – 48
Condenser – 133
Connecting rod
Bearings – 84
Specifications – 52, 54
Contact breaker points
Check and replacement – 39
Gap – 26
Conversion factors – 195
Crankcases
Inspection – 93
Reassembly – 98
Splitting – 82
Crankshaft bearings – 84
Cylinder barrels
Inspection – 85
Installation – 107
Removal – 62
Cylinder compression
Check – 48
Specifications – 26
Cylinder head
Disassembly, inspection and reassembly – 86
Installation – 108
Removal – 62
Specifications – 51, 52, 53
Valve job – 86

D

Disc brake
Bleeding – 48
Fluid level check – 30
Fluid replacement – 49
Front caliper
    Overhaul – 159
    Removal and installation – 158
Front disc
    Inspection – 154
    Removal and installation – 158
Front disc brake pad replacement – 154
Front master cylinder – 159
General information – 152

Pad inspection – 33
Rear caliper – 165
Rear disc
    Inspection – 163
    Removal and installation – 168
Rear disc brake pad replacement – 163
Rear master cylinder – 165
Specifications – 27, 151
Distributor – 136
Drive belt (final) – 49
Drive chain (final)
Automatic oiler – 37
Check, adjustment and lubrication – 32
Drum brakes
Front
    Cable lubrication – 35
    Inspection and shoe replacement – 153
Rear – 162
Shoe check – 33
Specifications – 27, 151

E

Electrical system
General information – 172
Specifications – 171
Troubleshooting – 25, 172
Wiring diagrams – 184
Electronic ignition system check – 134
Engine
Break-in procedure – 115
Component inspection – 84
Disassembly – 62
General information – 56
Initial start-up after overhaul – 115
Inspection and repair – 82
Installation – 111
Oil and filter change – 37
Oil level check – 30
Reassembly – 62, 98
Removal – 57
Specifications – 50 through 56
Evaporative emission control system – 130, 183
Exhaust system
Installation – 114
Removal – 58

F

Fast idle speed – 27
Fasteners
Check – 35
Types – 10
Final drive
Belt – 49
Chain – 32, 37
Sprocket (engine) – 73, 106
Sprocket (rear wheel) – 168
Float level – 127, 128, 129
Fluid level checks – 30
Forks
Disassembly, inspection and reassembly – 144
Oil change – 45
Removal and installation – 142
Specifications – 28, 138

Index

Frame – inspection and repair – 138
Front brake components – see Brakes
Front wheel
    Bearings – 45
    Removal and refitting – 152
Fuel system
    Carburetor – 36, 119, 120, 127, 128, 129
    Check – 34
    Control valve – 118
    Filter – 43
    General information – 118
    Specifications – 116
    Tank removal and installation – 118

G

Gearshift and gearbox – see *Transmission*
Generator
    Check – 175
    Installation – 178
    Overhaul – 176
    Removal – 176

H

Handlebar switches – 180
Head
    Disassembly, inspection and reassembly – 86
    Installation – 108
    Removal – 62
    Specifications – 51, 52, 53
    Valve job – 86
Headlight
    Adjustment – 48
    Bulb replacement – 174
    Specifications – 172
Horn – 181

I

Identification numbers – 8
Idle speed – 26, 36
Ignition system
    Air gap – 133
    Coil – 133
    Component removal and installation – 135
    Condenser – 133
    Contact breaker points – 39
    Distributor – 136
    Electronic (check) – 134
    General information – 133
    Spark plugs – 39
    Specifications – 26, 132
    Switch – 181
    Timing check and adjustment – 40
    VOES (Vacuum Operated Electric Switch) – 135
Ignition components (engine mounted)
    Installation – 101
    Removal – 71
Introduction to the Harley-Davidson Sportster – 7

K

Kickstart mechanism
    Inspection – 97
    Installation – 103
    Removal – 71

L

Levels (fluid) – 30
Light switch – 181
Lubricants and fluids – 28
Lubrication
    Cables – 35
    Final drive chain – 32
    General – 35
    Oil pump – 75

M

Maintenance intervals – 29
Maintenance techniques, tools and working facilities – 10
Master cylinder
    Front – 159
    Rear – 165
Motorcycle chemicals and lubricants – 16

O

Oil
    Change
        Engine – 37
        Transmission – 43
    Fork – 28, 45
    Level check
        Engine – 30
        Transmission – 31
    Pump – 75
    Seals – 97

P

Piston
    Inspection – 85
    Installation – 107
    Removal – 66
    Rings – 107
    Specifications – 51, 52, 53, 54
Points (contact breaker) – 39
Primary drive
    Chain adjustment – 27, 43
    Inspection – 96
    Installation – 101
    Removal – 68

R

Rear brake components – see *Brakes*

Index

Rear wheel
    Bearings – 46,
    Removal and installation – 161
Rear wheel sprocket – 168
Recommended lubricants and fluids – 28
Regulator check – 178
Rings (piston) – 107
Rocker boxes/rocker arms
    Inspection – 91
    Installation – 110
    Removal – 62
Routine maintenance
    Intervals – 29
    Introduction – 30

S

Safety first! – 17
Seals – 97
Seat – 149
Shock absorbers – 139
Sidestand check and maintenance – 48
Spark plugs
    Check and replacement – 39
    Specifications – 26
Spokes – 36
Sprocket (final drive)
    Engine – 73, 106
    Rear wheel – 168
Starter drive assembly – 97
Starter motor
    Disassembly, inspection and reassembly – 179
    Installation – 105
    Removal – 66
Starter solenoid – 180
Steering head bearings
    Check – 44
    Maintenance – 149
Suspension inspection – 44
Swingarm – 140
Switches
    Brake light – 181, 182
    Handlebar – 180
    Ignition and light – 181

T

Taillight bulb – 174
Tappets
    Inspection – 92
    Installation – 101

    Removal – 66
    Specifications – 51, 53
Throttle
    Cable adjustment – 131
    Check and lubrication – 44
Timing gears and bearings – 92
Tires
    Check – 36
    Pressures – 27
    Removal and installation – 168
Tools – 11
Transmission
    Dismantling and reassembly – 93
    Inspection – 93
    Oil change – 43
    Oil level check – 31
    Installation – 101
    Removal – 79
    Specifications – 55
Troubleshooting – 18
Tubes (tire) – 170
Tune-up and routine maintenance – 26
Turn signal
    Bulbs – 174
    Relay /cancel unit – 183

V

Valves
    Adjustment (pre-Evolution engine only) – 36
    Disassembly, inspection and reassembly – 86
    Servicing – 86
    Specifications – 26, 51, 52, 53
VOES (Vacuum Operated Electric Switch) – 136

W

Wheel bearings
    End play – 150, 153, 162
    Repacking procedure – 45
Wheel sprocket – 168
Wheels
    General information – 152
    Inspection and repair – 152
    Removal and installation
        Front – 152
        Rear – 161
    Specifications – 150
Wiring diagrams – 184
Working facilities – 15